# Innovative Methods for Rare Disease Drug Development

# Chapman & Hall/CRC Biostatistics Series

Series Editors

Shein-Chung Chow
*Editor-in-Chief, Duke University School of Medicine, USA*

Byron Jones
*Novartis Pharma AG, Switzerland*

Jen-pei Liu
*National Taiwan University, Taiwan*

Karl E. Peace
*Georgia Southern University, USA*

Bruce W. Turnbull
*Cornell University, USA*

## RECENTLY PUBLISHED TITLES

For more information about this series, please visit: https://www.routledge.
com/Chapman--Hall-CRC-Biostatistics-Series/book-series/CHBIOSTATIS

# Innovative Methods for Rare Disease Drug Development

Shein-Chung Chow, PhD

Duke University School of Medicine, Durham, North Carolina

## CRC Press
Taylor & Francis Group
Boca Raton London New York

CRC Press is an imprint of the
Taylor & Francis Group, an **informa** business

A CHAPMAN & HALL BOOK

First edition published 2020
by CRC Press
6000 Broken Sound Parkway NW, Suite 300, Boca Raton, FL 33487-2742

and by CRC Press
2 Park Square, Milton Park, Abingdon, Oxon, OX14 4RN

First issued in paperback 2022

**Visit the Taylor & Francis Web site at
http://www.taylorandfrancis.com**

**and the CRC Press Web site at
http://www.crcpress.com**

*Library of Congress Cataloging-in-Publication Data*
Names: Chow, Shein-Chung, 1955- author.
Title: Innovative methods for rare disease drug development / Shein-Chung
Chow, PhD, Duke University School of Medicine, Durham, North Carolina.
Description: Boca Raton, FL : CRC Press, 2021. |
Series: Chapman & Hall/CRC biostatistics series |
Includes bibliographical references and index.
Identifiers: LCCN 2020030145 | ISBN 9780367502102 (hardback) |
ISBN 9781003049364 (ebook)
Subjects: LCSH: Drug development. | Rare diseases.
Classification: LCC RM301.25 C46 2021 | DDC 615.1/9—dc23
LC record available at https://lccn.loc.gov/2020030145

ISBN 13: 978-0-367-50290-4 (pbk)
ISBN 13: 978-0-367-50210-2 (hbk)
ISBN 13: 978-1-00-304936-4 (ebk)

DOI: 10.1201/9781003049364

Typeset in Palatino
by codeMantra

# Contents

# Preface

In the United States, a rare disease is defined by the Orphan Drug Act of 1983 as a disorder or condition that affects less than 200,000 persons, while an orphan drug is a drug product developed for treating patients with rare diseases. As indicated in US Food and Drug Administration (FDA) guidance, most rare diseases are genetically related and thus are present throughout the person's entire life, even if symptoms do not immediately appear (FDA, 2015b, 2019a). To encourage the development of rare disease drug products, FDA provides several incentives to expedite development programs for rare diseases. These expedited programs include (i) fast-track designation, (ii) breakthrough therapy designation, (iii) priority review designation, and (iv) accelerated approval for regulatory review and approval of rare disease drug products. For the approval of drug products, FDA requires that substantial evidence regarding effectiveness and safety of the drug products be provided. Substantial evidence can only be obtained through the conduct of adequate and well-controlled investigations (21 CFR 314.126(a)). As indicated in its recent guidance, FDA does not have the intention to create a statutory standard for the approval of orphan drugs that is different from the standard for the approval of drugs in common conditions (FDA, 2015b, 2019a).

For the approval of drug products, a traditional approach is to conduct a randomized clinical trial by testing a null hypothesis of no treatment effect (i.e., ineffectiveness) against an alternative hypothesis that there is treatment effect (i.e., effectiveness). Effectiveness is concluded if we reject the null hypothesis at a pre-specified level of significance (say 5%). To ensure there is high probability of correctly detecting a clinically meaningful difference (or treatment effect), a pre-study power calculation (i.e., power analysis for sample size calculation) is often performed to ensure that there are sufficient number of subjects for achieving a desired power for correctly detecting a clinically meaningful treatment effect if such a treatment effect truly exists. In rare disease clinical trials, however, power calculation may not be feasible because there are only limited number of subjects available for the rare diseases under study. In this case, innovative thinking and approach for rare diseases drug development are needed for providing substantial evidence meeting the same standard as drugs in common conditions with certain statistical assurance.

This book is intended to be the first book entirely devoted to the discussion of innovative design and analysis for rare diseases drug development. The scope of this book will focus on some innovative thinking that (i) there is demonstrating *not-ineffectiveness* rather than demonstrating effectiveness in placebo-controlled studies with a limited number of subjects available (Chow and Huang, 2019) and (ii) sample size requirement is based on probability monitoring procedure (Huang and Chow, 2019). Along this line,

Chow (2020) proposed an innovative approach utilizing two-stage seamless adaptive trial design in conjunction with the concept of real-world data and real-world evidence, which is proposed for obtaining substantial evidence that will meet the same standard for the approval of rare drug products. This book consists of 14 chapters concerning regulatory requirement, innovative design, and analysis for rare diseases drug development.

Chapter 1 provides some background regarding rare diseases drug development. Also included in this chapter are regulatory perspectives, including regulatory incentives/guidance and brief introduction of the out-of-the-box innovative thinking for rare diseases drug development. Chapter 2 gives some basic considerations concerning rare diseases drug development. Chapter 3 focuses on hypotheses testing for clinical evaluation of rare diseases drug products. Chapter 4 discusses the endpoint selection in rare diseases clinical trials. Also included in this chapter is the potential use of biomarker and development of therapeutic index. Chapter 5 provides clinical strategy for non-inferiority and equivalence margin selection based on risk assessment. Chapter 6 evaluates the probability of inclusiveness after not-ineffectiveness has been established. Chapter 7 discusses sample size requirement for rare diseases drug development based on a newly proposed probability monitoring procedure. Chapter 8 reviews FDA's recent clinical initiative for potential use of real-world data/evidence in support of regulatory approval of new indications and/or labeling change. Chapter 9 provides an innovative approach for rare diseases drug development. Chapters 10–12 cover some useful complex innovative trial designs, including the *n*-of-1 trial design, a two-stage adaptive trial design, and master protocols (the platform trial design) for rare diseases drug development, respectively. Chapters 13–14 provide discussions on case studies for gene therapy for rare diseases and NASH (nonalcoholic steatohepatitis) study for liver diseases with unmet medical need, respectively.

From Taylor & Francis, we would like to thank Mr. David Grubbs for providing me the opportunity to work on this book. I would like to thank colleagues from Duke University School of Medicine and Office of Biostatistics (OB)/ Office of Translational Science (OTS), Center for Drug Evaluation and Research (CDER) at Food and Drug Administration (FDA) for their support during the preparation of this book. I wish to express my gratitude to many friends from the academia, the pharmaceutical industry, and regulatory agencies such as FDA, European Medicines Agency (EMA), and National Medical Products Administration (NMPA) of China for their encouragement and support.

Finally, the views expressed are those of the author and not necessarily those of Duke University School of Medicine. I am solely responsible for the contents and errors of this book. Any comments and suggestions that will lead to the improvement of revision of this book are very much appreciated.

**Shein-Chung Chow, PhD**
*Duke University School of Medicine*
*Durham, North Carolina*

# Author

**Shein-Chung Chow, PhD,** is currently a Professor at the Department of Biostatistics and Bioinformatics, Duke University School of Medicine, Durham, North Carolina, the United States. Between 2017 and 2019, Dr. Chow was on-leave for the US Food and Drug Administration (FDA) as an Associate Director at the Office of Biostatistics (OB), Center for Drug Evaluation and Research (CDER) at the FDA. Dr. Chow is also a Special Government Employee (SGE) appointed by the FDA as an Oncologic Drug Advisory Committee (ODAC) voting member and Statistical Advisor to the FDA. Prior to joining Duke University School of Medicine, Dr. Chow was the Director of TCOG (Taiwan Cooperative Oncology Group) Statistical Center and the Executive Director of National Clinical Trial Network Coordination Center in Taiwan. Prior to that, Dr. Chow also held various positions in the pharmaceutical industry such as Vice President, Biostatistics, Data Management, and Medical Writing at Millennium Pharmaceuticals, Inc., Cambridge, Massachusetts; Executive Director, Statistics and Clinical Programming at Covance, Inc., Director and Department Head at Bristol-Myers Squibb Company, Plainsboro, New Jersey. Dr. Chow was the Editor-in-Chief of the *Journal of Biopharmaceutical Statistics* (2002–2020). Dr. Chow is also the Editor-in-Chief of the *Biostatistics Book Series* at Chapman and Hall/CRC Press of Taylor & Francis Group. He was elected Fellow of the American Statistical Association and an elected member of the ISI (International Statistical Institute). Dr. Chow is the author or co-author of over 310 methodology papers and 31 books, including *Design and Analysis of Bioavailability and Bioequivalence Studies, Sample Size Calculations in Clinical Research, Adaptive Design Methods in Clinical Trials, Biosimilars: Design and Analysis of Follow-on Biologics., Analytical Similarity Assessment in Biosimilar Drug Development*, and most recently *Innovative Methods for Rare Disease Drug Development.*

Dr. Chow received a BS in mathematics from National Taiwan University, Taiwan, and a PhD in statistics from the University of Wisconsin, Madison, Wisconsin.

# 1

## Introduction

## 1.1 What Is Rare Disease?

In the United States, a rare disease is defined as a condition that affects fewer than 200,000 people (ODA, 1983). Under this definition, there may be as many as 7,000 rare diseases in the United States. The total number of Americans living with a rare disease is estimated at between 25 and 30 million. This estimate has been used by the rare disease community for several decades to highlight that while individual diseases may be rare, the total number of people with a rare disease is large. In the United States, however, only a few types of rare diseases are tracked when a person is diagnosed, including certain infectious diseases, birth defects, and cancers, as well as the diseases on state newborn screening tests. Because most rare diseases are not tracked, it is hard to determine the exact number of rare diseases or how many people are affected by those diseases. Rare diseases are also known as orphan diseases because drug companies were not interested in developing treatments under economic (or return of investment) consideration. To overcome this dilemma, in 1983, the United States Congress passed the Orphan Drug Act, which created several financial incentives to encourage pharmaceutical companies in order to develop new drugs for rare diseases.

Other countries have their own official definitions of a rare disease. For example, in the European Union (EU), a disease or disorder is defined as rare in Europe when it affects fewer than 1 in 2,000 people. Under this definition, there are more than 6,000 rare diseases. On the whole, rare diseases may affect 30 million EU citizens. Eighty percent of rare diseases are of genetic origin, and are often chronic and life-threatening. On the other hand, in Japan, according to the National Programme on Rare and Intractable Diseases (NPRID) launched in 1972, rare diseases are defined as those with a prevalence of less than 50,000, or 1 in 2,500, and with no known cause or cure. Under this definition, 123 diseases have been identified/specified by an Expert Advisory Board on the basis of research priorities, which include Behçet disease, multiple sclerosis, and amyotrophic lateral sclerosis. In China, thus far, rare diseases have not been officially defined. The definition that is commonly considered is based on a consensus of experts reached by the

1

Genetics Branch of the Chinese Medical Association in May 2010. According to this definition, a rare disease is a disease with a prevalence of less than 1/500,000, or a neonatal morbidity of less than 1/10,000 (Ma et al., 2011; He et al., 2018). Since epidemiological data on rare diseases are lacking in China, the current list of rare diseases is based on actual conditions, and the list is mainly derived from the professional opinions of the Expert Committee on the Diagnosis, Treatment, and Care for Rare Diseases established by the Medical Administration Bureau of the former National Health and Family Planning Commission.

For orphan drug designation, FDA considers using the mechanism of action (MOA) of the drug to determine what distinct disease or condition the drug is intended to treat, diagnose, or prevent. Whether a given medical condition constitutes a distinct disease or condition for the purpose of orphan drug designation depends on a number of factors, assessed cumulatively, including pathogenesis of the disease or condition, course of the disease or condition, prognosis of the disease or condition, and resistance to treatment. These factors are analyzed in the context of the specific drug for which designation is requested. During the course of reviewing a request for orphan drug designation, equipped with the most current scientific literature about a particular disease or condition, FDA may come to a new understanding about the nature of that disease or condition. Below is a list of some diseases or conditions for which FDA's views on how it categorizes or otherwise understands the disease or condition have evolved. This is not a comprehensive list of orphan disease determinations, but reflective of some of the more common questions we receive. FDA will update this list as appropriate when it makes orphan drug designation determinations that change how we approach the disease or condition in question. For a complete list of orphan drug designations and approvals, see the searchable Orphan Products Designation Database (Table 1.1).

As most rare diseases may affect far fewer persons, one of the major concerns of rare disease clinical trials is that often there are only a small number of subjects available. However, FDA does not have the intention to create a statutory standard for the approval of orphan drugs that is different from the standard for the approval of drugs in common conditions. Thus, in rare disease clinical trials, power calculation for the required sample size may not be feasible. In this case, innovative thinking and approach are necessary for achieving the same standard with a limited number of patients available. In this chapter, some innovative thinking such as (i) probability monitoring procedure for the justification of the selected small sample size, (ii) the concept of demonstrating not-ineffectiveness (non-inferiority) rather than demonstrating effectiveness (superiority) with a limited number of patients available, (iii) the combined use of data from randomized clinical trial (RCT) and real-world data and real-world evidence (RWD/RWE) in support of rare diseases drug development, and (iv) innovative two-stage adaptive seamless clinical trial design are proposed.

**TABLE 1.1**

List of Orphan Disease Determination

| Disease or Condition | FDA's Perspectives |
|---|---|
| Ovarian, fallopian tube, and primary peritoneal cancer | FDA considers ovarian cancer, fallopian tube cancer, and primary peritoneal cancer to be one distinct disease or condition |
| Metastatic brain cancer | FDA considers any primary tumor type that has metastasized to the brain to be its own distinct disease or condition. For example, breast cancer that has metastasized to the brain is a distinct disease from breast cancer |
| Pulmonary hypertension | FDA recognizes the five WHO classifications of pulmonary hypertension as distinct diseases or conditions |
| Scleroderma | FDA considers systemic sclerosis to be a different disease or condition than localized scleroderma |
| Lymphoma | FDA recognizes the WHO classifications of lymphoma as distinct diseases or conditions |
| Familial adenomatous polyposis | FDA recognizes familial adenomatous polyposis as a distinct disease or condition from sporadic adenomatous polyps |
| Medication-induced dyskinesia in Parkinson's disease (PD) | FDA recognizes medication-induced dyskinesia in PD as the disease or condition. Levodopa-induced dyskinesia in PD is considered to be a subset of medication-induced dyskinesia in PD |

Section 1.2 outlines the regulatory perspectives (including regulatory incentives and regulatory guidance) regarding rare disease drug development. Section 1.3 discusses sponsors' strategy for rare disease drug development; includes the unique and remarkable story of Velcade® drug development; and posts the practical and challenging issues that are commonly encountered in rare disease drug development. Section 1.4 describes and discusses some innovative designs. Section 1.5 reveals the aim and scope of this book.

## 1.2 Regulatory Perspectives

### 1.2.1 Regulatory Incentives

As indicated in FDA (2015b), most rare diseases are genetically related and thus are present throughout the person's entire life, even if symptoms do not immediately appear. Many rare diseases appear early in life, and about 30% of children with rare diseases will die before reaching their fifth birthday. FDA is to advance the evaluation and development of products, including drugs, biologics, and devices that demonstrate a promise for the diagnosis and/or treatment of rare diseases or conditions. Along this line, FDA evaluates the scientific and clinical data submissions from sponsors to identify and designate products as promising for rare diseases and to further advance the scientific development of such promising medical products. To encourage the development of rare disease drug products, FDA provides several incentive (expedited) programs, including (i) fast-track designation, (ii) breakthrough therapy designation, (iii) priority review designation, and (iv) accelerated approval for regulatory review and approval of rare disease drug products, which are briefly described below.

**Fast-Track Designation** – Fast track is a designation of an investigational drug for expedited review to facilitate the development of drugs, which (i) treat a serious or life-threatening condition and (ii) fill an unmet medical need. Fast-track designation must be requested by the sponsors. The request can be initiated at any time during the drug development process. FDA will review the request and make a decision within 60 days.

Fast-track designation is designed to aid in the development and expedite the review of drugs, which show a promise in treating a serious or life-threatening disease and address an unmet medical need. Serious condition is referred to as the determination of the seriousness of a disease. The determination is a matter of judgment, but is generally based on whether the drug will affect the factors such as survival, day-to-day functioning, or the likelihood that the disease (if left untreated) will progress from a less severe condition to a more serious one. For a drug to address an unmet medical need, the drug must be developed as a treatment or preventative measure for a disease that has no current therapy. If there are existing therapies, a fast-track eligible drug must show some advantages over the available treatment, e.g., (i) showing superior effectiveness, (ii) avoiding serious side effects of an available treatment, (iii) improving the diagnosis of a serious disease where early diagnosis results in an improved outcome, (iv) decreasing a clinically significant toxicity of an available treatment, and (v) addressing an expected public health need.

Once a drug receives fast-track designation, early and frequent communication between the FDA and a drug sponsor is encouraged throughout the entire drug development and review process. The frequency of communication assures that questions and issues are resolved quickly, often leading to earlier drug approval and access by patients. Note that the drug sponsor may appeal to the division responsible for reviewing the application within the Center for Drug Evaluation and Research (CDER) at the FDA if its request for fast-track designation is not granted or any other general dispute. The drug sponsor can subsequently utilize the agency's procedures for internal review or dispute resolution if necessary.

**Breakthrough Therapy Designation** – Breakthrough therapy designation is a process designed to expedite the development and review of drugs that are intended to treat a serious condition, and preliminary clinical evidence indicates that the drug may demonstrate a substantial improvement over the available therapy on a clinically significant endpoint. To determine whether the improvement over available therapy is substantial is a matter of judgment and depends on both the magnitude of the treatment effect, which could include the duration of the effect, and the importance of the observed clinical outcome. In general, the preliminary clinical evidence should show a clear advantage over the available therapy.

For the purposes of breakthrough therapy designation, clinically significant endpoint generally refers to an endpoint that measures an effect on irreversible morbidity or mortality (IMM) or on symptoms that represent serious consequences of the disease. A clinically significant endpoint can also refer to the findings that suggest an effect on IMM or serious symptoms, which includes, but are not limited to, (i) an effect on an established surrogate endpoint, (ii) an effect on a surrogate endpoint or intermediate clinical endpoint considered reasonably likely to predict a clinical benefit (i.e., the accelerated approval standard), (iii) an effect on a pharmacodynamic biomarker that does not meet criteria for an acceptable surrogate endpoint, but strongly suggests the potential for a clinically meaningful effect on the underlying disease, and (iv) a significantly improved safety profile compared to the available therapy (e.g., less dose-limiting toxicity (DLT) for an oncology agent), with an evidence of similar efficacy.

A drug that receives breakthrough therapy designation is eligible for the following: (i) all fast-track designation features, (ii) intensive guidance on an efficient drug development program, beginning as early as phase 1, and (iii) organizational commitment involving senior managers. Similar to fast-track designation, breakthrough therapy designation is requested by the drug sponsor. If a sponsor has not requested breakthrough therapy designation, FDA may suggest that the sponsor consider submitting a request provided that (i) after reviewing submitted data and information (including preliminary clinical evidence), the agency thinks the drug development program may meet the criteria for breakthrough therapy designation and (ii) the remaining drug development program can benefit from the designation. It should be noted that FDA does not anticipate that breakthrough therapy designation requests will be made after the submission of an original BLA (Biologics License Application) or NDA (New Drug Application) or a supplement. FDA will respond to breakthrough therapy designation requests within 60 days of receipt of the request.

**Priority Review** – Prior to approval, each drug marketed in the United States must go through a detailed FDA review process. In 1992, under the Prescription Drug User Fee Act (PDUFA), FDA agreed to the specific goals for improving the drug review time and created a two-tiered system of review times – *standard review* and *priority review*. A priority review designation means FDA's goal is to take action on an application within 6 months (compared to 10 months under the standard review).

A *priority review* designation will direct the overall attention and resources to the evaluation of applications for drugs that, if approved, would be significant improvements in the safety or effectiveness of the treatment, diagnosis, or prevention of serious conditions when compared to the standard applications. Significant improvement may be demonstrated by the following: (i) evidence of increased effectiveness in treatment, prevention, or diagnosis of condition; (ii) elimination or substantial reduction of a treatment-limiting drug reaction; (iii) documented enhancement of patient compliance that is expected to lead to an improvement in serious outcomes; or (iv) evidence of safety and effectiveness in a new subpopulation.

FDA decides on the review designation for every application. However, an applicant may expressly request priority review as described in the Guidance for Industry Expedited Programs for Serious Conditions – Drugs and Biologics. It does not affect the length of the clinical trial period. FDA informs the applicant of a priority review designation within 60 days of the receipt of the original BLA, NDA, or efficacy supplement. Designation of a drug as "priority" does not alter the scientific/medical standard for approval or the quality of evidence necessary.

**Accelerated Approval** – In 1992, FDA initiated the FDA Accelerated Approval Program to allow faster approval of drugs for serious conditions that fill an unmet medical need. The faster approval relies on the use of surrogate endpoints. Drug approval typically requires clinical trials with endpoints that demonstrate a clinical benefit, like increased survival for cancer patients. Drugs with an accelerated approval can initially be tested in clinical trials that use a surrogate endpoint, or something that is thought to predict the clinical benefits. Surrogate endpoints typically require less time, and in the case of a cancer patient, it is much faster to measure a reduction in tumor size, e.g., than the overall patient survival.

### 1.2.2 Regulatory Guidance

In January 2019, FDA published a draft guidance on *Rare Diseases: Common Issues in Drug Development* to assist sponsors of drug and biological products for the treatment or prevention of rare diseases (FDA, 2019a). The purpose of this draft guidance is to assist sponsors in conducting more efficient and successful drug development programs. As indicated in the draft guidance, the statutory requirements for marketing approval for drugs to treat rare and common diseases are the *same*. FDA does not have the intention to create a statutory standard for the approval of orphan drugs that is different from the standard for the approval of drugs in common conditions. For the approval of drug products, FDA requires that substantial evidence regarding effectiveness and safety of the drug products be provided. Substantial evidence is based on the results of adequate and well-controlled investigations (21 CFR 314.126(a)).

FDA recognized that issues that are encountered in rare disease drug development are frequently more difficult to be addressed because there is often limited medical and scientific knowledge, natural history data, and drug development experience. This draft guidance addresses the importance of the following elements in development programs for rare diseases: (i) adequate description and understanding of the disease's natural history;

(ii) adequate understanding of the pathophysiology of the disease and the drug's MOA; (iii) nonclinical pharmacotoxicological and human toxicological considerations to support the proposed clinical investigation or investigations; (iv) selection or development of outcome assessments and endpoints; (v) evidence to establish safety and effectiveness; (vi) drug manufacturing considerations during drug development (e.g., pharmaceutical quality system considerations); (vii) participation of patients, caretakers, and advocates in development programs; and most importantly, (viii) interactions with the agency (FDA, 2015b, 2019a). As FDA pointed out, early consideration of these issues gives sponsors the opportunity to efficiently and effectively address the issues and to have productive meetings with FDA.

**Natural History Studies** – Since the natural history of rare diseases is often poorly understood, there is a need for prospectively designed, protocol-driven natural history studies initiated in the earliest drug development planning stages. Although it is not required, FDA encourages sponsors to evaluate early the depth and quality of existing natural history knowledge to determine if it is sufficient to inform their drug development programs (FDA, 2019a). A natural history study initiated early may run in parallel with early stages of drug development and may allow updating of drug development strategies as new learning emerges. In general, natural history study designs can be characterized as (i) retrospective or prospective and (ii) cross-sectional or longitudinal. Retrospective and prospective studies differ with respect to when patient data are collected. The information to be collected in the study is typically set forth in a protocol or procedure manual. On the other hand, cross-sectional and longitudinal natural history studies collect data from the cohorts of patients. Note that cross-sectional and longitudinal studies may be retrospective or prospective.

As the draft guidance pointed out, each type of natural history study has advantages and disadvantages (FDA, 2019a). In general, retrospective studies may be conducted more quickly than prospective studies. However, retrospective studies are limited in that they can only obtain data elements available in the existing records. Retrospective studies are also limited by many factors, including, but not limited to, inconsistent measurement procedures, irregular time intervals, and unclear use of terms that may limit the completeness and generalizability of the information. These limitations often preclude the use of such studies as an external control group for drug trials if it is not possible to match the characteristics of patients in the drug trial with the historical controls. Prospective studies provide systematically and comprehensively captured data using consistent medical terms and methodologies relevant to future clinical trials. For a prospective design, a cross-sectional study may be conducted more quickly than a longitudinal study. However, cross-sectional studies are unable to provide a comprehensive description of the course of progressive or recurrent disease. Cross-sectional studies may be helpful to inform the design of a longitudinal natural history study. Longitudinal studies typically yield the most comprehensive information about a disease, can characterize the course of disease within patients, and can help distinguish different phenotypes.

**Endpoint Selection** – For many rare diseases, the well-established efficacy endpoints for the disease may not be available. Thus, FDA suggested that a sponsor should define a trial endpoint by selecting a patient assessment to be used as an outcome measure and define when in the trial the patient would be assessed (FDA, 2019a). As indicated in the draft guidance, endpoint selection in a clinical trial involves the knowledge and understanding of the following: (i) the range and course of clinical manifestations associated with the disease; (ii) the clinical characteristics of the specific target population, which may be a subset of the total population with a disease; (iii) the aspects of the disease that are meaningful to the patient and that could be assessed to evaluate the drug's effectiveness; and (iv) the possibility of using the accelerated approval pathway. Despite continuing efforts to develop novel surrogate endpoints, FDA emphasized that only the usual clinical endpoints for the adequate and well-controlled trials can provide the substantial evidence of effectiveness supporting marketing approval of the drug (FDA, 2019a). Thus, it is suggested that sponsors should select endpoints considering the objectives of each trial in the context of the overall clinical development program.

More details and discussions regarding the study endpoint selection are provided in Chapter 4.

**Remarks** – FDA draft guidance has addressed some important issues that are commonly encountered in rare disease drug development, such as the use of natural history data and endpoint selection. In addition, this draft also indicated that (i) exploratory evidence from early-phase trials helps inform the choice of dose and timing of endpoints, and (ii) the use of adaptive seamless trial design may allow early evidence to be used later in a study, especially helpful when there are limited numbers of patients to study. If an adaptive design is under consideration, a thorough statistical analysis plan, including the key features of the trial design and preplanned analyses, should be discussed with the review division before the trial initiation. This draft guidance, however, has little discussion of the general issues of sample size requirement and statistical analysis, which play important roles for the success of rare diseases drug development.

## 1.3 Sponsor's Perspectives

Despite regulatory incentives for rare diseases drug development, it is of great interest to the sponsors regarding how to increase the probability of success with a limited number of subjects available and at the same time fulfill with regulatory requirement for review and approval of rare diseases regulatory submissions.

### 1.3.1 Practical Difficulties and Challenges

For rare diseases drug development, despite FDA's incentive programs, some practical difficulties and challenges are inevitably encountered at the planning stage of rare diseases clinical trials. These practical difficulties and challenges include, but are not limited to, (i) insufficient power due to small sample size available; (ii) little or no prior information regarding dose finding; (iii) the potential use of AI (artificial intelligence) machine learning; and (iv) inflexibility in study design, which have impact on the probability of success of the intended clinical trials.

**Insufficient Power** – In practice, for rare diseases drug development, it is expected that the intended clinical trial may not have a desired power (i.e., the probability of correctly detecting a clinically meaningful difference or treatment effect when such a difference or treatment effect truly exists) for confirming efficacy of the test treatment under investigation at the 5% level of significance due to small sample available. Thus, the commonly considered power calculation for sample size is not feasible for rare diseases clinical trials. In this case, the sponsor must seek for alternative methods for sample size calculation for achieving certain statistical assurance for the intended rare diseases clinical trials.

As indicated in Chow, Shao et al. (2017), in addition to power analysis, other methods such as precision analysis, reproducibility analysis, and probability monitoring procedure could be used for the calculation of sample size in order to achieve certain statistical assurance in clinical trials. The precision analysis is to select a sample size that controls type I error rate within a desired precision, while the reproducibility analysis is to select a sample size that will achieve a desired probability of reproducibility. The probability monitoring procedure is to justify a selected sample size based on the control of the probability across efficacy/safety boundaries.

More details and discussion regarding the sample size calculation for rare diseases clinical trials are provided in Chapter 7.

**Inefficient Dose Finding** – Regarding dose finding for maximum tolerable dose (MTD), a traditional "3+3" dose escalation design is often considered. The traditional "3+3" escalation design is to enter three patients at a new dose level and then enter another three patients when a dose limiting toxicity (DLT) is observed. The assessment of the six patients is then made to determine whether the trial should be stopped at the level or to escalate to the next dose level. Note that DLT is referred to as unacceptable or unmanageable safety profile which is pre-defined by some criteria such as Grade 3 or greater hematological toxicity according to the US National Cancer Institute's Common Toxicity Criteria (CTC). This dose-finding design, however, suffers the following disadvantages: (i) inefficient, (ii) often underestimation of the MTD especially when the starting dose is too low, (iii) depending upon the DLT rate at MTD, and (iv) the low probability of correctly identifying the MTD.

Alternatively, it is suggested that a continued reassessment method (CRM) should be considered (Song and Chow, 2015). For the method of CRM, the dose–response relationship is continually reassessed based on the accumulative data collected from the trial. The next patient who enters the trial is then assigned to the potential MTD level. Thus, the CRM involves (i) dose toxicity modeling, (ii) dose-level selection, (iii) reassessment of model parameters, and (iv) assignment of next patient. Chang and Chow (2005) considered the CRM method in conjunction with a Bayesian approach for dose–response trials which substantially improve the CRM for dose finding.

To select a more efficient dose-finding design between the "3+3" escalation design and the CRM design, FDA recommends the following criteria for design selection: (i) number of patients expected, (ii) number of DLT expected, (iii) toxicity rate, (iv) probability of observing DLT prior to MTD, (v) probability of correctly achieving the MTD, and (vi) probability of overdosing. Song and Chow (2015) compared the "3+3" dose escalation design and the CRM design in conjunction with a Bayesian approach for a radiation therapy dose-finding trial based on a clinical trial simulation study. The results indicated that (i) CRM has an acceptable probability of correctly reaching the MTD, (ii) the "3+3" dose escalation design is always under-estimating the MTD, and (iii) CRM generally performs better than the "3+3" dose escalation design.

**Inflexible Study Design** – Under the restriction of only small sample available, the usual parallel-group design is considered not flexible. Instead, it is suggested that some complex innovative designs (CIDs) such as $n$-of-1 trial design, adaptive trial design, master protocol study design, and Bayesian sequential design should be considered.

In recent years, the $n$-of-1 trial design has become a very popular design for the evaluation of the difference in treatment effect within the same individual when $n$ treatments are administered at different dosing periods. In general, as compared to the parallel-group design, $n$-of-1 trial design requires less subjects for the evaluation of the test treatment under investigation. On the other hand, adaptive trial design has the flexibility for modifying the study protocol as it continues after the review of interim data. Clinical trial utilizing adaptive design methods can not only increase the probability of success of drug development, but also shortens the development process. More details regarding the $n$-of-1 trial design and the adaptive trial design are provided in Chapter 10 and Chapter 11, respectively.

In addition, the concept of master protocol and the use of Bayesian sequential design have received much attention lately, which also provide some flexibility for the evaluation of treatment effect in rare diseases drug development. More details and discussions regarding the application of master protocol trial design in clinical development are given in Chapter 12.

## 1.3.2 The Remarkable Story of the Development of Velcade®

**Fast-Track Destination** – Velcade® (bortezomib) is an antineoplastic agent (a proteasome inhibitor) indicated for the treatment of multiple myeloma and mantle cell lymphoma (see Figure 1.1). Multiple myeloma is the second most common cancer (an incurable cancer) of the blood, representing approximately 1% of all cancers and 2% of all cancer deaths. It is estimated that approximately 45,000 Americans have multiple myeloma with about 15,000 new cases diagnosed each year. Only about percent of multiple myeloma patients survive longer than 5 years with the disease. Although the disease is predominantly a cancer of the elderly (the average age at diagnosis is 70 years of age), recent statistics indicate both increased incidence and younger age of onset. Thus, multiple myeloma is considered a rare disease and meets the requirement for fast-track designation for the expedited review.

Velcade® was co-developed by Millennium Pharmaceuticals, Inc. and Johnson & Johnson Pharmaceutical Research & Development. On May 13, 2003, the FDA approved Velcade® under fast-track application for the

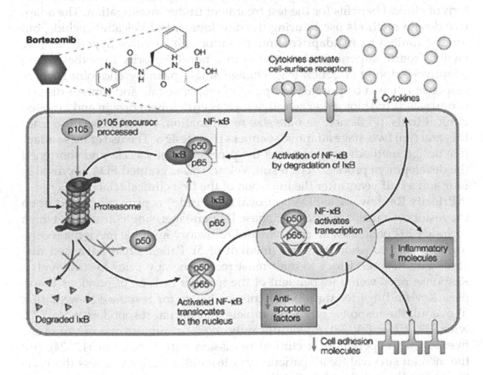

**Nature Reviews | Drug Discovery**

**FIGURE 1.1**
Velcade® (bortezomib), a proteasome inhibitor.

treatment of multiple myeloma. As indicated by Sánchez-Serrano (2017), the successful story behind the development of this drug is quite unique and remarkable, and primarily due to the adaption of core model. The core model, as epitomized by the success story of bortezomib, emphasizes the potential power of maximizing collaborative approaches and is useful in providing insights to policy makers, scientists, investors, and the public on how the process of drug development can be optimized, which should eventually lead to lower drug discovery and development costs; the creation of better, safer, and more effective medicines; and affordable access to the best drugs, including those for neglected and orphan diseases, not only in the developed countries, but also worldwide. Only then will we be able to build a better global system of health care, one which is not only more egalitarian, but also more humane.

**Innovative Science** – Under the core model, adaptive methods were frequently developed and used in clinical trials not only to (i) increase the probability of success of the intended clinical trials, but also to (ii) shorten the development process. The adaptive design methods gave the principal investigator (PI) the flexibility to identify any sign, signal, and trend/pattern of clinical benefits for the test treatment under investigation. The adaptive design methods used during the development of Velcade® include, but are not limited to, (i) adaptive randomization, (ii) adaptive hypotheses (i.e., switch from a superiority hypothesis to a non-inferiority hypothesis), (iii) adaptive-endpoint selection (e.g., change single primary endpoint such as response rate to a co-primary endpoint of response rate and time-to-disease progression), (iv) biomarker adaptive such as enrichment design and targeted clinical trials, (v) flexible sample size re-estimation, (vi) adaptive dose finding, and (vii) two-stage adaptive seamless trial design. The use of these adaptive design methods has increased the probability of success and shortened the development process. As a result, Velcade® was granted FDA approval in four and a half years after the initiation of the first clinical trial.

**Priority Review** – The FDA approval of Velcade® is primarily based upon the results of a major multicenter phase II open-label, single-arm trial, which included 202 patients with relapsed and refractory multiple myeloma receiving at least two prior therapies (median of 6). Patients had advanced disease, with 91% refractory to their most recent therapy prior to study entry. Response rates were independent of the number or type of previous therapies. Key findings for the 188 patients evaluable for response showed that (i) overall, the response rate for complete and partial responders was 27.7% with 95% CI of (21, 35), (ii) significantly, 17.6% or almost one out of every five patients experienced a clinical remission with 95% CI of (12, 24), (iii) the median survival for all patients was 16 months (range was less than one to greater than 18 months), (iv) the median duration of response for complete and partial responders was 12 months with 95% CI of (224 days; NE), and (v) side effects were generally predictable and manageable. For safety assessment, in 228 patients who were treated with Velcade® in two phase II studies of multiple myeloma, the most commonly reported adverse events were asthenic conditions (including fatigue, malaise, and weakness) (65%), nausea (64%), diarrhea (51%), decreased appetite (including anorexia) (43%), constipation (43%), thrombocytopenia (43%), peripheral neuropathy (including peripheral sensory neuropathy and peripheral neuropathy aggravated) (37%), pyrexia (36%), vomiting (36%), and anemia (32%). Fourteen percent of patients experienced at least one episode of grade four toxicity, with the most common toxicity being thrombocytopenia (3%) and neutropenia (3%).

In May 2003, Velcade® was launched for the treatment of relapsed and refractory multiple myeloma – a cancer of the blood. At the time, the FDA granted approval for the treatment of multiple myeloma for patients who had not responded to at least two other therapies for the disease. Velcade®, the first FDA-approved proteasome inhibitor, reached the market in record time and represented the first treatment in more than a decade to be approved for patients with multiple myeloma. In late December 2007, Millennium successfully submitted a supplemental new drug application (SNDA) to the FDA for Velcade® for previously untreated multiple myeloma. The SNDA submitted to the FDA for this indication included data from a phase III study, a large, well-controlled international clinical trial, comparing a Velcade®-based regimen to a traditional standard of care. Priority review was granted by the FDA in January 2008. On June 20, 2008, the FDA approved Velcade® in combination for patients with previously untreated multiple myeloma. This means that Millennium can market Velcade® to patients who have not had any prior therapies for multiple myeloma (a first-line therapy).

**Lessons Learn** – Millennium used innovative science to develop novel products that would address the unmet medical needs of patients. Millennium's success in bringing Velcade® to patients so rapidly reflects the high level of collaboration among many partners, both internally and externally. Moving forward, the sponsors of rare diseases drug development should adopt the innovative science in developing breakthrough products that make a difference in patients' lives. The FDA priority review and rapid approval of Velcade® represents a major advance in our fight against rare diseases. With its new and unique MOA of inhibiting the proteasome, Velcade® is different from traditional chemotherapies and represents a new treatment option for patients. Thus, innovative thinking and approach like CIDs are necessarily implemented for rare disease drug development.

## 1.4 Innovative Thinking

For rare disease drug development, one of the major challenges is that there are only limited subjects available for clinical trials. FDA (2019a), however, indicated that the agency does not have an intention to create a statutory standard for rare diseases drug development. In this case, some out-of-the-box

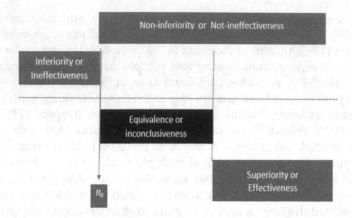

**FIGURE 1.2**
Demonstrating not-ineffectiveness or effectiveness in active-controlled (top line) and placebo-controlled (bottom line) studies.

innovative thinking designs are necessarily applied for obtaining substantial evidence for the approval of rare disease drug product (Chow, 2020). The out-of-the-box innovative thinking designs include (i) probability monitoring procedure for sample size requirement, (ii) the concept of demonstrating not-ineffectiveness rather than demonstrating effectiveness, (iii) borrowing RWD in support of regulatory approval of rare diseases drug products, and (iv) the use of CID to shorten the process of drug development. Along this line, Chow (2020) and Chow and Huang (2020) proposed an innovative approach for rare diseases drug development by first demonstrating not-ineffectiveness with limited subjects available and then utilizing (borrowing) RWD to rule out the probability of inconclusiveness for the demonstration of effectiveness under a two-stage adaptive seamless trial design. The proposed innovative approach can not only overcome the problem of small patient population for rare diseases, but also achieve the same standard for the evaluation of drug products with common conditions.

**Probability Monitoring Procedure for Sample Size** – For rare disease clinical development, it is recognized that a pre-study power analysis for sample size calculation is not feasible due to the fact that there are only limited number of subjects available for the intended trial, especially when the anticipated treatment effect is relatively small and/or the variability is relatively large. In this case, alternative methods such as precision analysis (or confidence interval approach), reproducibility analysis, and probability monitoring approach may be considered for providing substantial evidence with certain statistical assurance (Chow et al., 2017). It, however, should be noted that the resultant sample sizes from these different analyses could be very different with different levels of statistical assurance achieved. Thus, for rare disease clinical trials, it is suggested that an appropriate sample size should be selected for achieving certain statistical assurance under a valid trial design. To overcome the problem, Huang and Chow (2019) proposed a probability monitoring procedure for sample size calculation/justification, which can substantially reduce the required sample size for achieving certain statistical assurance.

As an example, an appropriate sample size may be selected based on a probability monitoring approach such that the probability of crossing safety boundary is controlled at a pre-specified level of significance. Suppose an investigator plans to monitor the safety of a rare disease clinical trial sequentially at several times, $t_i$, $i = 1, ..., K$. Let $n_i$ and $P_i$ be the sample size and the probability of observing an event at time $t_i$. Thus, an appropriate sample size can be selected such that the following probability of crossing safety stopping boundary is less than a pre-specified level of significance

$$p_k = P\{\text{across safety stopping boundry} \mid n_k, P_k\} < \alpha, k = 1,..., K. \quad (1.1)$$

Note that the concept of the probability monitoring approach should not be mixed up the concepts with those based on power analysis, precision analysis, and reproducibility analysis. Statistical methods for data analysis should reflect the desired statistical assurance under the trial design.

**Demonstrating Not-Ineffectiveness Versus Demonstrating Effectiveness** – For the approval of a new drug product, the sponsor is required to provide substantial evidence regarding safety and efficacy of the drug product under investigation. In practice, a typical approach is to conduct adequate and well-controlled (placebo-controlled) clinical studies, and test the following point hypotheses:

$$H_0: \textit{in}\text{effectiveness versus} \quad H_a: \text{effectiveness.} \quad (1.2)$$

The rejection of the null hypothesis of *in*effectiveness is in favor of the alternative hypothesis of effectiveness. Most researchers interpret that the rejection of the null hypothesis is the demonstration of the alternative hypothesis of effectiveness. It, however, should be noted that "in favor of effectiveness" does not imply "the demonstration of effectiveness." In practice, hypotheses (1.2) should be

$$H_0: \textit{in}\text{effectiveness versus} \quad H_a: \text{not-}\textit{in}\text{effectiveness.} \quad (1.3)$$

In other words, the rejection of $H_0$ would lead to the conclusion of "*not* $H_0$," which is $H_a$, i.e., "not-ineffectiveness" as given in (1.3). As can be seen from $H_a$ in (1.2) and (1.3), the concept of *effectiveness* (1.2) and the concept of *not-ineffectiveness* (1.3) are not the same. Not-ineffectiveness does not imply effectiveness in general. Thus, the traditional approach for clinical evaluation of the drug product under investigation can only demonstrate "*not-ineffectiveness*" but not *"effectiveness."* The relationship between demonstrating "effective-ness" (1.2) and demonstrating "not-ineffectiveness" (1.3) is illustrated in Figure 1.2. As it can be seen from Figure 1.2, in a placebo-controlled study, "not-ineffectiveness" consists of two parts, namely, the portion of "incon-clusiveness" and the portion of "effectiveness." As a result, the rejection of the null hypothesis of ineffectiveness cannot directly imply that the drug product is effective unless the probability of inconclusiveness, denoted by $p_{IC}$, is negligible, i.e.,

$$p_{IC} = P\{\text{inconclusiveness}\} < \varepsilon, \tag{1.4}$$

where $\varepsilon$ is a pre-specific number which is agreed upon between clinician and regulatory reviewer (Chow and Huang, 2019a).

Note that in active-controlled studies, the concept of demonstrating "not-ineffectiveness" is similar to that of establishing non-inferiority of the test treatment as compared to the active control agent. One can test for superiority (i.e., effectiveness) once the non-inferiority has been estab-lished without paying any statistical penalties. More details regarding demonstrating not-ineffectiveness for rare diseases drug development are provided in Chapter 6.

**The Use of RWD/RWE** – The 21st-century Cures Act passed by the United States Congress in December 2016 requires that the FDA shall establish a program to evaluate the potential use of RWE, which is derived from RWD to (i) support the approval of new indication for a drug approved under sec-tion 505 (c) and (ii) satisfy post-approval study requirements. RWD refers to data relating to patient's health status and/or the delivery of health care routinely collected from a variety of sources. RWD sources include, but are not limited to, electronic health record (EHR), administrative claims and enrolment, personal digital health applications, public health databases, and emerging sources. In practice, RWE offers the opportunities to develop robust evidence using high-quality data and sophisticated methods for pro-ducing causal-effect estimates regardless randomization method/model is used. In this chapter, we have demonstrated that the assessment of treat-ment effect (RWE) based on RWD could be biased due to potential selec-tion and information biases of RWD. Although the fit-for-purpose RWE may meet regulatory standards under certain assumptions, it is not the same as substantial evidence (current regulatory standard). In practice, it is then suggested that when there are gaps between the fit-for-purpose RWE and substantial evidence, we should make efforts to fill the gaps for an accurate and reliable assessment of the treatment effect.

In order to map RWE to substantial evidence (current regulatory stan-dard), we need to have good understanding of the RWD in terms of data relevancy/quality and its relationship with substantial evidence so that a fit-for-regulatory purpose RWE can be derived to map to regulatory standard.

As indicated by Corrigan-Curay (2018), there is a value of using RWE to support regulatory decisions in drug review and approval process. However, incorporating RWE into evidence generation, many factors must be considered at the same time before we can map RWE to substantial evidence (current regulatory standard) for regulatory review and approval. These factors include, but are not limited to, (i) efficacy or safety; (ii) relationship to available evidence; (iii) clinical context, e.g., rare, severe, life-threatening, or unmet medical need; and (iv) natural of endpoint/concerns about bias. In addition, leveraging RWE to support new indications and label revisions can help accelerate high-quality RWE earlier in the product lifecycle, providing more relevant evidence to support higher-quality and higher-value care for patients. Incorporating RWE into product labeling can lead to better-informed patient and provider decisions with more relevant information. For this purpose, it is suggested characterizing RWD quality and relevancy for regulatory purposes. Ultimate regulatory acceptability, however, will depend upon how robust these studies can be. That is, how well they minimize the potential for bias and confounding.

More details regarding the use of RWD/RWE for rare diseases drug development are provided in Chapter 8.

**Innovative Trial Design** – As indicated earlier, small patient population is a challenge to rare disease clinical trials. Thus, there is a need for innovative trial designs in order to obtain substantial evidence with a small number of subjects available for achieving the same standard for regulatory approval. In this sub-section, several innovative trial designs, including *n*-of-1 trial design, an adaptive trial design, master protocols, and a Bayesian design, are discussed.

One of the major dilemmas for rare diseases clinical trials is the unavailability of patients with the rare diseases under study. In addition, it is unethical to consider a placebo control in the intended clinical trial. Thus, it is suggested an *n*-of-1 crossover design be considered. An *n*-of-1 trial design is to apply *n* treatments (including placebo) in an individual at different dosing periods with sufficient washout in between dosing periods. A complete *n*-of-1 trial design is a crossover design consisting of all possible combinations of treatment assignment at different dosing periods.

Another useful innovative trial design for rare disease clinical trials is an adaptive trial design. In its draft guidance on adaptive clinical trial design, FDA defines an adaptive design as a study that includes a *prospectively* planned opportunity for the modification of one or more specified aspects of the study design and hypotheses based on the analysis of (usually interim) data from subjects in the study (FDA, 2010, 2019c). The FDA guidance has been served as an official document describing the potential use of adaptive designs in clinical trials since it was published in 2019. It, however, should be noted that the FDA draft guidance on adaptive clinical trial design is currently being revised in order to reflect pharmaceutical practice and FDA's current thinking.

Woodcock and LaVange (2017) introduced the concept of master protocol for studying multiple therapies, multiple diseases, or both in order to answer more questions in a more efficient and timely fashion. Master protocols include the following types of trials: umbrella, basket, and platform. The type of umbrella trial is to study multiple targeted therapies in the context of a single disease, while the type of basket trial is to study a single therapy in the context of multiple diseases or disease subtypes. The platform is to study multiple targeted therapies in the context of a single disease in a perpetual manner, with therapies allowed us to enter or leave the platform on the basis of decision algorithm. As indicated by Woodcock and LaVange (2017), if designed correctly, master protocols offer a number of benefits, including streamlined logistics, improved data quality, collection and sharing, as well as the potential to use innovative statistical approaches to study design and analysis. Master protocols may be a collection of sub-studies or a complex statistical design or platform for rapid learning and decision-making.

Under the assumption that historical data (e.g., previous studies or experience) are available, Bayesian methods for borrowing information from different data sources may be useful. These data sources could include, but are not limited to, natural history studies and expert's opinion regarding prior distribution about the relationship between endpoints and clinical outcomes. The impact of borrowing on results can be assessed through the conduct of sensitivity analysis. One of the key questions of particular interest to the investigator and regulatory reviewer is that how much to borrow in order to (i) achieve the desired statistical assurance for substantial evidence, and (ii) maintain the quality, validity, and integrity of the study.

**Innovative Approach** – Combining the out-of-the-box innovative thinking regarding rare disease drug development described in the previous section, Chow and Huang (2019b) and Chow (2020) proposed the following innovative approach utilizing a two-stage adaptive approach in conjunction with the use of RWD/RWE for rare diseases drug development. This innovative approach is briefly summarized below.

*Step 1.* Select a small sample size $n_1$ at stage 1 as deemed appropriate by the PI based on both medical and nonmedical considerations. Note that $n_1$ may be selected based on the probability monitoring procedure.

*Step 2.* Test hypotheses (3) for not-ineffectiveness at the $\alpha_1$ level, a prespecified level of significance. If fails to reject the null hypothesis of ineffectiveness, then stop the trial due to futility. Otherwise, proceed to the next stage.

Note that an appropriate value of $\alpha_1$ can be determined based on the evaluation of the trade-off with the selection of $\alpha_2$ for controlling the overall type I error rate at the significance level of $\alpha$. The goal of this step is to establish non-inferiority (i.e., not-ineffectiveness) of the test treatment with a limited number of subjects available at the $\alpha_1$-level of significance based on the concept of probability monitoring procedure for sample size justification and performing a non-inferiority (not-ineffectiveness) test with a significance level of $\alpha_1$.

*Step 3a*. Recruit additional $n_2$ subjects at stage 2. Note that $n_2$ may be selected based on the probability monitoring procedure. Once the non-inferiority (not-ineffectiveness) has been established at stage 1, sample size re-estimation may be performed for achieving the desirable statistical assurance (say 80% power) for the establishment of effectiveness of the test treatment under investigation at the second stage (say $N_2$, sample size required at stage 2).

*Step 3b*. Obtain (borrow) $N_2 - n_2$ data from previous studies (RWD) if the sample size of $n_2$ subjects is not enough for achieving the desirable statistical assurance (say 80% power) at stage 2. Note that data obtained from the $n_2$ subjects are from RCT, while data obtained from the other $N_2 - n_2$ are from RWD.

*Step 4*. Combined data from both Step 3a (data obtained from RCT) and Step 3b (data obtained from RWD) at stage 2, perform a statistical test to eliminate the probability of inconclusiveness. That is, perform a statistical test to determine whether the probability of inconclusiveness has become negligible at the $\alpha_2$-level of significance. For example, if the probability of inconclusiveness is less than a pre-specified value (say 5%), we can then conclude that the test treatment is effective.

In summary, for review and approval of rare diseases drug products, Chow and Huang (2020) proposed first to demonstrate not-ineffectiveness with limited information (patients) available at a pre-specified level of significance of $\alpha_1$. Then, after the not-ineffectiveness of the test treatment has been established, collect additional information (RWD) to rule out the probability of *inconclusiveness* for the demonstration of effectiveness at a pre-specified level of significance of $\alpha_2$ under the two-stage adaptive seamless trial design.

More details regarding innovative approach for the assessment of rare diseases drug development are provided in Chapter 9.

**Remarks** – In this chapter, some out-of-the-box innovative thinking designs regarding rare disease drug development are described. These innovative thinking designs include (i) probability monitoring procedure for sample size calculation/justification for certain statistical assurance, (ii) the concept of testing non-inferiority (i.e., demonstrating not-ineffectiveness) with a limited number of subjects available, (iii) utilizing (borrowing) RWD from various data sources in support of regulatory approval of rare diseases drug products, and (iv) the use of a two-stage adaptive seamless trial design to shorten the process of drug development. Combining these innovative thinking designs, under a two-stage adaptive seamless trial design, Chow and Huang (2019b) and Chow (2020) proposed an innovative approach for rare diseases drug development by first demonstrating not-ineffectiveness based on limited subjects available and then utilizing (borrowing) the RWD to rule out the probability of inconclusiveness for the demonstration of effectiveness. Chow and Huang's proposed innovative approach can not only overcome the problem of small patient population for rare diseases, but also achieve the same standard for the evaluation of drug products with common conditions.

## 1.5 Aim and Scope of This Book

This book is intended to be the first book entirely devoted to the discussion of innovative design and analysis for rare diseases drug development. The scope of this book will focus on some out-of-the-box innovative thinking and approach for rare diseases drug development. Along this line, an innovative approach based on these out-of-the-box innovative thinking designs is proposed for obtaining substantial evidence that will meet the same standard for the approval of rare drug products (Chow and Huang, 2019b; Chow, 2020; Chow and Huang, 2020).

This book consists of 14 chapters concerning regulatory requirement, innovative design, and analysis for rare diseases drug development. This chapter provides some background regarding rare diseases drug development. Also included in this chapter are regulatory perspectives, including regulatory incentives/guidance and brief introduction of the out-of-the-box innovative thinking for rare diseases drug development. Chapter 2 provides some basic considerations concerning rare diseases drug development. Chapter 3 focuses on hypotheses testing for the clinical evaluation of rare diseases drug products. Chapter 4 discusses the endpoint selection in rare diseases clinical trials. Also included in these chapters is the potential use of biomarker and development of therapeutic index. Chapter 5 provides the clinical strategy for non-inferiority and equivalence margin selection based on risk assessment. Chapter 6 evaluates the probability of inclusiveness after not-ineffectiveness has been established. Chapter 7 discusses the sample size requirement for rare diseases drug development based on a newly proposed probability monitoring procedure. Chapter 8 reviews FDA's recent clinical initiative for potential use of RWD/RWE in support of regulatory approval of new indications and/or labeling change. Chapter 9 gives an innovative approach for rare diseases drug development. Chapters 10–12 cover some useful complex innovative trial designs, including the $n$-of-1 trial design, a two-stage adaptive trial design, and master protocols (the platform trial design) for rare diseases drug development, respectively. Chapters 13–14 provide discussions on case studies for gene therapy for rare diseases and NASH (Non-Alcoholic SteatoHepatitis) study for liver diseases with unmet medical need, respectively.

# 2

# *Basic Considerations*

## 2.1 Introduction

As indicated in Chapter 1, in the United States, most rare diseases are genetically related (FDA, 2015b, 2019a). FDA is to advance the evaluation and development of products, including drugs, biologics, and devices that demonstrate a promise for the diagnosis and/or treatment of rare diseases or conditions. Along this line, FDA evaluates scientific and clin-ical data submissions from sponsors to identify and designate products as promising for rare diseases and to further advance scientific development of such promising medical products. Following the Orphan Drug Act, FDA also provides incentives for sponsors to develop the products for rare diseases. These incentive (expedited) programs include (i) fast-track designation, (ii) breakthrough therapy designation, (iii) priority review designation, and (iv) accelerated approval for regulatory review and approval of rare disease drug products.

For rare diseases drug development, one of the most challenges for rare disease clinical trials is probably that there is only a small patient population available. It is then a great concern how to conduct clinical trials with the limited number of subjects available for obtaining substantial evidence regarding the safety and effectiveness for approval of the drug product under investigation. FDA, however, emphasizes that they do not have an intention to create a statutory standard for the approval of rare diseases (orphan) drugs that is different from the standard for the approval of drugs in common conditions. Thus, it is suggested that innovative trial designs such as an adaptive trial design or a complete *n*-of-1 trial design should be used for an accurate and reliable and a more effective assessment of rare disease drug products under investigation.

For the approval of drug products, FDA requires that substantial evidence regarding effectiveness and safety of the drug products be provided. Substantial evidence is based on the results of adequate and well-controlled investigations (21 CFR 314.126(a)). For rare diseases drug development, with the limited number of subjects available, it is expected that there may not have sufficient power for detecting a clinically meaningful difference (treatment effect). In rare disease clinical trials, power calculation for the required sample size may not be feasible. In this case, alternative methods such as precision analysis, reproducibility analysis, or probability monitoring approach may be considered for providing substantial evidence with certain statistical assurance (Chow, Shao et al., 2017). In practice, however, small patient population is a challenge to rare disease clinical trials for obtaining substantial evidence through the conduct of adequate and well-controlled investigations. Data collection from adequate and well-controlled clinical investigation is essential for obtaining substantial evidence for the approval of the drug products. Data collection is a key to the success of the intended trial. Thus, utilizing out-of-the-box innovative trial designs and statistical methods in rare disease setting is extremely important to the success of rare diseases drug development.

Section 2.2 outlines some basic considerations in rare disease clinical trials. Section 2.3 introduces several innovative trial designs such as a complete *n*-of-1 trial design, an adaptive design, and a Bayesian design. Section 2.4 derives and discusses statistical methods for data analysis under these innovative trial designs. Section 2.5 proposes several criteria for the evaluation of rare disease clinical trials. Section 2.6 gives some concluding remarks.

## 2.2 Basic Considerations

For the approval of drug products, FDA requires that substantial evidence regarding the effectiveness and safety of the drug products under development be provided. This, however, is a major challenge for rare disease drug product development due to the fact that there are only a limited number of subjects available. Thus, some basic principles are necessarily considered (FDA, 2015b, 2019a). These basic considerations are described below.

**Historical Data** – In its draft guidance, FDA encourages the sponsors to evaluate the existing natural history knowledge in drug development programs (FDA, 2015b). Natural history studies help in defining the disease population, selecting clinical endpoints (sensitive and specific measures), and developing new or optimized biomarkers in early rare disease drug development. Most importantly, natural history studies provide an external control group for clinical trials. This is important especially for rare disease clinical trials with small patient population.

In practice, natural history studies could be conducted either prospectively or retrospectively (e.g., based on the existing medical records like patient charts). Data could be collected from the cohorts of patients under either a cross-sectional study or a longitudinal study. As indicated by the FDA, for a prospective design, a cross-sectional study may be conducted more quickly than a longitudinal study. However, cross-sectional studies are unable to provide a comprehensive description of the course of progressive or recurrent diseases (FDA, 2015b).

**Ethical Consideration** – As indicated by Grady (2017), the principles that contribute to ethical research include collaborative partnership with stakeholders, social value, scientific validity, fair subject evaluation, a balance of risk and benefit, informed consent, respect for participants, and independent review (see also, Coors et al., 2017). Independent review is usually conducted by institutional review board (IRB). IRB ensures that ethical requirements are fulfilled, including biases are eliminated, ethical issues between investigators and participants are balanced, and the research does not exploit individuals or groups. In practice, critical and emerging ethical issues are challenges to rare disease clinical research. These challenges include relatively few participants, the need for multi-site studies, innovative designs, and the need to protect privacy in a contained research environment.

As an example, like most clinical trials, it may not be ethical to conduct placebo-controlled trials for rare diseases when effective treatments exist. Placebo-controlled trials, however, are generally considered acceptable when risks are minimized and there is no increased risk of serious harm to those in the placebo arm. In case of placebo-controlled trials, an unequal treatment allocation, e.g., 2 (test treatment):1 (placebo), may be considered to minimize possible ethical concern. For another example, since there are a relatively small number of subjects to enroll, an innovative trial design such as $n$-of-1 crossover design and adaptive trial design may be useful for obtaining substantial evidence regarding the effectiveness and safety of the test treatment under investigation.

**The Use of Biomarkers** – In clinical trials, a biomarker is usually used to select *right* patient population (i.e., patient population who are most likely to respond to the test treatment). This process is usually referred to as an enrichment process assuming that the biomarker is predictive of clinical outcomes. A biomarker is also often used in the development of diagnostic procedure for the early detection of disease. It can also help in achieving personalized medicine (individualized medicine or precision medicine).

As compared to a hard (or gold standard) endpoint like survival, a biomarker often has the following characteristics: (i) It can be measured earlier, easier, and more frequently; (ii) there are less subjects to competing risks; (iii) it is less affected by other treatment modalities; (iv) it can detect a larger effect size (i.e., a smaller sample size is required); and (v) it is a predictive of clinical endpoint. The use of biomarker has the following advantages: (i) It can lead to better target population; (ii) it can detect a larger effect size (clinical benefit) with a smaller sample size; and (iii) it allows an early and faster decision-making.

Thus, under the assumption that there is a well-established relationship between a biomarker and clinical outcomes, the use of biomarker in rare disease clinical trials can not only allow screening for possible responders at enrichment phase, but also provides the opportunity to detect a signal of potential safety concerns early and provides supportive evidence of efficacy with a small number of patients available.

**Generalizability** – As indicated in the FDA draft guidance, about half of the people affected by rare diseases are children. Thus, conducting clinical trials with pediatric patients is critical in rare disease drug development so that the drug can be properly labeled for pediatric use. Note that a pediatric population is defined as the group of subjects whose ages are between birth to 17 years. FDA encourages sponsors to develop formulations for pediatric population.

In clinical development, after the drug product has been shown to be effective and safe with respect to the targeted patient population, it is often of interest to determine how likely the clinical results can be reproducible to a *different but similar* patient population with the same disease. For example, if an approved drug product is intended for the adult patient population with a certain rare disease, it is often of interest to study the effect of the drug product on a different but similar patient population, such as pediatric or elderly patient population with the same rare disease. In addition, it is also of interest to determine whether the clinical results can be generalized to from one patient population to patient with/without ethnic differences. Shao and Chow (2002) proposed the concept of the probability of generalizability, which is the measurement of reproducibility probability of a future trial with the population slightly deviated from the targeted patient population of the current trials. The assessment of generalizability probability can be used to determine whether the clinical results can be generalized from the targeted patient population to a different but similar patient population with the same rare disease.

**Sample Size Requirement** – For rare disease drug development, the Orphan Drug Act provides incentives associated with orphan drug designation to make the development of rare disease drug products with a small number of patients financially viable (FDA, 2015b). However, FDA does not have an intention to create a statutory standard for the approval of orphan drugs that is different from the standard for the approval of drugs in common conditions. Thus, sample size requirement has become one of the most challenging issues in rare disease clinical trials.

In rare disease clinical trials, power calculation for the required sample size may not be feasible due to a limited number of subjects available, especially when the anticipated treatment effect is relatively small. In this case, alternative methods such as precision analysis (or confidence interval approach), reproducibility analysis, or probability monitoring approach, and Bayesian approach may be considered for providing substantial evidence with certain statistical assurance (Chow et al., 2017). It, however, should be noted that the resultant sample sizes from these different analyses could be very different with different levels of statistical assurance achieved. Thus, for rare disease clinical trials, it is suggested that an appropriate sample size should be selected for achieving certain statistical assurance under a valid trial design.

In practice, it is suggested that statistical methods used for data analysis should be consistent with statistical methods used for sample size estimation for scientific validity of the intended clinical trial. The concepts of power analysis, precision analysis, reproducibility analysis, probability monitoring approach, and Bayesian approach should not be mixed up with, while the statistical methods for data analysis should reflect the desired statistical assurance under the trial design.

## 2.3 Complex Innovative Design

As indicated earlier, small patient population is a challenge to rare disease clinical trials. Thus, there is a need for innovative trial designs in order to obtain substantial evidence with a small number of subjects available for achieving the same standard for regulatory approval. In this section, several innovative trial designs, including $n$-of-1 trial design, an adaptive trial design, master protocols, and a Bayesian design, are discussed.

**Complete $n$-of-1 Trial Design** – One of the major dilemmas for rare diseases clinical trials is the unavailability of patients with the rare diseases under study. In addition, it is unethical to consider a placebo control in the intended clinical trial. Thus, it is suggested an $n$-of-1 crossover design be considered. An $n$-of-1 trial design is to apply $n$ treatments (including placebo) in an individual at different dosing periods with sufficient washout in between dosing periods. A complete $n$-of-1 trial design is a crossover design consisting of all possible combinations of treatment assignment at different dosing periods.

Suppose there are $p$ dosing periods and two test treatments, e.g., a test ($T$) treatment and a reference ($R$) product, to be compared. A complete $n$-of-1 trial design for comparing two treatments consists of $\Pi_{i=1}^{p} 2$, where $p \geq 2$, sequences of $p$ treatments (either $T$ or $R$ at each dosing period). If $p = 2$, then the $n$-of-1 trial design is a $4 \times 2$ crossover design, i.e., ($RR$, $RT$, $TT$, and $TR$), which is a typical Balaam's design. When $p = 3$, the $n$-of-1 trial design becomes an $8 \times 3$ crossover design, while the complete $n$-of-1 trial design with $p = 4$ is a $16 \times 4$ crossover design, which is illustrated in Table 2.1.

As indicated in a recent FDA draft guidance, a two-sequence dual design, i.e., ($RTR$ and $TRT$) and a $4 \times 2$ crossover designs, i.e., ($RTRT$ and $RRRR$), recommended by the FDA are commonly considered switching designs for assessing interchangeability in biosimilar product development (FDA, 2017a). These two switching designs, however, have the limitation for fully characterizing relative risk (i.e., a reduction in efficacy or an increase in the incidence rate of adverse event rate). On the other hand, these two trial designs are special case of the complete $n$-of-1 trial design with three or four dosing periods, respectively. Under the complete $n$-of-1 crossover design with four

**TABLE 2.1**

Complete *n*-of-1 Trial Design with $p = 4$

| Group | Period 1 | Period 2 | Period 3 | Period 4 |
|:---:|:---:|:---:|:---:|:---:|
| 1 | R | R | R | R |
| 2 | R | T | R | R |
| 3 | T | T | R | R |
| 4 | T | R | R | R |
| 5 | R | R | T | R |
| 6 | R | T | T | T |
| 7 | T | R | T | R |
| 8 | T | T | T | T |
| 9 | R | R | R | T |
| 10 | R | R | T | T |
| 11 | R | T | R | T |
| 12 | R | T | T | R |
| 13 | T | R | R | T |
| 14 | T | R | T | T |
| 15 | T | T | R | T |
| 16 | T | T | T | R |

*Note:* The first block (a 4 × 2 crossover design) is a complete *n*-of-1 design with two periods, while the second block is a complete *n*-of-1 design with three periods.

dosing periods, all possible switching and alternations can be assessed, and the results can be compared within the same group of patients and between different groups of patients.

A complete *n*-of-1 trial design has the following advantages: (i) each subject is at his/her own control; (ii) it allows a comparison between the test product and the placebo if the intended trial is a placebo-controlled study (this has raised the ethical issue of using placebo on the patients with critical conditions); (iii) it allows estimates of intra-subject variability; (iv) it provides estimates for treatment effect in the presence of possible carry-over effect, and most importantly, (v) it requires less subjects for achieving the study objectives of the intended trial design. However, the *n*-of-1 trial design suffers from the drawbacks of (i) possible dropouts or missing data and (ii) patients' disease status that may change at each dosing period prior to dosing.

More details regarding the potential use of complete $n$-of-1 trial design for rare diseases drug development are further discussed in Chapter 10.

**Adaptive Trial Design** – Another useful innovative trial design for rare disease clinical trials is an adaptive trial design. In its draft guidance on adaptive clinical trial design, FDA defines an adaptive design as a study that includes a *prospectively* planned opportunity for the modification of one or more specified aspects of the study design and hypotheses based on the analysis of (usually interim) data from subjects in the study (FDA, 2010). The FDA draft guidance has been served as an official document describing the potential use of adaptive designs in clinical trials since it was published in 2010. It, however, should be noted that the FDA draft guidance on adaptive clinical trial design is currently being revised in order to reflect pharmaceutical practice and FDA's current thinking.

In practice, a two-stage seamless adaptive trial design is probably the most commonly considered adaptive trial design in clinical trials. A seamless trial design is referred to as a program that addresses study objectives within a single trial that are normally achieved through separate trials in clinical development. An adaptive seamless design is a seamless trial design that would use data from patients enrolled before and after the adaptation in the final analysis. Thus, a two-stage seamless adaptive design consists of two phases (stages), namely, a learning (or exploratory) phase (stage 1) and a confirmatory phase (stage 2). The learning phase provides that opportunity for adaptations such as stop the trial early due to safety and/or futility/efficacy based on accrued data at the end of the learning phase. A two-stage seamless adaptive trial design reduces lead time between the learning (i.e., the first study for the traditional approach) and confirmatory (i.e., the second study for the traditional approach) phases. Data collected at the learning phase are combined with those data obtained at the confirmatory phase for final analysis. The specific design features of a two-stage seamless adaptive trial design can overcome the limitations and dilemma of rare disease clinical trials. Thus, a two-stage seamless adaptive trial design is not only feasible but also useful for rare disease clinical trials (Table 2.2).

In practice, two-stage seamless adaptive trial designs can be classified into the following four categories depending upon study objectives and

**TABLE 2.2**

Types of Two-Stage Seamless Adaptive Designs

| Study Objectives | Study Endpoint | |
|---|---|---|
| | Same (S) | Different (D) |
| Same (S) | I = SS | II = SD |
| Different (D) | III = DS | IV = DD |

study endpoints used at different stages. In other words, we have Category I (SS) – same study objectives and same study endpoints; Category II (SD) – same study objectives but different study endpoints; Category III (DS) – different study objectives but same study endpoints; and Category IV (DD) – different study objectives and different study endpoints. Note that different study objectives are usually referred to dose finding (selection) at the first stage and efficacy confirmation at the second stage, while different study endpoints are directed to biomarker versus clinical endpoint or the same clinical endpoint with different treatment durations.

Category I trial design is often viewed as a similar design to a group sequential design with one interim analysis although there are differences between a group sequential design and a two-stage seamless design. In this chapter, our emphasis will be placed on Category II designs. The results obtained can be similarly applied to Category III and Category IV designs with some modifications for controlling the overall type I error rate at a pre-specified level. In practice, typical examples for a two-stage adaptive seamless design include a two-stage adaptive seamless phase I/II design and a two-stage adaptive seamless phase II/III design. For the two-stage adaptive seamless phase I/II design, the objective at the first stage is for biomarker development and the study objective for the second stage is to establish early efficacy. For a two-stage adaptive seamless phase II/III design, the study objective is for treatment selection (or dose finding), while the study objective at the second stage is for efficacy confirmation.

More details regarding the potential use of two-stage adaptive seamless trial design for rare diseases drug development are provided in Chapter 11.

**Master Protocol** – Woodcock and LaVange (2017) introduced the concept of master protocol for studying multiple therapies, multiple diseases, or both in order to answer more questions in a more efficient and timely fashion (see also, Redman and Allegra, 2015). Master protocols include the following types of trials: umbrella, basket, and platform. The type of umbrella trial is to study multiple targeted therapies in the context of a single disease, while the type of basket trial is to study a single therapy in the context of multiple diseases or disease subtypes. The platform is to study multiple targeted therapies in the context of a single disease in a perpetual manner, with therapies allowed us to enter or leave the platform on the basis of decision algorithm. As indicated by Woodcock and LaVange (2017), if designed correctly, master protocols offer a number of benefits, including streamlined logistics, improved data quality, collection and sharing, as well as the potential to use innovative statistical approaches to study design and analysis. Master protocols may be a collection of sub-studies or a complex statistical design or platform for rapid learning and decision-making.

In practice, master protocol is intended for the addition or removal of drugs, arms, and study hypotheses. Thus, in practice, master protocols may or may not be adaptive, umbrella, or basket studies. Since master protocol has the ability to combine a variety of logistical, innovative, and correlative elements, it allows learning more from smaller patient populations. Thus, the concept of master protocols in conjunction with adaptive trial design described in the previous section may be useful for rare diseases clinical investigations although it has been most frequently implemented in oncology research.

A typical example for rare disease drug development is to utilize the concept of master protocol like platform trials. A platform trial is an exploratory multi-arm clinical trial evaluating one or more treatments on one or more cohorts (or populations) with an objective to screen and identify promising treatments in connection with some cohorts for further investigation (see Figure 2.1). A platform trial is typically followed by confirmative studies further investigating potential arms identified by the screening outcomes.

Figure 2.2 provides a typical platform trial for the development of treatments for Ebola. The platform trial is motivated by the need to rapidly evaluate multiple potential treatments. In other words, there are multiple treatment regimens (single agents or combinations) for treating patients

| Design type | Description |
|---|---|
| 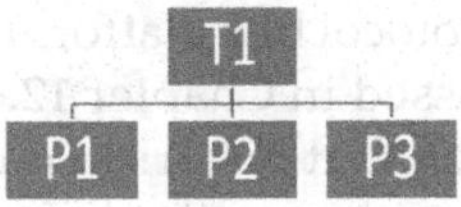 T1 — P1 P2 P3 | This design is to evaluate one new treatment for multiple cohorts or populations. |
| 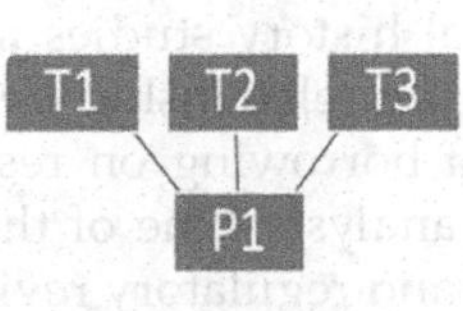 T1 T2 T3 — P1 | This design is to evaluate multiple new treatments for the same cohort or population. |
| 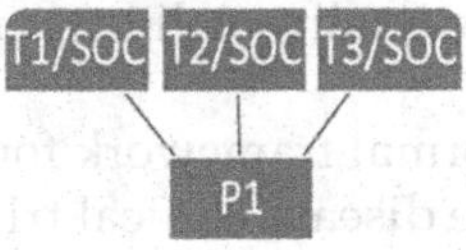 T1/SOC T2/SOC T3/SOC — P1 | This design is to evaluate one or more new treatments in combination with different standard of cares (SOCs) for the same cohort or population. |
| 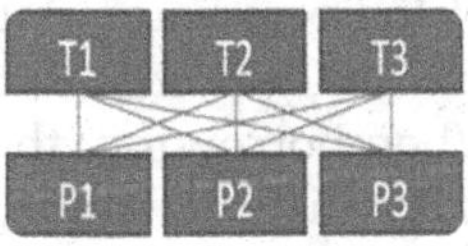 T1 T2 T3 — P1 P2 P3 | This design is to evaluate multiple new treatments for multiple cohorts or populations. |

**FIGURE 2.1**
Types of platform trials.

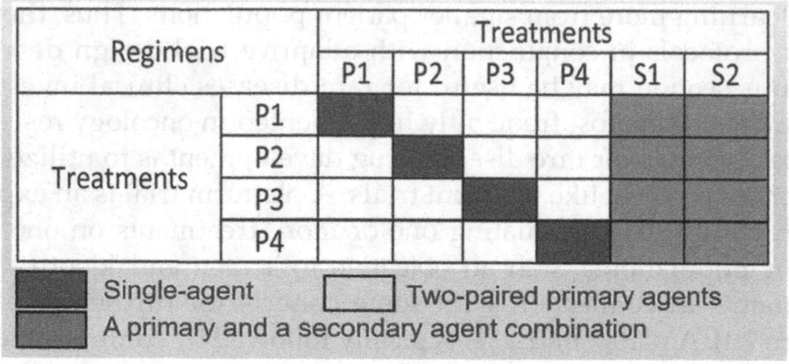

| Regimens | | Treatments | | | | | |
|----------|-----|-----|-----|-----|-----|-----|-----|
| | | P1 | P2 | P3 | P4 | S1 | S2 |
| Treatments | P1 | | | | | | |
| | P2 | | | | | | |
| | P3 | | | | | | |
| | P4 | | | | | | |

■ Single-agent          □ Two-paired primary agents
   A primary and a secondary agent combination

**FIGURE 2.2**
Ebola platform trial. (*Note*: Possible combinations assuming a hypothetical state of the trial with four primary agents (*P*) and two secondary agents (*S*).)

with Ebola virus disease, which is considered a rare disease according to the definition. Under the platform trial, it is suggested that response adaptive randomization and adaptive agent determination be used for achieving the study objectives for a rapid evaluation of multiple treatments under study.

More details regarding the use of master protocol like platform trial design for rare diseases drug development are discussed in Chapter 12.

**Bayesian Approach** – Under the assumption that historical data (e.g., previous studies or experience) are available, Bayesian methods for borrowing information from different data sources may be useful. These data sources could include, but are not limited to, natural history studies and expert's opinion regarding prior distribution about the relationship between endpoints and clinical outcomes. The impact of borrowing on results can be assessed through the conduct of sensitivity analysis. One of the key questions of particular interest to the investigator and regulatory reviewer is that how much to borrow in order to (i) achieve the desired statistical assurance for substantial evidence, and (ii) maintain the quality, validity, and integrity of the study.

Although Bayesian approach provides a formal framework for borrowing historical information, which is useful in rare disease clinical trials, borrowing can only be done under the assumption that there is a well-established relationship between patient populations (e.g., from previous studies to the current study). In practice, it is suggested not to borrow any data from previous studies whenever possible. The primary analysis should be relied on the data collected from the current study. When borrowing, the associated risk should be carefully evaluated for scientific/statistical validity of the final conclusion. It should be noted that Bayesian approach may not be feasible if there is no prior experience or study available. The determination of prior in Bayesian is always debatable because the primary assumption of the selected prior is often difficult, if not impossible, to be verified.

## 2.4 Statistical Methods for Data Analysis

Statistical methods used for data analysis should be consistent with statistical methods used for sample size calculation. Moreover, statistical methods should be able to overcome the issue of small sample size, and achieve certain statistical assurance. Statistical methods for data analysis should be innovative in order to confirm that the observed treatment effect has provided a substantial evidence to support the safety and efficacy of the test treatment under investigation.

**Analysis for Complete *n*-of-1 Trial Design** – As it can be seen from Table 2.1, the complete *n*-of-1 trial design with four dosing periods can be generally described as a *K*-sequence, *J*-period (i.e., $K \times J$) crossover design. In this section, consider the following model for comparing two treatments, i.e., a test (*T*) drug and a reference (*R*) drug:

$$Y_{ijk} = \mu + G_k + S_{ik} + P_j + D_{d_{(j,k)}} + C_{d_{(j-1,k)}} + e_{ijk}$$

$$i = 1, 2, \ldots, n_k; j = 1, 2, \ldots, J; k = 1, 2, \ldots, K; d = T \text{ or } R, \qquad (2.1)$$

where $\mu$ is the overall mean, $G_k$ is the fixed *k*th sequence effect, $S_{ik}$ is the random effect for the *i*th subject within the *k*th sequence with mean 0 and variance $\sigma_S^2$, $P_j$ is the fixed effect for the *j*th period, $D_{d_{(j,k)}}$ is the drug effect for the drug at the *k*th sequence in the *j*th period, $C_{d_{(j-1,k)}}$ is the carry-over effect, and $e_{ijk}$ is the random error with mean 0 and variance $\sigma_e^2$. Under the model, it is assumed that $S_{ik}$ and $e_{ijk}$ are mutually independent.

Under model (2.1), denote *P* as the following parameter vector:

$$(\mu, G_1, G_2, P_1, P_2, P_3, P_4, D_T, D_R, C_T, C_R)',$$

which contains all unknown parameters in the model. Thus, the moment estimator can be obtained in linear form of the observed cell means $\tilde{Y} = \beta' \bar{Y}$, where $\bar{Y}$ is the vector of observed mean vector. Then, $E(\tilde{Y}) = E(\beta' \bar{Y}) = L'P$, which is the parameter of interest based on the linear contrast *L*. Let $\omega_{jk}$ be the expected value at the *k*th sequence and *j*th period, then $\omega = X'P$ with *X* being the design matrix. Setting $E(\beta' \bar{Y}) = L'P$ implies that $\beta' \omega = L'P \Rightarrow \beta' X'P = L'P$. Thus, in order to have $E(\beta' \bar{Y}) = L'P$, we should set $\beta' X' = L' \Rightarrow L = X\beta$. Therefore, the method of moment estimation for $L'P$ is $\beta' \bar{Y}$ with $\beta = (X'X)^- X'L$. Thus, we would reject the null hypothesis if

$$T_D = \frac{\hat{D} - \theta}{\hat{\sigma}_e^2 \sqrt{\dfrac{1}{11n}}} > t\left[\frac{\alpha}{2}, 16n - 5\right].$$

The corresponding confidence interval can be obtained as follows:

$$\tilde{D} \pm t\left[\frac{\alpha}{2}, 16n - 5\right]\hat{\sigma}_e^2\sqrt{\frac{1}{11n}}.$$

More details regarding statistical tests in the presence/absence of carry-over effects under the complete $n$-of-1 trial design with $p = 4$ are provided in Chapter 10. Also included in Chapter 10 is the corresponding power analysis for sample size calculation.

**Analysis under an Adaptive Trial Design** – Statistical analysis for a two-stage seamless design (with the same study objectives and same study end-points at different stages) is similar to that of a group sequential design with one interim analysis. Thus, standard statistical methods for a group sequential design can be applied. For other kinds of two-stage seamless trial designs, standard statistical methods for a group sequential design are not appropriate and hence should not be applied directly. In this section, statistical methods for other types of two-stage adaptive seamless design are described. For illustration purpose, in this section, we will discuss $n$-stage adaptive design based on individual $p$-values from each stage (Chow and Chang, 2011).

Consider a clinical trial with $K$ interim analyses. The final analysis is treated as the $K$th interim analysis. Suppose that at each interim analysis, a hypothesis test is performed followed by some actions that are dependent on the analysis results. Such actions could be an early stopping due to futility/efficacy or safety, sample size re-estimation, modification of randomization, or other adaptations. In this setting, the objective of the trial can be formulated using a global hypothesis test, which is an intersection of the individual hypothesis tests from the interim analyses

$$H_0 : H_{01} \cap \ldots \cap H_{0K}, \tag{2.2}$$

where $H_{0i}, i = 1, \ldots, K$ is the null hypothesis to be tested at the $i$th interim analysis. Note that there are some restrictions on $H_{0i}$; i.e., rejection of any $H_{0i}, i = 1, \ldots, K$ will lead to the same clinical implication (e.g., drug is efficacious); hence, all $H_{0i}, i = 1, \ldots, K$ are constructed for testing the *same* endpoint within a trial. Otherwise, the global hypothesis cannot be interpreted.

In practice, $H_{0i}$ is tested based on a sub-sample from each stage, and without the loss of generality, assume that $H_{0i}$ is a test for the efficacy of a test treatment under investigation, which can be written as

$$H_{0i} : \eta_{i1} \geq \eta_{i2} \quad \text{versus} \quad H_{ai} : \eta_{i1} < \eta_{i2},$$

where $\eta_{i1}$ and $\eta_{i2}$ are the responses of the two treatment groups at the $i$th stage. It is often the case that when $\eta_{i1} = \eta_{i2}$, the $p$-value $p_i$ for the sub-sample at the $i$th stage is uniformly distributed on $[0, 1]$ under $H_0$ (Bauer and Kohne, 1994). This desirable property can be used to construct a test statistic for multiple-stage seamless adaptive designs. As an example, Bauer and Kohne (1994) used Fisher's combination of the $p$-values. Similarly, Chang (2007) considered a linear combination of the $p$-values as follows:

$$T_k = \sum_{i=1}^{K} w_{ki} p_i, \, i = 1, \dots, K, \tag{2.3}$$

where $w_{ki} > 0$ and $K$ is the number of analyses planned in the trial. For simplicity, consider the case where $w_{ki} = 1$. This leads to

$$T_k = \sum_{i=1}^{K} p_i, \, i = 1, \dots, K. \tag{2.4}$$

The test statistic $T_k$ can be viewed as cumulative evidence against $H_0$. The smaller the $T_k$ is, the stronger the evidence is. Equivalently, we can define the test statistic as $T_k = \sum_{i=1}^{K} p_i / K$, which can be viewed as an average of the evidence against $H_0$. The stopping rules are given by

$$\begin{cases} \text{Stop for efficacy} & \text{if } T_k \leq \alpha_k \\ \text{Stop for futility} & \text{if } T_k \geq \beta_k \, , \\ \text{Continue} & \text{otherwise} \end{cases} \tag{2.5}$$

where $T_k, \alpha_k,$ and $\beta_k$ are monotonic increasing functions of $k$, $\alpha_k < \beta_k$, $k = 1, \dots, K-1,$ and $\alpha_K = \beta_K$. Note that $\alpha_k$ and $\beta_k$ are referred to as the efficacy and futility boundaries, respectively. To reach the $k$th stage, a trial has to pass 1 to $(k-1)$th stages. Therefore, a so-called proceeding probability can be defined as the following unconditional probability:

$$\psi_k(t) = P\left(T_k < t, \alpha_1 < T_1 < \beta_1, \dots, \alpha_{k-1} < T_{k-1} < \beta_{k-1}\right)$$

$$= \int_{\alpha_1}^{\beta_1} \dots \int_{\alpha_{k-1}}^{\beta_{k-1}} \int_{-\infty}^{t} f_{T_1 \dots T_k}(t_1, \dots, t_k) \, dt_k \, dt_{k-1} \dots dt_1 \tag{2.6}$$

where $t \geq 0, t_i, i = 1, \dots, k$ is the test statistic at the $i$th stage, and $f_{T_1 \dots T_k}$ is the joint probability density function. The error rate at the $k$th stage is given by

$$\pi_k = \psi_k(\alpha_k). \tag{2.7}$$

When efficacy is claimed at a certain stage, the trial is stopped. Therefore, the type I error rates at different stages are mutually exclusive. Hence, the experiment-wise type I error rate can be written as follows:

$$\alpha = \sum_{k=1}^{K} \pi_k. \tag{2.8}$$

Note that the above method was derived based on individual p-values obtained at different stages under the assumption that the study objectives and study endpoints at different stages are the same. For other types of two-stage adaptive seamless trial designs, the above methods need to be modified in order to account for different study objectives and/or different study endpoints at different stages. More details regarding statistical methods for the analysis of other types of adaptive trial designs are discussed in Chapter 11.

## 2.5 Evaluation of Rare Disease Clinical Trials

Due to the small sample size in rare disease clinical trials, the conclusion drawn may not achieve a desired level of statistical inference (e.g., power or confidence interval). In this case, it is suggested that the following methods be considered for the evaluation of the rare disease clinical trial to determine whether substantial evidence of safety and efficacy has been achieved. Let $n_1$, $n_2$, and $N$ be the sample size of the intended trial at interim, sample size of the data borrowed from previous studies, and sample size required for achieving a desired power (say 80%), respectively.

**Predictive Confidence Interval (PCI)** – Let $\bar{T}_i$ and $\bar{R}_i$ be the sample mean of the $i$th sample for the test product and the reference product, respectively. Also, let $\hat{\sigma}_1$, $\hat{\sigma}_2$, and $\hat{\sigma}^{\cdot}$ be the pooled sample standard deviation of difference in sample mean between the test product and the reference product based on the first sample ($n_1$), the second sample ($n_1 + n_2$), and the third sample ($N$), respectively. Under a parallel design, the usual confidence interval of the treatment effect can be obtained based on the $i$th sample and $j$th sample as follows:

$$CI_i = \bar{T}_i - \bar{R}_i \pm z_{1-\alpha}\hat{\sigma}_i,$$

where $i = 1, 2,$ and $N$. In practice, for rare disease clinical trials, we can compare these confidence intervals in terms of their relative efficiency for a complete clinical picture. Relative efficiency of $CI_i$ as compared to $CI_j$ is defined as

$$R_{ij} = \hat{\sigma}_i / \hat{\sigma}_j,$$

where $i$ and $j$ represent the $i$th sample and $j$th sample, respectively.

**Probability of Reproducibility** – Although there will not be sufficient power due to the small sample size available in rare diseases clinical trials, alternatively, we may consider empirical power based on the observed treatment effect and the variability associated with the observed difference adjusted for the sample size required for achieving the desired power (i.e., $N$). The empirical power is also known as reproducibility probability of the clinical results for future studies if the studies shall be conducted under the similar experimental conditions. Shao and Chow (2002) studied how to evaluate the reproducibility probability using this approach under several study designs for comparing means with both equal and unequal variances.

When the reproducibility probability is used to provide substantial evidence of the effectiveness of a drug product, the estimated power approach may produce an optimistic result. Alternatively, Shao and Chow (2002) suggested that the reproducibility probability be defined as a lower confidence bound of the power of the second trial. The reproducibility probability can be used to determine the clinical results observed from the rare disease clinical trial that has provided substantial evidence for the evaluation of safety and efficacy of the test treatment under investigation.

In addition, they also suggested a more sensible definition of reproducibility probability using the Bayesian approach. Under the Bayesian approach, the unknown parameter $\theta$ is a random vector with a prior distribution, say $\pi(\theta)$, which is assumed known. Thus, the reproducibility probability can be defined as the conditional probability of $|T| > C$ in the future trial, given the data set $x$ observed from the previous trial(s), i.e.,

$$P\{|T| > C \mid x\} = \int P(|T| > C \mid \theta)\pi(\theta \mid x)d\theta,$$

where $T = T(y)$ is based on the data set $y$ from the future trial and $\pi(\theta \mid x)$ is the posterior density of $\theta$, given $x$.

To study the reproducibility probability, we need to specify the test procedure, i.e., the form of the test statistic $T$. For simplicity, consider two-samples design with equal variances. Suppose that a total of $n = n_1 + n_2$ patients are randomly assigned to two groups, a treatment group and a control group. In the treatment group, $n_1$ patients receive the treatment (or a test drug) and produce responses $x_{11}, \dots, x_{1n}$. In the control group, $n_2$ patients receive the placebo (or a reference drug) and produce responses $x_{21}, \dots, x_{2n_2}$. This design is a typical two-group parallel design in clinical trials. We assume that $x_{ij}$'s are independent and normally distributed with means $\mu_i, i = 1, 2$ and a common variance $\sigma^2$. Suppose that the hypotheses of interest are

$$H_0: \mu_1 - \mu_2 = 0 \text{ versus } H_a: \mu_1 - \mu_2 \neq 0. \tag{2.9}$$

The discussion for a one-sided $H_a$ is similar.

Consider the commonly used two-sample $t$-test, which rejects $H_0$ if and only if $|T| > t_{0.975,n-2}$, where $t_{0.975,n-2}$ is the 97.5th percentile of the $t$-distribution with $n - 2$ degrees of freedom

$$T = \frac{\bar{x}_1 - \bar{x}_2}{\sqrt{\frac{(n_1 - 1)s_1{}^2 + (n_2 - 1)s_2{}^2}{n - 2}}\sqrt{\left(\frac{1}{n_1} + \frac{1}{n_2}\right)}} \tag{2.10}$$

and $\bar{x}_i$ and $s_i{}^2$ are, respectively, the sample mean and variance based on the data from the $i$th treatment group. The power of $T$ for the second trial is

$$p(\theta) = P\left(|T(y)| > t_{0.975,n-2}\right)$$
$$= 1 - \Im_{n-2}\left(t_{0.975,n-2}|\theta\right) + \Im_{n-2}\left(-t_{0.975,n-2}|\theta\right) \tag{2.11}$$

where

$$\theta = \frac{\mu_1 - \mu_2}{\sigma\sqrt{\frac{1}{n_1} + \frac{1}{n_2}}} \tag{2.12}$$

and $\Im_{n-2}(\bullet|\theta)$ denotes the distribution function of the noncentral $t$-distribution with $n - 2$ degrees of freedom and the noncentrality parameter $\theta$. Note that $p(\theta) = p(|\theta|)$.

Values of $p(\theta)$ as a function of $|\theta|$ are provided in Table 2.3. Using the idea of replacing $\theta$ by its estimate $T(x)$, where $T$ is defined by (2.10), we obtain the following reproducibility probability:

$$\hat{P} = 1 - \Im_{n-2}\left(t_{0.975,n-2}|T(x)\right) + \Im_{n-2}\left(-t_{0.975,n-2}|T(x)\right), \tag{2.13}$$

which is a function of $|T(x)|$. When $|T(x)| > t_{0.975,n-2}$,

$$\hat{P} \approx \begin{cases} 1 - \Im_{n-2}\left(t_{0.975,n-2}|T(x)\right) & \text{if } T(x) > 0 \\ \Im_{n-2}\left(-t_{0.975,n-2}|T(x)\right) & \text{if } T(x) < 0 \end{cases}. \tag{2.14}$$

If $\Im_{n-2}$ is replaced by the normal distribution and $t_{0.975,n-2}$ is replaced by the normal percentile, then formula (25.8) is the same as that in Goodman (1992) who studied the case where the variance $\sigma^2$ is known. Table 2.4 can be used to find the reproducibility probability $\hat{P}$ with a fixed sample size $n$. For example, if $T(x) = 2.9$ was observed in a clinical trial with $n = n_1 + n_2 = 40$, then the reproducibility probability is 0.807. If $T(x) = 2.9$ was observed in a clinical trial with $n = 36$, then an extrapolation of the results in Table 2.3 (for $n = 30$ and 40) leads to a reproducibility probability of 0.803.

For the case where the two samples have unequal variances, $x_{ij}$'s are independently distributed as $N(\mu_i, \sigma^2_i), i = 1, 2$. When $\sigma^2_1 \neq \sigma^2_2$, there exists no

**TABLE 2.3**

Values of the Power Function $p(\theta)$ in (2.11)

| $|\theta|$ | 10 | 20 | Total 30 | Sample 40 | Size 50 | 60 | 100 | ∞ |
|---|---|---|---|---|---|---|---|---|
| 1.96 | 0.407 | 0.458 | 0.473 | 0.480 | 0.484 | 0.487 | 0.492 | 0.500 |
| 2.02 | 0.429 | 0.481 | 0.496 | 0.504 | 0.508 | 0.511 | 0.516 | 0.524 |
| 2.08 | 0.448 | 0.503 | 0.519 | 0.527 | 0.531 | 0.534 | 0.540 | 0.548 |
| 2.14 | 0.469 | 0.526 | 0.542 | 0.550 | 0.555 | 0.557 | 0.563 | 0.571 |
| 2.20 | 0.490 | 0.549 | 0.565 | 0.573 | 0.578 | 0.581 | 0.586 | 0.594 |
| 2.26 | 0.511 | 0.571 | 0.588 | 0.596 | 0.601 | 0.604 | 0.609 | 0.618 |
| 2.32 | 0.532 | 0.593 | 0.610 | 0.618 | 0.623 | 0.626 | 0.632 | 0.640 |
| 2.38 | 0.552 | 0.615 | 0.632 | 0.640 | 0.645 | 0.648 | 0.654 | 0.662 |
| 2.44 | 0.573 | 0.636 | 0.654 | 0.662 | 0.667 | 0.670 | 0.676 | 0.684 |
| 2.50 | 0.593 | 0.657 | 0.675 | 0.683 | 0.688 | 0.691 | 0.697 | 0.705 |
| 2.56 | 0.613 | 0.678 | 0.695 | 0.704 | 0.708 | 0.711 | 0.717 | 0.725 |
| 2.62 | 0.632 | 0.698 | 0.715 | 0.724 | 0.728 | 0.731 | 0.737 | 0.745 |
| 2.68 | 0.652 | 0.717 | 0.735 | 0.743 | 0.747 | 0.750 | 0.756 | 0.764 |
| 2.74 | 0.671 | 0.736 | 0.753 | 0.761 | 0.766 | 0.769 | 0.774 | 0.782 |
| 2.80 | 0.690 | 0.754 | 0.771 | 0.779 | 0.783 | 0.786 | 0.792 | 0.799 |
| 2.86 | 0.708 | 0.772 | 0.788 | 0.796 | 0.800 | 0.803 | 0.808 | 0.815 |
| 2.92 | 0.725 | 0.789 | 0.805 | 0.812 | 0.816 | 0.819 | 0.824 | 0.830 |
| 2.98 | 0.742 | 0.805 | 0.820 | 0.827 | 0.831 | 0.834 | 0.839 | 0.845 |
| 3.04 | 0.759 | 0.820 | 0.835 | 0.842 | 0.846 | 0.848 | 0.853 | 0.860 |
| 3.10 | 0.775 | 0.834 | 0.849 | 0.856 | 0.859 | 0.862 | 0.866 | 0.872 |
| 3.16 | 0.790 | 0.848 | 0.862 | 0.868 | 0.872 | 0.874 | 0.879 | 0.884 |
| 3.22 | 0.805 | 0.861 | 0.874 | 0.881 | 0.884 | 0.886 | 0.890 | 0.895 |
| 3.28 | 0.819 | 0.873 | 0.886 | 0.892 | 0.895 | 0.897 | 0.901 | 0.906 |
| 3.34 | 0.832 | 0.884 | 0.897 | 0.902 | 0.905 | 0.907 | 0.911 | 0.916 |
| 3.40 | 0.844 | 0.895 | 0.907 | 0.912 | 0.915 | 0.917 | 0.920 | 0.925 |
| 3.46 | 0.856 | 0.905 | 0.916 | 0.921 | 0.924 | 0.925 | 0.929 | 0.932 |
| 3.52 | 0.868 | 0.914 | 0.925 | 0.929 | 0.932 | 0.933 | 0.936 | 0.940 |
| 3.58 | 0.879 | 0.923 | 0.933 | 0.937 | 0.939 | 0.941 | 0.943 | 0.947 |
| 3.64 | 0.889 | 0.931 | 0.940 | 0.944 | 0.946 | 0.947 | 0.950 | 0.953 |
| 3.70 | 0.898 | 0.938 | 0.946 | 0.950 | 0.952 | 0.953 | 0.956 | 0.959 |
| 3.76 | 0.907 | 0.944 | 0.952 | 0.956 | 0.958 | 0.959 | 0.961 | 0.965 |
| 3.82 | 0.915 | 0.950 | 0.958 | 0.961 | 0.963 | 0.964 | 0.966 | 0.969 |
| 3.88 | 0.923 | 0.956 | 0.963 | 0.966 | 0.967 | 0.968 | 0.970 | 0.973 |
| 3.94 | 0.930 | 0.961 | 0.967 | 0.970 | 0.971 | 0.972 | 0.974 | 0.977 |

*Source:* Shao and Chow (2002).

**TABLE 2.4**

95% Lower Confidence Bound $|\hat{\theta}|_-$

|          |      |      | Total | Sample | Size |      |      |      |
| -------- | ---- | ---- | ----- | ------ | ---- | ---- | ---- | ---- |
| $|T(x)|$ | 10   | 20   | 30    | 40     | 50   | 60   | 100  | ∞    |
| 4.5      | 1.51 | 2.01 | 2.18  | 2.26   | 2.32 | 2.35 | 2.42 | 2.54 |
| 4.6      | 1.57 | 2.09 | 2.26  | 2.35   | 2.41 | 2.44 | 2.52 | 2.64 |
| 4.7      | 1.64 | 2.17 | 2.35  | 2.44   | 2.50 | 2.54 | 2.61 | 2.74 |
| 4.8      | 1.70 | 2.25 | 2.43  | 2.53   | 2.59 | 2.63 | 2.71 | 2.84 |
| 4.9      | 1.76 | 2.33 | 2.52  | 2.62   | 2.68 | 2.72 | 2.80 | 2.94 |
| 5.0      | 1.83 | 2.41 | 2.60  | 2.71   | 2.77 | 2.81 | 2.90 | 3.04 |
| 5.1      | 1.89 | 2.48 | 2.69  | 2.80   | 2.86 | 2.91 | 2.99 | 3.14 |
| 5.2      | 1.95 | 2.56 | 2.77  | 2.88   | 2.95 | 3.00 | 3.09 | 3.24 |
| 5.3      | 2.02 | 2.64 | 2.86  | 2.97   | 3.04 | 3.09 | 3.18 | 3.34 |
| 5.4      | 2.08 | 2.72 | 2.95  | 3.06   | 3.13 | 3.18 | 3.28 | 3.44 |
| 5.5      | 2.14 | 2.80 | 3.03  | 3.15   | 3.22 | 3.27 | 3.37 | 3.54 |
| 5.6      | 2.20 | 2.88 | 3.11  | 3.24   | 3.31 | 3.36 | 3.47 | 3.64 |
| 5.7      | 2.26 | 2.95 | 3.20  | 3.32   | 3.40 | 3.45 | 3.56 | 3.74 |
| 5.8      | 2.32 | 3.03 | 3.28  | 3.41   | 3.49 | 3.55 | 3.66 | 3.84 |
| 5.9      | 2.39 | 3.11 | 3.37  | 3.50   | 3.58 | 3.64 | 3.75 | 3.94 |
| 6.0      | 2.45 | 3.19 | 3.45  | 3.59   | 3.67 | 3.73 | 3.85 | 4.04 |
| 6.1      | 2.51 | 3.26 | 3.53  | 3.67   | 3.76 | 3.82 | 3.94 | 4.14 |
| 6.2      | 2.57 | 3.34 | 3.62  | 3.76   | 3.85 | 3.91 | 4.03 | 4.24 |
| 6.3      | 2.63 | 3.42 | 3.70  | 3.85   | 3.94 | 4.00 | 4.13 | 4.34 |
| 6.4      | 2.69 | 3.49 | 3.78  | 3.93   | 4.03 | 4.09 | 4.22 | 4.44 |
| 6.5      | 2.75 | 3.57 | 3.86  | 4.02   | 4.12 | 4.18 | 4.32 | 4.54 |

*Source:*  Shao and chow (2002).

exact testing procedure for the hypotheses in (25.3). When both $n_1$ and $n_2$ are large, an approximate 5% level test rejects $H_0$ when $|T| > z_{0.975}$, where

$$T = \frac{\bar{x}_1 - \bar{x}_2}{\sqrt{\dfrac{s_1^2}{n_1} + \dfrac{s_2^2}{n_2}}}. \tag{2.15}$$

Since $T$ is approximately distributed as $N(\theta, 1)$ with

$$\theta = \frac{\mu_1 - \mu_2}{\sqrt{\dfrac{\sigma_1^2}{n_2} + \dfrac{\sigma_2^2}{n_2}}}, \tag{2.16}$$

the reproducibility probability obtained by using the estimated power approach is given by

$$\hat{P} = \Phi(T(x) - Z_{0.975}) + \Phi(-T(x) - Z_{0.975}). \qquad (2.17)$$

When the variances under different treatments are different and the sample sizes are not large, a different study design, such as a matched-pair parallel design or a $2 \times 2$ crossover design, is recommended. A matched-pair parallel design involves $m$ pairs of matched patients. One patient in each pair is assigned to the treatment group, and the other is assigned to the control group. Let $x_{ij}$ be the observation from the $j$th pair and the $i$th group. It is assumed that the differences $x_{1j} - x_{2j}, j = 1, \ldots, m$ are independent and identically distributed as $N(\mu_1 - \mu_2, \sigma^2_D)$. Then, the null hypothesis $H_0$ is rejected at the 5% level of significance if $|T| > t_{0.975, m-1}$, where

$$T = \frac{\sqrt{m}\,(\bar{x}_1 - \bar{x}_2)}{\hat{\sigma}^2_D} \qquad (2.18)$$

and $\hat{\sigma}^2_D$ is the sample variance based on the differences $x_{1j} - x_{2j}, j = 1, \ldots, m$. Note that $T$ has the noncentral $t$-distribution with $m - 1$ degrees of freedom and the noncentrality parameter

$$\theta = \frac{\sqrt{m}\,(\mu_1 - \mu_2)}{\sigma^2_D}. \qquad (2.19)$$

Consequently, the reproducibility probability obtained by using the estimated power approach is given by (2.13) with $T$ defined by (2.18) and $n - 2$ replaced by $m - 1$.

Suppose that the study design is a $2 \times 2$ crossover design in which $n_1$ patients receive the treatment at the first period and the placebo at the second period, and $n_2$ patients receive the placebo at the first period and the treatment at the second period. Let $x_{ij}$ be the normally distributed observation from the $j$th patient at the $i$th period and $l$th sequence. Then, the treatment effect $\mu_D$ can be unbiasedly estimated by

$$\hat{\mu}_D = \frac{\bar{x}_{11} - \bar{x}_{12} - \bar{x}_{21} + \bar{x}_{22}}{2} \sim N\left(\mu_D, \frac{\sigma^2_D}{4}\left(\frac{1}{n_1} + \frac{1}{n_2}\right)\right),$$

where $\bar{x}_{ij}$ is the sample mean based on $x_{ij}, j = 1, \ldots, n_1$ and $\sigma^2_D = \text{var}(x_{l1j} - x_{l2j})$ An unbiased estimator of $\sigma^2_D$ is

$$\hat{\sigma}^2_D = \frac{1}{n_1 + n_2 - 2} \sum_{l=1}^{2} \sum_{j=1}^{m} (x_{l1j} - x_{l2j} - \bar{x}_{l1} + \bar{x}_{l2})^2,$$

which is independent of $\hat{\mu}_D$ and distributed as $\sigma_D^2/(n_1 + n_2 - 2)$ times the chi-square distribution with $n_1 + n_2 - 2$ degrees of freedom. Thus, the null hypothesis $H_0: \mu_D = 0$ is rejected at the 5% level of significance if $|T| > t_{0.975,n-2}$, where $n = n_1 + n_2$ and

$$T = \frac{\hat{\mu}_D}{\frac{\hat{\sigma}_D}{2}\sqrt{\left(\frac{1}{n_1} + \frac{1}{n_2}\right)}}. \tag{2.20}$$

Note that $T$ has the noncentral $t$-distribution with $n - 2$ degrees of freedom and the noncentrality parameter

$$\theta = \frac{\mu_D}{\frac{\sigma_D}{2}\sqrt{\left(\frac{1}{n_1} + \frac{1}{n_2}\right)}}. \tag{2.21}$$

Consequently, the reproducibility probability obtained by using the estimated power approach is given by (2.13) with $T$ defined by (2.20).

Other methods such as confidence interval approach and Bayesian approach for the assessment of the probability of reproducibility can be found in Shao and Chow (2002). Moreover, similar idea can be applied to assess the generalizability from one patient population (adults) to another (e.g., pediatrics or elderly). More details regarding the probability of reproducibility and generalizability can be found in Shao and Chow (2002).

## 2.6 Concluding Remarks

As discussed, for rare disease drug development, power analysis for sample size calculation may not be feasible due to the fact that there is a small patient population. FDA draft guidance emphasizes that the same standards for regulatory approval will be applied to rare diseases drug development despite the small patient population. Thus, often there is an insufficient power for rare disease drug clinical investigation. In this case, alternatively, it is suggested that sample size calculation or justification should be performed based on precision analysis, reproducibility analysis, or probability monitoring approach for achieving certain statistical assurance.

In practice, it is a dilemma for having the same standard with less subjects in rare disease drug development. Thus, it is suggested that innovative design and statistical methods should be considered and implemented for obtaining substantial evidence regarding effectiveness and safety in support of regulatory approval of rare disease drug products. In this chapter, several innovative trial designs such as complete $n$-of-1 trial design, adaptive seamless trial design, trial design utilizing the concept of master protocols, and Bayesian trial design are introduced. The corresponding statistical methods and sample size requirement under the respective study designs are derived. These study designs are useful in speeding up rare disease development process and identifying any signal, pattern or trend, and/or optimal clinical benefits of the rare disease drug products under investigation.

Due to the small patient population in rare disease clinical development, the concept of generalizability probability can be used to determine whether the clinical results can be generalized from the targeted patient population (e.g., adults) to a different but similar patient population (e.g., pediatrics or elderly) with the same rare disease. In practice, the generalizability probability can be evaluated through the assessment of sensitivity index between the targeted patient population and the different patient population (Lu et al., 2017). The degree of generalizability probability can then be used to judge whether the intended trial has provided substantial evidence regarding the effectiveness and safety for the different patient population (e.g., pediatrics or elderly).

In practice, although an innovative and yet complex trial design may be useful in rare disease drug development, it may introduce operational bias to the trial and consequently increase the probability of making errors. It is then suggested that quality, validity, and integrity of the intended trial utilizing an innovative trial design should be maintained.

# 3

# *Hypotheses Testing for Clinical Evaluation*

## 3.1 Introduction

In clinical trials, a typical approach for clinical evaluation of the safety and efficacy of a test treatment under investigation is to first test for the null hypothesis of no treatment difference in efficacy based on clinical data collected under a valid trial design. The investigator would reject the null hypothesis of no treatment difference and then conclude the alternative hypothesis that there is a difference in favor of the test treatment under investigation. As a result, if there is a sufficient power for correctly detecting a clinically meaningful difference if such a different truly exists, we claim that the test treatment is efficacious. The test treatment will be reviewed and approved by the regulatory agency if the recommended dose is well tolerated and there appear no safety concerns. In some cases, the regulatory agencies like FDA may issue an approvable letter pending a commitment for conducting a large scale of long-term safety surveillance trial.

In practice, the intended clinical trial is always powered for achieving the study objective with a desired power (say 80%) at a pre-specified level of significance (say 5%). However, powered the study based on a single primary endpoint (usually efficacy endpoint) may not be appropriate because one single primary efficacy endpoint may not be able to fully describe the performance of the treatment with respect to both efficacy and safety under study. Statistically, the traditional approach based on the single primary efficacy endpoint for the clinical evaluation of both safety and efficacy is a conditional approach (i.e., conditional on safety performance). It should be noted that under the traditional (conditional) approach, the observed safety profile may not be of any statistical meaning (i.e., the observed safety profile could be by chance alone and is not reproducible) and hence misleading. Thus, the traditional approach for the clinical evaluation of both efficacy and safety may have inflated the false-positive rate of the test treatment in treating the disease under investigation.

In the past several decades, the traditional approach is found to be inefficient because many drug products have been withdrawn from the market because of the risks to patients. Table 3.1 (reproduced from http://en.wikipedia.org/wiki/List_of_withdrawn_drugs) provides a list of (significant) withdrawn drugs between 1950 and 2010. As it can be seen from Table 3.1, most drugs withdrawn from the marketplace are due to safety concern (risks to the patients). Usually, this has been prompted by unexpected adverse effects that were not detected during phase III clinical trials and were only apparent from

**TABLE 3.1**

Significant Withdrawals of Drug Products between 1950 and 2010

| Drug Name | Withdrawn | Remarks |
|---|---|---|
| Thalidomide | 1950s–1960s | Withdrawn because of risk of teratogenicity; returned to market for use in leprosy and multiple myeloma under FDA orphan drug rules |
| Lysergic acid diethylamide (LSD) | 1950s–1960s | Marketed as a psychiatric cure-all; withdrawn after it became widely used recreationally |
| Diethylstilbestrol | 1970s | Withdrawn because of risk of teratogenicity |
| Phenformin and buformin | 1978 | Withdrawn because of risk of lactic acidosis |
| Ticrynafen | 1982 | Withdrawn because of risk of hepatitis |
| Zimelidine | 1983 | Withdrawn worldwide because of risk of Guillain–Barré syndrome |
| Phenacetin | 1983 | An ingredient in "A.P.C." tablet; withdrawn because of risk of cancer and kidney disease |
| Methaqualone | 1984 | Withdrawn because of risk of addiction and overdose |
| Nomifensine (Merital) | 1986 | Withdrawn because of risk of hemolytic anemia |
| Triazolam | 1991 | Withdrawn in the United Kingdom because of risk of psychiatric adverse drug reactions. This drug continues to be available in the United States. |
| Temafloxacin | 1992 | Withdrawn in the United States because of allergic reactions and cases of hemolytic anemia, leading to death of three patients |
| Flosequinan (Manoplax) | 1993 | Withdrawn in the United States because of an increased risk of hospitalization or death |
| Alpidem (Ananxyl) | 1996 | Withdrawn because of rare but serious hepatotoxicity |
| Fen-phen (popular combination of fenfluramine and phentermine) | 1997 | Phentermine remains on the market, dexfenfluramine and fenfluramine – later withdrawn because of heart valve disorder |
| Tolrestat (Alredase) | 1997 | Withdrawn because of risk of severe hepatotoxicity |
| Terfenadine (Seldane) | 1998 | Withdrawn because of risk of cardiac arrhythmias; superseded by fexofenadine |

*(Continued)*

**TABLE 3.1** (*Continued*)

Significant Withdrawals of Drug Products between 1950 and 2010

| Drug Name | Withdrawn | Remarks |
|---|---|---|
| Mibefradil (Posicor) | 1998 | Withdrawn because of dangerous interactions with other drugs |
| Etretinate | 1990s | Risk of birth defects; narrow therapeutic index |
| Temazepam (Restoril, Euhypnos, Normison, Remestan, Tenox, Norkotral) | 1999 | Withdrawn in Sweden and Norway because of diversion, abuse, and a relatively high rate of overdose deaths in comparison with other drugs of its group. This drug continues to be available in most of the world, including the United States, but under strict controls |
| Astemizole (Hismanal) | 1999 | Arrhythmias because of interactions with other drugs |
| Troglitazone (Rezulin) | 2000 | Withdrawn because of risk of hepatotoxicity; superseded by pioglitazone and rosiglitazone |
| Alosetron (Lotronex) | 2000 | Withdrawn because of risk of fatal complications of constipation; reintroduced 2002 on a restricted basis |
| Cisapride (Propulsid) | 2000s | Withdrawn in many countries because of risk of cardiac arrhythmias |
| Amineptine (Survector) | 2000 | Withdrawn because of hepatotoxicity, dermatological side effects, and abuse potential |
| Phenylpropanolamine (Propagest, Dexatrim) | 2000 | Withdrawn because of risk of stroke in women under 50 years of age when taken at high doses (75 mg twice daily) for weight loss |
| Trovafloxacin (Trovan) | 2001 | Withdrawn because of risk of liver failure |
| Cerivastatin (Baycol, Lipobay) | 2001 | Withdrawn because of risk of rhabdomyolysis |
| Rapacuronium (Raplon) | 2001 | Withdrawn in many countries because of risk of fatal bronchospasm |
| Rofecoxib (Vioxx) | 2004 | Withdrawn because of risk of myocardial infarction |
| Mixed amphetamine salts (Adderall XR) | 2005 | Withdrawn in Canada because of risk of stroke. See Health Canada press release. The ban was later lifted because the death rate among those taking Adderall XR was determined to be no greater than those not taking Adderall |
| Hydromorphone extended release (Palladone) | 2005 | Withdrawn because of a high risk of accidental overdose when administered with alcohol |
| Pemoline (Cylert) | 2005 | Withdrawn from US market because of hepatotoxicity |
| Natalizumab (Tysabri) | 2005–2006 | Voluntarily withdrawn from US market because of risk of progressive multifocal leukoencephalopathy (PML). Returned to market July 2006 |

(*Continued*)

**TABLE 3.1 (*Continued*)**

Significant Withdrawals of Drug Products between 1950 and 2010

| Drug Name | Withdrawn | Remarks |
|---|---|---|
| Ximelagatran (Exanta) | 2006 | Withdrawn because of risk of hepatotoxicity (liver damage) |
| Pergolide (Permax) | 2007 | Voluntarily withdrawn in the United States because of the risk of heart valve damage. Still available elsewhere |
| Tegaserod (Zelnorm) | 2007 | Withdrawn because of imbalance of cardiovascular ischemic events, including heart attack and stroke. Was available through a restricted access program until April 2008 |
| Aprotinin (Trasylol) | 2007 | Withdrawn because of increased risk of complications or death; permanently withdrawn in 2008 except for research use |
| Lumiracoxib | 2007–2008 | Progressively withdrawn around the world because of serious side effects, mainly liver damage |
| Rimonabant (Acomplia) | 2008 | Withdrawn around the world because of risk of severe depression and suicide |
| Efalizumab (Raptiva) | 2009 | Withdrawn because of increased risk of PML; to be completely withdrawn from market by June 2009 |
| Sibutramine (Reductil) | 2010 | Withdrawn in Europe because of increased cardiovascular risk |

*Source:* Wikipedia (2010). List of withdrawn drugs. http://en.wikipedia.org/wiki/List_of_withdrawn_drugs.

post-marketing surveillance data from the wider patient population. The list of withdrawn drugs given in Table 3.1 were approved by the regulatory agencies such as FDA and EMEA (European Medicines Evaluation Agency). Note that some of the drug products were approved to be marketed in Europe but had not yet been approved by the FDA for marketing in the United States.

In addition to drug withdrawals, drug products may be recalled due to the lack of good drug characteristics such as quality and stability. As an example, Table 3.2 summarizes the number of prescription and over-the-counter drugs got recalled between the fiscal years of 2004 and 2005. Most of the drug recalls are due to or related to safety issues although some of the causes for

**TABLE 3.2**

Summary of Drug Recalls between 2004 and 2005

| Fiscal Year | Prescription Drug | Over-the-Counter Drug |
|---|---|---|
| 2004 | 215 | 71 |
| 2005 | 401 | 101 |

*Source:* Report to the Nation issued by CDER/FDA.

recalls are due to failing to pass FDA inspection for stability testing and/or dissolution testing, which have an impact on the safety of the drug products currently on the marketplace. Thus, one of the controversial issues is that whether the traditional (conditional) hypotheses testing approach (based on efficacy alone) for the evaluation of the safety and efficacy of a test treatment under investigation is appropriate.

In clinical trials, clinical results are often reported by rounding up the number to certain decimal places. Statistical inference obtained based on data with different decimal places may lead to different conclusions. Therefore, the selection of the number of decimal places could be critical if the treatment effect is of marginally significance. Thus, how many decimal places should be used for reporting the clinical results has become an interesting question to the investigators who conduct clinical trials at various phases of clinical development. Chow (2000, 2011) introduced the concept of signal noise for determining the number of decimal places for the results obtained from the clinical trials. The idea is to select the minimum number of decimal places in such a way that there is no statistically significant difference between the data set presented by using the minimum decimal places and any other data sets with more decimal places.

In the next section, several composite hypotheses that will take both efficacy and safety into consideration are proposed. General concepts and principles for $\alpha$-adjustment for multiple comparisons for controlling the overall type I error rate are discussed in Section 3.3. In Section 3.4, for illustration purpose, statistical methods for testing the composite hypothesis that are derived, where represents testing for non-inferiority of the efficacy endpoint and is for superiority test of the safety endpoint. Section 3.5 studies the impact on power and sample size calculation when switching from testing for a single hypothesis to testing for a composite hypothesis. In Section 3.6, some statistical justification for Chow's proposal for the determination of appropriate decimal places in observations obtained from clinical research is provided.

## 3.2 Hypotheses for Clinical Evaluation

For the clinical evaluation of a test treatment under investigation, a placebo-controlled clinical trial or an active-controlled study is often conducted. A placebo-controlled trial is to demonstrate the superiority of the test treatment as compared to a placebo, while an active-controlled study is to show that the test treatment is not inferior or equivalent to an active control agent or a standard of care of the disease under study. Thus, hypotheses testing for superiority and non-inferiority/equivalence is often performed for the clinical evaluation of a test treatment under investigation.

### 3.2.1 Traditional Approach

For regulatory review and approval of a test treatment under investigation, a traditional approach is to conduct adequate and well-controlled clinical studies and test for hypotheses of superiority (S), non-inferiority (N), or (therapeutic) equivalence (E) on primary efficacy endpoint and at the same time closely monitor safety and tolerability of the test treatment under investigation. Clinical studies are often powered for generating substantial evidence regarding the effectiveness of the test treatment for regulatory review and approval. For safety assessment, the investigator usually examines the safety profile in terms of adverse events and other safety parameters to determine whether the test treatment is either better (superiority), non-inferior (non-inferiority), or similar (equivalence) as compared to the placebo or control. However, the clinical studies are generally not powered based on the primary safety endpoints.

This approach has been challenged for not being able to provide a complete clinical picture in terms of safety and effectiveness of the test treatment under investigation. As a result, some of drug products were withdrawn from the marketplace after approval by the regulatory agency. A typical example is the withdrawn of Vioxx in 2004. Vioxx is a COX-2 inhibitor drug for arthritis, approved by the FDA in 1999. It was withdrawn by the sponsor due to the safety concern of increased risk of heart attack and stroke. The sponsor's stock went down almost 30% in one day. In addition, the sponsor agreed to pay $950 million and plead guilty to a federal misdemeanor related to its marketing practices. The withdrawal of Vioxx indicated that powered a clinical study based on the primary efficacy endpoint may not be sufficient.

### 3.2.2 Assessing Both Safety and Efficacy Simultaneously

As an alternative to the traditional approach, Chow and Shao (2002) suggest testing composite hypotheses that take both safety and efficacy into consideration. For illustration purpose, Table 3.3 provides a summary of all possible scenarios of composite hypotheses for the clinical evaluation of safety and efficacy of a test treatment under investigation.

Statistically, we would reject the null hypothesis at a pre-specified level of significance and conclude the alternative hypothesis with a desired power.

**TABLE 3.3**

Composite Hypotheses for Clinical Evaluation

| Efficacy | Safety | | |
|---|---|---|---|
| | N | S | E |
| N | NN | NS | NE |
| S | SN | SS | SE |
| E | EN | ES | EE |

*Note:* N = non-inferiority; S = superiority; E = equivalence.

For example, the investigator may be interested in testing non-inferiority in efficacy and superiority in safety of a test treatment as compared to a control. In this case, we can consider testing the null hypothesis, in which N denotes the non-inferiority in efficacy and S represents the superiority of safety. We would reject the null hypothesis and conclude the alternative hypothesis that the test treatment is non-inferior to the active control agent and its safety is superior to the active control agent. To test the null hypothesis, appropriate statistical tests should be derived under the null hypothesis. The derived test statistics can then be evaluated for achieving the desired power under the alternative hypothesis. The selected sample size will ensure that the intended trial will achieve the study objectives of (i) establishing non-inferiority of the test treatment in efficacy and (ii) showing superiority of the safety profile of the test treatment at a pre-specified level of significance.

Note that the composite hypothesis problem described above is different from multiple comparisons. Multiple comparisons usually consist of a set of null hypotheses. The overall hypothesis is that all individual null hypotheses are true, and the alternative hypothesis is that at least one of the null hypotheses is not true. In contrast, when it comes to the composite hypothesis problem, say, the alternative hypothesis is that the test drug is non-inferior (N) in efficacy and superior (S) in safety. Then, the null hypothesis is not NS; i.e., the test drug is inferior in efficacy *or* the test drug is *not* superior in safety. In other words, the null hypothesis consists of three subsets of null hypothesis: First, the test drug is inferior in efficacy and superior in safety; second, the test drug is non-inferior in efficacy and *not* superior in safety; third, the test drug is inferior in efficacy and *not* superior in safety. It would be complicated to consider all these three subsets of null hypothesis. If the third subset of null hypothesis is considered, naturally the alternative hypothesis is that the test drug is either non-inferior in efficacy *or* superior in safety, which is different from the hypothesis that the test drug is non-inferior in efficacy *and* superior in safety.

### 3.2.3 Remarks

It also should be noted that in the interest of controlling the overall type I error rate at the $\alpha$-level, appropriate $\alpha$-levels (say $\alpha_1$ for efficacy and $\alpha_2$ for safety) should be chosen. By switching from a single hypothesis testing to a composite hypothesis testing, sample size increase is expected. As a result, test for a composite hypothesis of safety and efficacy may not be feasible for rare diseases drug development due to the fact that there are limited number of subjects available for clinical evaluation. In this case, it is suggested that some out-of-the-box innovative thinking regarding sample size requirement and the potential use of real-world data and real-word evidence should be considered. More details about innovative thinking regarding sample size requirement and the potential use of real-world data and real-world evidence can be found in Chapters 7 and 8, respectively.

## 3.3 Multiplicity in Clinical Evaluation

In clinical trials, *multiplicity* is usually referred to as multiple inferences that are made in simultaneous context (Westfall and Bretz, 2010). As a result, $\alpha$-adjustment for multiple comparisons is to make sure that the simultaneously observed differences are not by chance alone. In clinical trials, the most commonly seen multiplicity includes the comparison of (i) multiple treatments (dose groups), (ii) multiple endpoints, (iii) multiple time points, (iv) interim analyses, (v) multiple tests of the sample hypothesis, (vi) variable/model selection, and (vii) subgroup analyses. In clinical trials, it is often interest to determine that the observed differences are not by chance alone and they are reproducible. In this case, the level of significance is necessarily adjusted for controlling the *overall* type I error rate at a pre-specified level of significance for multiple endpoints. This has raised the critical issue of multiplicity in clinical evaluation. In this section, some general concepts and principles for $\alpha$-adjustment for multiple comparisons are briefly described.

### 3.3.1 General Concepts

By definition, the type I error rate is defined as the probability of rejecting the null hypothesis when the null hypothesis is true. Let $H_{0i}$: $H_i$ versus $H_{ai}$: $K_i$, $i = 1,\ldots, m$ be the $m$ hypotheses to be tested, where $H_i$ and $K_i$ are the null hypothesis and the alternative hypothesis of the $i$th hypothesis. Table 3.4 summarizes type I error in multiple comparisons.

Under the above structure, the following terms are defined. First, per-comparison error rate (PCER) can be defined as

$$\text{PCER} = \frac{E(V)}{m},$$

which is the ratio of the expected value of committing type I error. Second, the family-wise error rate (FWER) can be defined as

$$\text{FWER} = P(V > 0),$$

**TABLE 3.4**

Type I Error in Multiple Comparisons

| Null Hypothesis | Fail to Reject The Null Hypothesis | Reject The Null Hypothesis | Total |
|---|---|---|---|
| True | $U$ | $V$ | $m_0$ |
| False | $T$ | $S$ | $m - m_0$ |
| Total | $W$ | $R$ | $m$ |

$m_0$ is the number of times that the null hypotheses are rejected.

which indicates that at least one of the $m$ tests has been rejected. Similarly, we can define the generalized family-wise error rate (gFWER) as follows:

$$\text{gFWER} = P(V > k),$$

which indicates that at least $k$ of the $m$ tests have been rejected. According to the above discussion, the false discovery rate (FDR) can be evaluated as follows:

$$\text{FDR} = E(Q) = E\left(\frac{V}{R} \middle| R > 0\right) P(R > 0),$$

which is the expected value of the ratio, i.e., $E(Q)$ between rejection and wrong rejection of the null hypothesis, where $Q = V/R$.

It should be noted that in practice, depending upon the study objectives, the users may want to control certain kinds of type I error rate.

## 3.3.2 General Principles

In this section, the concepts of union–intersection test (UIT) and intersection–union test (IUT) will be described and discussed.

For UIT, the following hypotheses are tested:

$$H_0: H_I = \bigcap_{i=1}^{m} H_i \text{ versus } H_a: K_U = \bigcup_{i=1}^{m} K_i. \tag{3.1}$$

Under the union–intersection hypotheses (3.1), we would reject $H_0$ if one or more $H_{0i}: H_i$ versus $H_{ai}: K_i$ is rejected. In this case, $\alpha$-adjustment is not necessary.

On the other hand, for IUT, the following hypotheses are tested:

$$H_0: H_U = \bigcup_{i=1}^{m} H_i \text{ versus } H_a: K_I = \bigcap_{i=1}^{m} K_i. \tag{3.2}$$

Under the intersection–union hypotheses (3.2), we would reject $H_0$ if all $H_{0i}: H_i$ versus $H_{ai}: K_i$ are rejected. In this case, $\alpha$ is necessarily adjusted.

For an effective alpha adjustment for multiplicity, several methods are available in the literature. For more details, the readers are encouraged to review the references of Finner and Strassburger (2002), Senn and Bretz (2007), Dmitrienko et al. (2010), and Bretz et al. (2010).

## 3.4 Statistical Methods for Testing Composite Hypothesis of $NS$

For illustration purpose, consider the composite hypotheses that $H_0$: not $NS$ versus $H_a$: $NS$ in clinical evaluation of a test treatment under investigation, where $N$ represents the hypothesis for testing non-inferiority in efficacy and $S$ stands for the hypothesis for testing superiority in safety. Let $X$ and $Y$ be the efficacy and safety endpoint, respectively. Assume that $(X, Y)$ follows a bivariate normal distribution with mean $(\mu_X, \mu_Y)$ and variance–covariance matrix $\Sigma$, i.e., where

$$\Sigma = \begin{pmatrix} \sigma^2_x & \rho\sigma_X\sigma_Y \\ \rho\sigma_X\sigma_Y & \sigma^2_Y \end{pmatrix}.$$

Suppose that the investigator is interested in testing non-inferiority in efficacy and superiority in safety of a test treatment as compared to a control (e.g., an active control agent). The corresponding composite hypotheses may be considered:

$$H_0 : \mu_{X1} - \mu_{X2} \leq -\delta_X \text{ or } \mu_{Y1} - \mu_{Y2} \leq \delta_Y \text{ versus}$$

$$H_1 : \mu_{X1} - \mu_{X2} > -\delta_X \text{ and } \mu_{Y1} - \mu_{Y2} > \delta_Y \qquad ,$$

where $(\mu_{X1} - \mu_{Y1})$ and $(\mu_{X2} - \mu_{Y2})$ are the means of $(X, Y)$ for the test treatment and the control, respectively, and $\delta_X$ and $\delta_Y$ are the corresponding non-inferiority margin and superiority margin, respectively. Note that $\delta_X$ and $\delta_Y$ are the positive constants. If the null hypothesis is rejected based on a statistical test, we conclude that the test treatment is non-inferior to the control in efficacy endpoint $X$, and is superior over the control in safety endpoint $Y$.

To test the above composite hypotheses, suppose that a random sample of $(X, Y)$ is collected from each treatment arm. In particular, $(X_{11}, Y_{11}), \ldots, (X_{1n}, Y_{1n})$ are i.i.d. $N((\mu_{X1}, \mu_{Y1}), \Sigma)$, which is the random sample from the test treatment, and $(X_{21}, Y_{21}), \ldots, (X_{2n_2}, Y_{2n_2})$ are i.i.d. $N((\mu_{X2}, \mu_{Y2}), \Sigma)$, which is the random sample from the control treatment. Let $\bar{X}_1$ and $\bar{X}_2$ be the sample means of $X$ in the test treatment and the control, respectively. Similarly, $\bar{Y}_1$ and $\bar{Y}_2$ are the sample means of $Y$ in the test treatment and the control, respectively. It can be verified that the sample mean vector $(\bar{X}_i, \bar{Y}_i)$ follows a bivariate normal distribution. In particular, $(\bar{X}_i, \bar{Y}_i)$ follows $N((\mu_{Xi}, \mu_{Yi}), n_i^{-1}\Sigma)$. Since $(\bar{X}_1, \bar{Y}_1)$ and $(\bar{X}_2, \bar{Y}_2)$ are independent bivariate normal vectors, it follows that $(\bar{X}_1 - \bar{X}_2)$ $(\bar{Y}_1 - \bar{Y}_2)$ is also normally distributed as $N((\mu_{X1} - \mu_{X2}, \mu_{Y1} - \mu_{Y2}), (n_1^{-1} + n_2^{-1})\Sigma)$. For simplicity, we assume $\Sigma$ is known, i.e., the values of parameters $\sigma_X^2, \sigma_Y^2$, and $\rho$ are known. To test the composite hypothesis $H_0$ for both efficacy and safety, we may consider the following test statistics:

$$T_X = \frac{\bar{X}_1 - \bar{X}_2 + \delta_X}{\sqrt{(n_1^{-1} + n_2^{-1})\sigma_X^2}},$$

$$T_Y = \frac{\bar{Y}_1 - \bar{Y}_2 + \delta_Y}{\sqrt{(n_1^{-1} + n_2^{-1})\sigma_Y^2}}.$$

Thus, we would reject the null hypothesis $H_0$ for large values of $T_X$ and $T_Y$. Let $C_1$ and $C_2$ be the critical values for $T_X$ and $T_Y$, respectively. Then, we have

$$P(T_X > C_1, T_Y > C_2) = P\left( U_X > C_1 - \frac{\mu_{X1} - \mu_{X2} + \delta_X}{\sqrt{(n_1^{-1} + n_2^{-1})\sigma_X^2}}, U_y > C_2 - \frac{\mu_{Y1} - \mu_{Y2} + \delta_Y}{\sqrt{(n_1^{-1} + n_2^{-1})\sigma_Y^2}} \right),$$

(3.3)

where $(U_X, U_Y)$ is the standard bivariate normal random vector, i.e., a bivariate normal random vector with zero means, unit variances, and a correlation coefficient of $\rho$.

Under the null hypothesis $H_0$ that $\mu_{X1} - \mu_{X2} \le -\delta_X$ or $\mu_{Y1} - \mu_{Y2} \le \delta_Y$, it can be shown that the upper limit of $P(T_X > C_1, T_Y > C_2)$ is the maximum of the two probabilities, i.e., $\max\{1 - \Phi(C_1), 1 - \Phi(C_2)\}$, where $\Phi$ is the cumulative distribution function of the standard normal distribution. A brief proof is given below.

For given constants $a_1$ and $a_2$, which are a standard bivariate normal vector

$$(U_X, U_Y) \sim N\left( (0,0), \begin{pmatrix} 1 & \rho \\ \rho & 1 \end{pmatrix} \right),$$

we have

$$P(U_X > a_1, U_Y > a_2) = \frac{1}{2\pi\sqrt{1-\rho^2}} \int_{a_1}^{+\infty} \int_{a_2}^{+\infty} \exp-\left\{ \frac{x^2 + y^2 - 2\rho xy}{2(1-\rho^2)} \right\} dy\, dx$$

$$= \frac{1}{\sqrt{2\pi}} \int_{a_1}^{+\infty} \exp\left\{ -\frac{x^2}{2} \right\} \int_{a_2}^{+\infty} \frac{1}{\sqrt{2\pi(1-\rho^2)}} \exp\left\{ -\frac{(y-\rho x)^2}{2(1-\rho^2)} \right\} dy\, dx$$

$$= 1 - \Phi(a_1) - \frac{1}{\sqrt{2\pi}} \int_{a_1}^{+\infty} \Phi\left( \frac{a_2 - \rho x}{\sqrt{1-\rho^2}} \right) \exp\left\{ -\frac{x^2}{2} \right\} dx.$$

(3.4)

Since the joint distribution of $(U_X, U_Y)$ is symmetric, (3.3) is also equal to

$$1 - \Phi(a_2) - \frac{1}{\sqrt{2\pi}} \int_{a_2}^{+\infty} \Phi\left(\frac{a_2 - \rho y}{\sqrt{1-\rho^2}}\right) \exp\left\{-\frac{y^2}{2}\right\} dy. \qquad (3.5)$$

Based on (3.3), $p(T_X > C_1, T_Y > C_2)$ can be expressed by (3.4) and (3.5) with $a_1$ and $a_2$ replaced by

$$D_1 = C_1 - \frac{\mu_{X1} - \mu_{X2} + \delta_X}{\sqrt{(n_1^{-1} + n_2^{-1})\sigma_X^2}},$$

and

$$D_2 = C_2 - \frac{\mu_{Y1} - \mu_{Y2} + \delta_Y}{\sqrt{(n_1^{-1} + n_2^{-1})\sigma_Y^2}},$$

respectively. Under the null hypothesis $H_0$ that $\mu_{X1} - \mu_{X2} + \delta_X$ or $\mu_{Y1} - \mu_{Y2} + \delta_Y$, it's true that either $D_1 \geq C_1$ or $D_2 \geq C_2$. Since integrals in (3.4) and (3.5) are positive, it follows that

$$P(T_X > C_1, T_Y > C_2 | H_0) < \max(1 - \Phi(C_2)).$$

To complete the proof, we need to show that for any $\varepsilon > 0$, $\delta_X$ and $\delta_Y (> 0)$ and given values of other parameters, there exist values of $\mu_{X1} - \mu_{X2}$ and $\mu_{Y1} - \mu_{Y2}$ such that (3.4) is larger than $1 - \Phi(C_1) - \varepsilon$ and $1 - \Phi(C_2) - \varepsilon$. Let $\mu_{X2} - \mu_{X2} = -\delta_X$. Then, (3.4) becomes

$$1 - \Phi(C_1) - \frac{1}{\sqrt{2\pi}} \int_{C_1}^{+\infty} \Phi\left(\frac{D_2 - \rho x}{\sqrt{1-\rho^2}}\right) \exp\left\{-\frac{x^2}{2}\right\} dx. \qquad (3.6)$$

For $\rho > 0$, there exists a negative value $K$ such that when $D_2 < K$, for any $x$ in $[C_1, +\infty)$,

$$\Phi\left(\frac{D_2 - \rho x}{\sqrt{1-\rho^2}}\right) < \varepsilon.$$

For sufficiently large, $\mu_{Y1} - \mu_{Y2}$, it can happen that $D_2 < K$. Therefore, for sufficiently large, $\mu_{Y1} - \mu_{Y2}$, (3.4) $> 1 - \Phi(C_1) - \varepsilon$. For $\rho \leq 0$, express the integral in (3.6) as $I_1 + I_2$, where

$$I_1 = \int_{C_1}^{E} \Phi\left(\frac{D_2 - \rho x}{\sqrt{1-\rho^2}}\right) \exp\left\{-\frac{x^2}{2}\right\} dx \text{ and } I_2 = \int_{E}^{+\infty} \Phi\left(\frac{D_2 - \rho x}{\sqrt{1-\rho^2}}\right) \exp\left\{-\frac{x^2}{2}\right\} dx.$$

$\varepsilon$ is chosen such that $I_2 \leq \int_E^{+\infty} \exp\left\{-\frac{x^2}{2}\right\} dx < 0.5 \, \varepsilon$ The first inequality holds as the cumulative distribution is always $\leq 1$. For a chosen value of $\varepsilon$, the argument for $\rho > 0$ can be applied to prove $I_1 < 0.5\varepsilon$ for sufficiently large $\mu_{Y1} - \mu_{Y2}$. Hence, $P(T_X > C_1, T_Y > C_2 | H_0)$ is greater than $1 - \Phi(C_1) - \varepsilon$ for $\mu_{X1} - \mu_{X2} = -\delta_X$ and sufficiently large $\mu_{Y1} - \mu_{Y2}$. Similarly, it can be shown that $P(T_X > C_1, T_Y > C_2 | H_0)$ is greater than $1 - \Phi(C_2) - \varepsilon$ for $\mu_{Y1} - \mu_{Y2} = \delta_Y$ and sufficiently large $\mu_{X1} - \mu_{X2}$. This completes the proof.

Therefore, the type I error of the test based on $T_X$ and $T_Y$ can be controlled at the level of $\alpha$ by appropriately choosing corresponding critical values of $C_1$ and $C_2$. Denote by $Z_\alpha$ the upper $\alpha$-percentile of the standard normal distribution. Then, the power function of the above test is $P(T_x > Z_{\alpha_1}, T_Y > Z_{\alpha_2})$, which can be calculated from (3.3) and the cumulative distribution function of the standard bivariate distribution.

## 3.5 The Impact on Power and Sample Size Calculation

### 3.5.1 Fixed Power Approach

As indicated earlier, when switching from testing a single hypothesis (i.e., based on a single study endpoint like the efficacy endpoint in clinical trials) to testing a composite hypothesis (i.e., based on two study endpoints such as both efficacy and safety endpoints in clinical trials), increase in sample size is expected. Let $X$ be the efficacy endpoint in clinical trials. Consider testing the following single non-inferiority hypothesis with a non-inferiority margin of $\delta_X$:

$$H_{01}: \mu_{X1} - \mu_{X2} \leq -\delta_X \quad \text{versus} \quad H_{11}: \mu_{X1} - \mu_{X2} > -\delta_X.$$

Then, a commonly used test is to reject the null hypothesis $H_{01}$ at the $\alpha$ level of significance if $T_X > Z_\alpha$. The total sample size for concluding the test treatment is non-inferior to the control with $1 - \beta$ power if the difference of mean $\mu_{X1} - \mu_{X2} > -\delta_X$ is

$$N_X = \frac{(1+r)^2 (Z_\alpha + Z_\beta)^2 \sigma_X^2}{r(\mu_{X1} - \mu_{X2} + \delta_X)^2},$$

where $r = n_2 / n_1$ is the sample size allocation ratio between the control and the test treatment. Table 3.5 gives the total sample size ($N_X$) for the test of non-inferiority based on efficacy endpoint $X$ and total sample size ($N$) for testing composite hypothesis based on both efficacy endpoint $X$ and safety endpoint $Y$, for various scenarios. In particular, we calculated the sample sizes for

$\alpha = 0.05, \beta = 0.20, \mu_{Y1} - \delta_Y = 0.3, = 1$, and several values of $\Delta = \mu_{X1} - \mu_{X2} + \delta_X$ and other parameters. For a hypothesis of superiority of the test treatment in safety, i.e., the component with respect to safety in the composite hypothesis, the preceding specified values of type I error rate, power, and $\mu_{Y1} - \mu_{Y2} - \delta_Y$ and $\sigma_Y$ require a total sample size $N_Y = 275$.

For many scenarios in Table 3.5, the total sample size $N$ for test of the composite hypothesis is much larger than the sample size for test of non-inferiority in efficacy ($N_X$). However, it happens in some cases that they are the same or their difference is quite small. Actually, $N$ is associated with the sample sizes for individual test of non-inferiority in efficacy ($N_X$) and of superiority in safety ($N_Y$), and the correlation coefficient ($\rho$) between $X$ and $Y$. When large difference exists between $N_X$ and $N_Y$, $N$ is quite close to the larger of $N_X$ and $N_Y$, and has little change along with change in $\rho$. In this numerical study, for $N_X = 69$ and 39 ($\ll 275$), $N$ is mostly equal to 275; for $N_X = 1392$ and 619 ($\gg 275$), the difference between $N$ and $N_X$ is 0 or negligible compared with the size of $N$. At the preceding four scenarios, a change in correlation coefficient between $X$ and $Y$ has little impact on $N$. On the one hand, the larger value of $N_X$ and $N_Y$ is not always close to $N$, especially when $N_X$ and $N_Y$ are close to each other. For example, in Table 3.4, when $N_X$ is equal to 275 ($= N_Y$), $N$ is 352 for $\rho = 0.5$, and 373 for $\rho = 0$. In addition, the results in Table 3.5 suggest that the correlation coefficient between $X$ and $Y$ is unlikely to have great influence on $N$, especially when the difference between $N_X$ and $N_Y$ is quite substantial. The above findings are consistent with the underlying "rule": When the two sample sizes are substantially different, taking $N$ as the larger of $N_X$ and $N_Y$ will ensure the

**TABLE 3.5**

Comparison of Sample Size between Tests for Multiple Endpoints and Single Endpoint

| $\sigma_X$ | $\rho$ | $\Delta = 0.2$ | | | $\Delta = 0.3$ | | | $\Delta = 0.4$ | | |
|---|---|---|---|---|---|---|---|---|---|---|
| | | $N_X$ | $N$ | $N/N_X$ | $N_X$ | $N$ | $N/N_X$ | $N_X$ | $N$ | $N/N_X$ |
| 0.5 | −1.0 | 155 | 304 | 1.96 | 69 | 276 | 4.00 | 39 | 275 | 7.05 |
| | −0.5 | 155 | 303 | 1.95 | 69 | 276 | 4.00 | 39 | 275 | 7.05 |
| | 0.0 | 155 | 300 | 1.94 | 69 | 276 | 4.00 | 39 | 275 | 7.05 |
| | 0.5 | 155 | 289 | 1.86 | 69 | 275 | 3.99 | 39 | 275 | 7.05 |
| | 1.0 | 155 | 275 | 1.77 | 69 | 275 | 3.99 | 39 | 275 | 7.05 |
| 1.0 | −1.0 | 619 | 647 | 1.05 | 275 | 381 | 1.39 | 155 | 304 | 1.96 |
| | −0.5 | 619 | 646 | 1.04 | 275 | 381 | 1.39 | 155 | 303 | 1.95 |
| | 0.0 | 619 | 642 | 1.04 | 275 | 373 | 1.36 | 155 | 300 | 1.94 |
| | 0.5 | 619 | 629 | 1.02 | 275 | 352 | 1.28 | 155 | 289 | 1.86 |
| | 1.0 | 619 | 619 | 1.00 | 275 | 275 | 1.00 | 155 | 275 | 1.77 |
| 1.5 | −1.0 | 1392 | 1392 | 1.00 | 619 | 647 | 1.05 | 348 | 433 | 1.24 |
| | −0.5 | 1392 | 1392 | 1.00 | 619 | 646 | 1.04 | 348 | 432 | 1.24 |
| | 0.0 | 1392 | 1392 | 1.00 | 619 | 642 | 1.04 | 348 | 424 | 1.22 |
| | 0.5 | 1392 | 1392 | 1.00 | 619 | 629 | 1.02 | 348 | 402 | 1.16 |
| | 1.0 | 1392 | 1392 | 1.00 | 619 | 619 | 1.00 | 348 | 348 | 1.00 |

powers of two individual tests for efficacy and safety are essentially 1 and $1 - \beta$, "resulting" in a power of $1 - \beta$ for test of the composite hypotheses; when $N_X$ and $N_Y$ are close to each other, taking $N$ as the larger of $N_X$ and $N_Y$ will ensure the power of the test of composite hypotheses is at about $(1 - \beta)^2$. Therefore, a significant increment in $N$ is required for achieving a power of $1 - \beta$.

## 3.5.2 Fixed Sample Size Approach

Based on the sample size in Table 3.4, power of the test of composite hypothesis $H_0$ was calculated with the results presented in Table 3.5, where $P$ is the power of test of composite hypothesis with $N_X$ in Table 3.4. $P_M$ is the power of the same test with max $(N_X, 275)$. With sample size $N_X$, the power of test of composite hypothesis is always not greater than the target value 80% as $N_X$ is always not larger than N, as presented in Table 3.6. In some cases that $\sigma_X = 1.5 > \sigma_Y = 1.0$, $N_X = N$. Hence, the corresponding $P = 80\%$. However, $P$ is less than 60% for many cases in our numerical study. The worst scenario is $P = 4.3\%$ when $N_X = 39$ for $\sigma_X = 0.5$, $\rho = -1$, and $\Delta = 0.4$. Therefore, the test of composite hypothesis of both efficacy and safety using sample size $N_X$, which is for achieving a certain power in testing hypothesis of efficacy only, may not have enough power to reject the null hypothesis. Interestingly, testing the composite hypothesis with max($N_X, 275$), the power $P_M$ is close to the target value 80% in most scenarios. Some exceptions happen when $N_X$ is close to 275 (corresponding to ($\Delta = 0.3$, $\sigma_X = 1.0$), and ($\Delta = 0.4$, $\sigma_X = 1.5$)) such that a significant

**TABLE 3.6**

Power (%) of Test of Composite Hypothesis

| $\sigma_X$ | $\rho$ | $\Delta = 0.2$ | | $\Delta = 0.3$ | | $\Delta = 0.4$ | |
|---|---|---|---|---|---|---|---|
| | | $P$ | $P_M$ | $P$ | $P_M$ | $P$ | $P_M$ |
| 0.5 | −1.0 | 38.9 | 75.3 | 14.7 | 80.0 | 4.3 | 80.0 |
| | −0.5 | 41.9 | 75.4 | 22.0 | 80.0 | 14.2 | 80.0 |
| | 0.0 | 47.1 | 76.2 | 27.7 | 80.0 | 19.2 | 80.0 |
| | 0.5 | 52.9 | 78.1 | 32.3 | 80.0 | 22.8 | 80.0 |
| | 1.0 | 58.8 | 80.0 | 34.5 | 80.0 | 23.9 | 80.0 |
| 1.0 | −1.0 | 78.2 | 78.2 | 60.1 | 60.1 | 38.9 | 75.3 |
| | −0.5 | 78.2 | 78.2 | 60.9 | 60.9 | 41.9 | 75.4 |
| | 0.0 | 78.6 | 78.6 | 64.0 | 64.0 | 47.1 | 76.2 |
| | 0.5 | 79.4 | 79.4 | 68.8 | 68.8 | 52.9 | 78.1 |
| | 1.0 | 80.0 | 80.0 | 80.0 | 80.0 | 58.8 | 80.0 |
| 1.5 | −1.0 | 80.0 | 80.0 | 78.2 | 78.2 | 67.6 | 67.6 |
| | −0.5 | 80.0 | 80.0 | 78.2 | 78.2 | 68.0 | 68.0 |
| | 0.0 | 80.0 | 80.0 | 78.6 | 78.6 | 70.1 | 70.1 |
| | 0.5 | 80.0 | 80.0 | 79.4 | 79.4 | 73.7 | 73.7 |
| | 1.0 | 80.0 | 80.0 | 80.0 | 80.0 | 80.0 | 80.0 |

increment in sample size from $\max(N_X, 275)$ to N is required. This suggest taking $N$ as the larger of the two sample sizes $N_X$ and $N_Y$ for testing hypothesis of individual endpoint when either of the two is much larger, say onefold larger than the other.

### 3.5.3 Remarks

Traditional approach for the clinical evaluation of a test treatment under investigation is to power the study based on the efficacy endpoint. The test treatment is considered approvable if its safety and tolerability are acceptable provided that the efficacy has been established. In practice, in the interest of controlling the overall type I error rate at a pre-specified level of significance, the type I error rate may be adjusted for multiple comparisons. It, however, should be noted that the overall type I error rate may be controlled at the risk of (i) decreasing the power and (ii) increasing the sample size when switching from testing a single hypothesis (for efficacy) to testing a composite hypothesis (for both efficacy and safety).

In this chapter, for illustration purpose, we assume that the two study endpoints follow a bivariate normal distribution. In practice, both efficacy and safety endpoints could be either a continuous variable, a binary response, or time-to-event data. Similar idea can be applied to determine the impact on power and sample size calculation when switching from testing a single hypothesis to testing a composite hypothesis. It, however, should be noted that closed forms for the relationships of powers and formulas for sample size calculation between the single hypothesis and the composite hypothesis may not exist. In this case, clinical trial simulation may be useful.

## 3.6 Significant Digits

As indicated earlier, statistical inference obtained based on data with different decimal places may lead to different conclusions. As an example, consider a parallel bioequivalence (BE) study. Suppose that there are 24 subjects in the group of test drug and 24 subjects in the group of reference drug. The data is given in Table 3.7. From the BE study results given in Table 3.8, it can be seen that keeping different number of decimal digits can lead to different conclusions. Thus, the selection of the number of decimal places could be critical if the treatment effect is of marginally significance. Chow (2000) introduced the concept of signal noise for determining the number of decimal places for the results obtained from the clinical trials. The idea is to select the minimum number of decimal places in such a way that there is no statistically significant difference between the data set presented by using the minimum decimal places and any other data sets with more decimal places. In what follows, Chow's proposal is briefly described.

**TABLE 3.7**

BE Example Data

| X | $X_0$ | $X_1$ | $X_2$ | Y | $Y_0$ | $Y_1$ | $Y_2$ |
|---|---|---|---|---|---|---|---|
| 1.169577 | 1 | 1.2 | 1.17 | 1.0722791 | 1 | 1.1 | 1.07 |
| 1.251990 | 1 | 1.3 | 1.25 | 1.0348811 | 1 | 1.0 | 1.03 |
| 1.449081 | 1 | 1.4 | 1.45 | 0.9020537 | 1 | 0.9 | 0.90 |
| 1.205818 | 1 | 1.2 | 1.21 | 1.1196368 | 1 | 1.1 | 1.12 |
| 1.355457 | 1 | 1.4 | 1.36 | 0.9736662 | 1 | 1.0 | 0.97 |
| 1.285863 | 1 | 1.3 | 1.29 | 1.1360977 | 1 | 1.1 | 1.14 |
| 1.519270 | 2 | 1.5 | 1.52 | 0.8531594 | 1 | 0.9 | 0.85 |
| 1.230438 | 1 | 1.2 | 1.23 | 1.1239591 | 1 | 1.1 | 1.12 |
| 1.374791 | 1 | 1.4 | 1.37 | 1.0642288 | 1 | 1.1 | 1.06 |
| 1.302860 | 1 | 1.3 | 1.30 | 0.9156539 | 1 | 0.9 | 0.92 |
| 1.396263 | 1 | 1.4 | 1.40 | 0.9044889 | 1 | 0.9 | 0.90 |
| 1.507581 | 2 | 1.5 | 1.51 | 0.9894644 | 1 | 1.0 | 0.99 |
| 1.337749 | 1 | 1.3 | 1.34 | 1.0281070 | 1 | 1.0 | 1.03 |
| 1.222744 | 1 | 1.2 | 1.22 | 0.8584933 | 1 | 0.9 | 0.86 |
| 1.235640 | 1 | 1.2 | 1.24 | 1.0074020 | 1 | 1.0 | 1.01 |
| 1.302359 | 1 | 1.3 | 1.30 | 0.9131539 | 1 | 0.9 | 0.91 |
| 1.379500 | 1 | 1.4 | 1.38 | 0.9563392 | 1 | 1.0 | 0.96 |
| 1.295147 | 1 | 1.3 | 1.30 | 1.2159481 | 1 | 1.2 | 1.22 |
| 1.376740 | 1 | 1.4 | 1.38 | 1.1442079 | 1 | 1.1 | 1.14 |
| 1.376414 | 1 | 1.4 | 1.38 | 1.0128952 | 1 | 1.0 | 1.01 |
| 1.321817 | 1 | 1.3 | 1.32 | 0.9561896 | 1 | 1.0 | 0.96 |
| 1.222626 | 1 | 1.2 | 1.22 | 0.8718494 | 1 | 0.9 | 0.87 |
| 1.140910 | 1 | 1.1 | 1.14 | 0.9620998 | 1 | 1.0 | 0.96 |
| 1.169492 | 1 | 1.2 | 1.17 | 0.9487145 | 1 | 0.9 | 0.95 |

*Note:* X: the original data from test drug; $X_i$: the original data with $i$ decimal digit; Y: the data from reference drug; $Y_i$: the data with $i$ decimal digit.

**TABLE 3.8**

BE Study

| Significant Digits | Confidence Interval | BE Limit | BE Result (Y/N) |
|---|---|---|---|
| 0 | (−0.013, 0.180) | (−0.2, 0.2) | Y |
| 1 | (0.261, 0.356) | (−0.2, 0.2) | N |
| 2 | (0.263, 0.362) | (−0.2, 0.2) | N |

## 3.6.1 Chow's Proposal

The number of significant decimal digits of a given data set obtained from an analytical experiment is defined as the minimum number of decimal places of the data set which satisfies the following two conditions: First, the data set with the minimum number of decimal places will achieve the desired accuracy and precision. Second, the data set with the minimum number of decimal places is not statistically distinguishable with those data sets with more decimal places than the minimum number of decimal places. In other words, the data set with significant decimal digits is not significantly different from those data sets with the number of decimal places that exceed the number of significant decimal digits.

Let $X$ be a continuous random variable and $X^*$ be its truncated value with $d$ decimal digits. We would claim that $X^*$ is not statistically different from $X$ if we fail to reject the following null hypothesis at the $\alpha$-level of significance

$$H_0: \mu_X = \mu_{X^*} \text{ versus } H_a: \mu_X \neq \mu_{X^*}, \tag{3.7}$$

where $\mu_X$ and $\mu_{X^*}$ are the population means for $X$ and $X^*$, respectively. When $X$ and $X^*$ are not statistically distinguishable, the $d$ decimal digits are considered significant decimal digits. Suppose $X$ is a continuous random variable with standard deviation $\sigma$ and $X^*$ is its truncated value after rounding up to the $d$th decimal place. Then, the maximum possible error due to the truncation would be less than $10^{1-d}$. As an example, if $d = 3$, the smallest and largest values for a given number with three decimal places are $a.bc0$ and $a.bc9$, respectively. Hence, the maximum possible error is less than 0.01, which is $10^{-2}$. Here, $-2$ is obtained as $-2 = 1 - d = 1-3$ intuitively; if this worst-case error is small *enough*, the distortion of the distribution due to rounding error would be negligible. But the question is how small would be considered *enough*? An idea is to apply the concept of signal noise in quality control and assurance to compare this error with $X$'s standard deviation $\sigma$. The significant digits can then be chosen by taking the first $d$ digits such that

$$\frac{10^{1-d}}{\sigma} < \delta' \text{ if and only if } \frac{10^{-d}}{\sigma} < \delta'/10 = \delta,$$

where $\delta$ is a constant, which is to be chosen such that the truncated observation $X^*$ is not statistically different from $X$ at the $\alpha$-level of significance. In practice, a convention choice of $\delta$ is $\delta = 10\%$. To provide a better understanding of the proposed procedure, the results for various choices of given choices of $\delta$ given $\sigma$ are summarized in Table 3.9. As it can be seen from Table 3.9, smaller would require more decimal places to be used in order to achieve a desired accuracy and precision. Table 3.9 also indicates that more decimal places are needed for a smaller $\sigma$-value.

## 3.6.2 Statistical Justification

Without the loss of generality, we assume $X$ follows a normal distribution with mean $\mu_x$ and variance $\sigma^2$, i.e., $X \sim N(\mu_X, \sigma^2)$. By properly truncation, $X^*$ is still approximately normally distributed with mean $\mu_{X^*}$ and variance $\sigma^2$,

**TABLE 3.9**

Significant Decimal Digits for Various Selection of $\delta$ Given $\sigma$

| | $\delta$ | | | | |
|---|---|---|---|---|---|
| $\sigma$ | 1% | 5% | 10% | 15% | 20% |
| 0.01 | 4 | 4 | 3 | 3 | 3 |
| 0.10 | 3 | 3 | 2 | 2 | 2 |
| 0.50 | 3 | 2 | 2 | 2 | 1 |
| 1.00 | 2 | 2 | 1 | 1 | 1 |
| 2.00 | 2 | 1 | 1 | 1 | 1 |

where $\mu_{X^*}$ may be different from $\mu_X$ due to rounding error. The following two-sample $t$ can be used to test the null hypothesis given in (3.5)

$$T = \frac{\sqrt{n}(\bar{X} - \bar{X}^*)}{\sqrt{s_X^2 + s_{X^*}^2}},$$

where $s_X^2$ and $s_{X^*}^2$ are the sample standard deviations of $X$ and $X^*$, respectively. Under the null hypothesis that $H_0: \mu_X = \mu_{X^*}$, the two-sample $T$ statistic follows a $t$-distribution with $2(n-1)$ degrees of freedom. We reject the null hypothesis $|T| > t_{\alpha/2,2(n-1)}$, where $t_{\alpha/2,2(n-1)}$ is the $(1-\alpha/2)$th quantile for a distribution with $2(n-1)$ degrees of freedom. Under the alternative hypothesis that $H_a: \mu_X \neq \mu_{X^*}$, the $t$ statistic can be written as

$$T = \frac{\sqrt{n}(\bar{X} - \bar{X}^*)}{\sqrt{s_X^2 + s_{X^*}^2}} = \frac{\sqrt{n/2[(\bar{X} - \mu_X)/\sigma - (\bar{X}^* - \mu_{X^*})/\sigma]} + \sqrt{n/2(\mu_X - \mu_{X^*})/\sigma}}{\sqrt{s_X^2/2\sigma^2 + s_{X^*}^2/2\sigma^2}}$$

$$= \frac{N(0,1) + \delta}{\chi^2_{2(n-1)}/2(n-1)} \sim t_{2(n-1)}(\delta),$$

where $t_{2(n-1)}(\delta)$ denotes a $t$-distribution with the noncentrality parameter of

$$\delta = \sqrt{\frac{n}{2}}\left(\frac{\mu_X - \mu_{X^*}}{\sigma}\right). \qquad (3.8)$$

When $|\delta|$ is smaller, there is a lower probability that $X^*$ will be different from $X$ under $t$-test. On the other hand, since $X^*$ is rounded at the $d$th decimal places, the maximum possible error due to truncation would be less than $10^{1-d} \geq |\mu_X - \mu_{X^*}|$. So a small value of $10^{-d}/\sigma$ would guarantee that $X^*$ is not significantly different from $X$. The above argument can be applied similarly to a more general situation where a transformation is performed. Let $f(x)$ be the function of transformation of X. In this case, the hypotheses of interest become

$$H_0: f(\mu_X) = f(\mu_{X^*}) \text{ versus } H_a: f(\mu_X) \neq f(\mu_{X^*}).$$

By Taylor's expansion, we have

$$\sqrt{n}\left(f(\bar{x}) - f\left(\bar{X}^*\right)\right) \approx f'(\mu_X)\sqrt{n}\left(\bar{X} - \bar{X}^*\right),$$

which approximately follows a normal distribution with mean $\sqrt{n}f'(\mu_X)(\mu_X - \mu_{X^*})$ and variance $2f'^2(\mu_X)\sigma^2$. As a result, the above null hypothesis can be tested by the following statistic:

$$T_f = \frac{\sqrt{n}f\left(\left(\bar{X}\right) - f\left(\bar{X}^*\right)\right)}{f'(\mu_X)\sqrt{s_{\bar{X}}^2 + s_{X^*}^2}}.$$

Under the null hypothesis, $T_f$ approximately follows a $t$-distribution with $2(n-1)$ degrees of freedom. Under the alternative hypothesis, $T_f$ can be written as

$$T_f \approx \frac{N\sqrt{n}f'(\mu_X)(\mu_X - \mu_{X^*}), 2f'^2(\mu_X)\sigma^2}{f'(\mu_X)\sqrt{s_{\bar{X}}^2 + s_{X^*}^2}}$$

$$= \frac{N\left(\dfrac{\sqrt{n}(\mu_X - \mu_{X^*})}{\sqrt{2}\sigma}, 1\right)}{\sqrt{\dfrac{s_{\bar{X}}^2}{2\sigma^2} + \dfrac{s_{X^*}^2}{2\sigma^2}}}$$

$$= \frac{N(0,1) + \delta}{\chi_{2(n-1)}^2/2(n-1)} \sim t_{2(n-1)}(\delta),$$

where $\delta$ is still the same noncentrality parameter as defined in (3.8). So if we choose significant digits properly, we can guarantee $\delta$ will be small and the probability that $X^*$ is statistically different from X will be small as well. This shows that the proposed procedure works as well for data after transformation. To illustrate the use of the proposed procedure for transformed data, consider a log-transformation, i.e., $f(x) = \log(x)$. Thus, the hypotheses become

$$H_0: \log(\mu_X) = \log(\mu_{X^*}) \text{ versus } H_a: \log(\mu_X) \neq \log(\mu_{X^*}).$$

Then, $f'(\mu_X) = 1/\mu_X$ and the test statistic is given by

$$T_f = \frac{\sqrt{n}\left[\log(\bar{X}) - \log(\bar{X}^*)\right]}{\frac{1}{\mu_X}\sqrt{s_X^2 + s_{X^*}^2}} \sim t_{2(n-1)}(\delta).$$

A numerical study is conducted to demonstrate the use of the proposed procedure. Thirty analytical results were generated from $N(\pi, 0.01)$, which are given in Table 3.9. For convenience's sake, we keep six decimal digits as the original values. If we choose $\delta$ to be equal to 10%, we have

$$\frac{10^{-d}}{\sigma} = \frac{10^{-d}}{0.01} \leq 0.1.$$

It can be seen that the minimum number of $d$ that satisfies the above expression is $d = 3$. Therefore, the number of significant decimal digits is chosen to be 3. Now consider four data sets $X_{ij}$, $j = 1, 2, 3, 4$, which are truncated at the $j$th decimal places, respectively. Then, a two-sample $t$-test is performed to test if $X_{1i}, X_{2i}, X_{3i}, X_{4i}$ are significantly different from one another and are significantly different from the original $X_i$. The results are summarized in Tables 3.10 and 3.11, from which we can see that $X_{1i}$ are significantly different from the rest of data sets. This shows that the rounding error can alter the distribution significantly. The results also indicate that $X_{3i}$ is not significantly different from $X_{4i}$. It shows the proposed procedure works well. It, however, should be noted that $X_{1i}$ is also not significantly different from $X_{3i}$ and $X_{4i}$. This indicates that the conventional choice of $\delta = 10\%$ may be conservative in this case.

**TABLE 3.10**

Pairwise Comparisons

| Comparison | $t$-statistic | $p$-value |
|---|---|---|
| $X_i$ versus $X_{1i}$ | 4.138 | <0.001 |
| $X_i$ versus $X_{2i}$ | 0.116 | 0.908 |
| $X_i$ versus $X_{3i}$ | 0.008 | 0.994 |
| $X_i$ versus $X_{4i}$ | 0.003 | 0.997 |
| $X_{1i}$ versus $X_{2i}$ | 4.072 | <0.001 |
| $X_{1i}$ versus $X_{2i}$ | 4.140 | <0.001 |
| $X_{1i}$ versus $X_{3i}$ | 4.137 | <0.001 |
| $X_{2i}$ versus $X_{3i}$ | 0.123 | 0.603 |
| $X_{2i}$ versus $X_{4i}$ | 0.112 | 0.911 |
| $X_{3i}$ versus $X_{4i}$ | 0.011 | 0.991 |

**TABLE 3.11**

Simulation Data Set for Two-Sample *t*-Test

| $i$ | $X_i$ | $X_{1i}$ | $X_{2i}$ | $X_{3i}$ | $X_{4i}$ |
|---|---|---|---|---|---|
| 1 | 3.145714 | 3.1 | 3.15 | 3.146 | 3.1457 |
| 2 | 3.140959 | 3.1 | 3.14 | 3.141 | 3.1410 |
| 3 | 3.141432 | 3.1 | 3.14 | 3.141 | 3.1414 |
| 4 | 3.127617 | 3.1 | 3.13 | 3.128 | 3.1276 |
| 5 | 3.142035 | 3.1 | 3.14 | 3.142 | 3.1420 |
| 6 | 3.146685 | 3.1 | 3.15 | 3.147 | 3.1467 |
| 7 | 3.146124 | 3.1 | 3.15 | 3.146 | 3.1461 |
| 8 | 3.138408 | 3.1 | 3.14 | 3.138 | 3.1384 |
| 9 | 3.125891 | 3.1 | 3.13 | 3.126 | 3.1259 |
| 10 | 3.136696 | 3.1 | 3.14 | 3.137 | 3.1367 |
| 11 | 3.133587 | 3.1 | 3.13 | 3.134 | 3.1336 |
| 12 | 3.158443 | 3.2 | 3.16 | 3.158 | 3.1584 |
| 13 | 3.140589 | 3.1 | 3.14 | 3.141 | 3.1406 |
| 14 | 3.128415 | 3.1 | 3.13 | 3.128 | 3.1284 |
| 15 | 3.149534 | 3.1 | 3.15 | 3.150 | 3.1495 |
| 16 | 3.153279 | 3.2 | 3.15 | 3.153 | 3.1532 |
| 17 | 3.147673 | 3.1 | 3.15 | 3.148 | 3.1477 |
| 18 | 3.140493 | 3.1 | 3.14 | 3.140 | 3.1405 |
| 19 | 3.150542 | 3.2 | 3.15 | 3.151 | 3.1505 |
| 20 | 3.123488 | 3.1 | 3.12 | 3.123 | 3.1235 |
| 21 | 3.161004 | 3.2 | 3.16 | 3.161 | 3.1610 |
| 22 | 3.140658 | 3.1 | 3.14 | 3.141 | 3.1407 |
| 23 | 3.151263 | 3.1 | 3.15 | 3.151 | 3.1512 |
| 24 | 3.124985 | 3.1 | 3.12 | 3.125 | 3.1250 |
| 25 | 3.140625 | 3.1 | 3.14 | 3.141 | 3.1406 |
| 26 | 3.168811 | 3.2 | 3.17 | 3.169 | 3.1688 |
| 27 | 3.159006 | 3.2 | 3.16 | 3.159 | 3.1590 |
| 28 | 3.143139 | 3.1 | 3.14 | 3.143 | 3.1431 |
| 29 | 3.123467 | 3.1 | 3.12 | 3.123 | 3.1235 |
| 30 | 3.146950 | 3.1 | 3.14 | 3.147 | 3.1470 |

### 3.6.3 Remarks

Statistical justification of the proposed procedure for the determination of the number of significant decimal digits in observations obtained from studies conducted in analytical research was made under the assumption of normality. In practice, the observed analytical results may be described better by other distributions like Weibull distribution for dissolution results of oral solid dosage form of a drug product. In this case, similar concept can be carried out to provide a valid statistical justification. In many cases, log-transformation is often considered for a better description or interpretation of the analytical results. For example, AUC (area under the plasma concentration–time curve) and $C_{max}$ (time to achieve maximum concentration) in the study of bioavailability and BE studies are known to be skewed to the right. As a result, a log-transformation is recommended. In this case, the proposed procedure is useful for the determination of the number of significant decimal digits to maintain certain degree of accuracy and precision for the assessment of BE. For the presentation of the analytical results, descriptive statistics such as mean, standard deviation, minimum, maximum, range, relative standard deviation (RSD) or coefficient of variation (CV), and statistical inferences such as confidence intervals and p-values are usually obtained. In practice, it is always a concern as to how many significant decimal digits should be used for descriptive statistics and statistical inferences to maintain the desired degree of accuracy and precision. In the interest of consistency, it is recommended that the same number of significant decimal digits be used for descriptive statistics and statistical inferences obtained from the analytical results.

In some cases, the analytical results may be expressed in a scientific form (e.g., $1.32 \times 10^5$ or $9.2 \times 10^{-7}$). The proposed procedure can be applied to its significant part (i.e., 1.32 for $1.32 \times 10^5$ or 9.2 for $9.2 \times 10^{-7}$) or its log (base 10)-transformation. When analytical results involve different data sets, it is suggested that each data set keeps its own significant decimal digits as determined by its standard deviation to maintain the same degree of accuracy and precision. A typical example is a dose proportionality study. The purpose of a dose proportionality study is usually to show that there is a linear relationship between dose and AUC within a given range. In other words, doubled the dose, AUC value is expected to be doubled. However, high dose will generally produce a large variability in AUC values. As a result, low dose, median dose, and high dose are expected to have different number of significant decimal digits to achieve the same degree of accuracy and precision. In the interest of keeping the same number of significant decimal digits, we may consider the AUC values adjusted for dose and then apply the proposed procedure to determine the number of significant decimal digits.

## 3.7 Concluding Remarks

In clinical trials, hypotheses testing is often performed for the clinical evaluation of the safety and efficacy of the test treatment under investigation. The traditional approach is to test for the null hypothesis of no treatment difference in efficacy and then evaluate the tolerability and safety of the test treatment under investigation. In other words, the intended clinical trial is powered based on the primary efficacy endpoint by ignoring the safety endpoint. This has led to the withdrawal of several drug products due to safety concern after regulatory approval. Thus, it is suggested that future clinical trials should be powered based on both efficacy and safety endpoints. This chapter also introduced general concepts and principles for $\alpha$-adjustment for multiplicity when performing composite hypotheses testing such as IUT and UIT. For illustration purpose, statistical test for testing composite hypotheses of (i) non-inferiority in efficacy and (ii) superiority in safety is derived in this chapter. Also included in this chapter is the impact on the sample size requirement when conducting composite hypotheses testing for the clinical evaluation of both efficacy and safety.

In practice, it is not uncommon to encounter the situation where the test results are marginal. In this case, it is critical to determine how many decimal places should be used for an accurate and reliable assessment of the treatment effect with proper scientific justification. This chapter introduced Chow's proposal for determining significant digits based on the concept of hypotheses testing for achieving statistical significance. Chow's proposal is found useful not only for clinical data presentation (up to certain decimal places) but also for scientific justification of statistical significance.

# 4

## Endpoint Selection in Clinical Trials

### 4.1 Introduction

In pharmaceutical/clinical development of a test treatment for treating patients with certain diseases, clinical trials are often conducted to evaluate the safety and effectiveness of the test treatment under investigation. In clinical trials, there may be multiple endpoints available for the evaluation of disease status and/or therapeutic effect of the test treatment under study (Williams et al., 2004; Filozof et al., 2017). In practice, it is usually not clear which study endpoint can best inform the disease status and should be used to evaluate the treatment effect. Thus, it is difficult to determine which endpoint should be used as the primary endpoint for the intended clinical trials; especially, these multiple endpoints may be correlated with some unknown correlation structures. Once the primary study endpoint has been selected, power calculation for sample size can be performed for achieving a desired power of correctly detecting the treatment effect if such a treatment effect truly exists. It, however, should be noted that different study endpoints may be of different data types (e.g., continuous, binary response, or time-to-event data) and may not be translated one another although they might be highly correlated with one another. Besides, it is very likely that different study endpoints will result in different sample size requirements. As a result, in a given clinical trial with a pre-selected sample size, it is not uncommon that some study endpoints may be achieved, but some may not. In this case, it is of interest to know which study endpoint is telling the truth or which endpoint can best inform the disease status and/or treatment effect posttreatment.

In this chapter, a therapeutic index (function) proposed by Filozof et al. (2017), which was developed based on a utility function to combine and utilize information collected from all study endpoints, is discussed. The developed therapeutic index is able to fully utilize all the information collected from the available study endpoints for an overall assessment of the safety and effectiveness of the test treatment under investigation. Statistical properties and performances of the proposed therapeutic index are evaluated both theoretically and via a numerical example.

In the next section, examples for clinical trials with multiple study endpoints are discussed. A therapeutic index function proposed by Filozof et al. (2017) that combines all available study endpoints in a clinical trial is described in Section 4.3. Statistical properties of the developed therapeutic index and its relative performances with respect to single endpoints and co-primary endpoints in terms of false-positive and false-negative rates are studied in Section 4.4. A numerical example is given in Section 4.5. Some concluding remarks are given in Section 4.6.

## 4.2 Clinical Trials with Multiple Endpoints

In clinical trials, a primary study endpoint is often chosen for a pre-study power analysis for sample size estimation. In practice, however, it is not uncommon to have multiple study endpoints (e.g., co-primary endpoints) for the evaluation of the safety and effectiveness of the test treatment under investigation. In this section, a couple of examples concern cancer clinical trials and clinical trials for the clinical evaluation of drug products in treating patients with nonalcoholic steatohepatitis (NASH).

### 4.2.1 Cancer Clinical Trials

One of the most typical examples for clinical trials with multiple endpoints would be cancer clinical trials. In cancer clinical trials, overall survival (OS), response rate (RR), and/or time-to-disease progression (TTP) are usually considered as primary clinical endpoints for the evaluation of effectiveness of the test treatment under investigation in regulatory submissions (see, e.g., Williams et al., 2004; Zhou et al., 2019). Williams et al. (2004) provided a list of oncology drug products approved by the United States Food and Drug Administration (FDA) based on single endpoint, co-primary endpoints, and/ or multiple endpoints between 1990 and 2002 (Table 4.1). As it can be seen from Table 4.1, a total of 57 oncology drug submissions were approved by the FDA between 1990 and 2002. Of the 57 applications, 18 were approved based on survival endpoint alone, while 18 were approved based on RR and/or TTP alone. About 9 submissions were approved based on RR plus tumor-related signs and symptoms (co-primary endpoints).

More recently, Zhou et al. (2019) provided a list of oncological and hematological drug approved by the FDA between 2008 and 2016 (Figure 4.1). As it can be seen from Figure 4.1, a total of 12 drugs were approved based on multiple endpoints. Both Table 4.1 and Figure 4.1 do not indicate that which study endpoint (including single endpoint, a co-primary endpoint, or a composite endpoint) should be used for evaluation and regulatory approval of the drug product under investigation.

**TABLE 4.1**

Endpoints Supporting Regulatory Approval of Oncology Drug Marketing
Application, January 1, 1990, to November 1, 2002

| | |
|---|---|
| Total | 57 |
| Survival | 18 |
| RR and/or TTP alone (predominantly hormone treatment of breast cancer or hematologic malignancies) | 18 |
| Tumor-related signs and symptoms | 13 |
|   RR + tumor-related signs and symptoms | (9) |
|   Tumor-related signs and symptoms alone | (4) |
| Disease-free survival (adjuvant setting) | 2 |
| Recurrence of malignant pleural effusion | 2 |
| Decreased incidence of new breast cancer occurrence | 2 |
| Decreased impairment creatinine clearance | 1 |
| Decreased xerostomia | 1 |

*Source:* Williams et al. (2004)

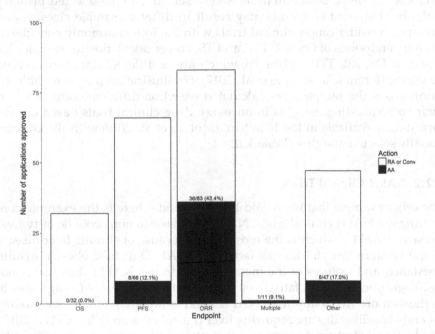

**FIGURE 4.1**

Number of applications approved by endpoint. The number of applications approved for a new indication in an oncology product by the US FDA during 2008 and 2016, grouped by primary endpoint of the trial that supported the application. Endpoints are abbreviated as follows: overall survival (OS), progression-free survival (PFS), objective response rate (ORR), relapse-free survival (RFS), event-free survival (EFS), multiple endpoints other than a co-primary endpoint of OS and PFS (Multiple), and other endpoints not included in the previous categories (Other). Types of approvals are abbreviated as follows: regular approval (RA), conversion to regular approval (Conv), and accelerated approval (AA). Data used were taken from the package inserts of the approved products and FDA records. (Zhou et al., 2019.)

In oncology drug development, commonly considered study endpoints include, but are not limited to, (i) OS, (ii) RR, (iii) TTP, and (iv) tumor size (TS). Note that in addition to OS, disease-free survival (DFS), progression-free survival (PFS), and relapse-free survival (RFS) are often considered in the review and approval process, especially in immunotherapy cancer clinical trials. Based on these four key study endpoints, there are a total of 15 possible endpoints: four single endpoints, i.e., {OS, RR, TTP, TS}; six co-primary endpoints (that combine two single endpoints), i.e., {(OS, RR), (OS, TTP), (OS, TS), (RR, TTP), (RR, TS), (TTP, TS)}; four composite endpoints (that combine three single endpoints), i.e., {(OS, RR, TTP), (OS, RR, TS), (OS, TTP, TS), (RR, TTP, TS)}; and one overall composite endpoint that combines all single endpoints. In practice, it is usually not clear which of the 15 study endpoints can best inform the disease status and/or evaluate the treatment effect. Moreover, different study endpoints may not translate one another although they may be highly correlated with one another.

In clinical trials, power calculation is often performed for sample size calculation. However, power analysis is very sensitive to the selected primary endpoint. Different endpoints may result in different sample sizes. As an example, consider cancer clinical trials with the four commonly considered primary endpoints of OS, RR, TTP, and TS. Power calculation for sample size based on OS, RR, TTP, and/or TS, which are of different data types, could be very different (Chow, Shao et al., 2017). For illustration purpose, Table 4.2 summarizes the sample sizes calculated based on different endpoints and their corresponding margins in oncology drug clinical trials based on historical data available in the literature (Motzer et al., 2019) with the conventionally selected margins (Table 4.2).

## 4.2.2 NASH Clinical Trials

The other example that we would like to consider here is the example concerning the NASH clinical trials. NASH is related to nonalcoholic fatty liver disease (NAFLD), which is the most common cause of chronic liver disease in the Western World. The risk factors for NAFLD include obesity, insulin resistance, and features of the metabolic syndrome. NAFLD has two major histologic phenotypes: a fatty liver and steatohepatitis. NASH progresses to cirrhosis more frequently than NAFLD and is rapidly rising as an etiology for end-stage liver disease requiring liver transplantation (Filozof et al., 2017). The histologic feature that appears to best predict mortality in NASH is the presence of significant fibrosis. There is currently no approved drug therapy for NASH in the United States. The development of such therapeutics is considered a public health priority.

For the assessment of therapeutic effect of NASH, there are three types of surrogate endpoints. Candidate surrogates are biomarkers that are under consideration for use in clinical trials but have not been accepted by the regulatory agency. Reasonably likely to predict clinical benefit are surrogates that have

**TABLE 4.2**

Sample Size Calculation Based on Different Endpoints

| Endpoint | OS | RR | TTP |
|---|---|---|---|
| $H_a$ | $\theta > \delta$ | $p_2 - p_1 > \delta$ | $\theta > \delta$ |
| Formula | $\dfrac{1}{\pi_1 \pi_2 p_o} \dfrac{\left(z_\alpha + z_\beta\right)^2}{\left(\ln\theta^* - \ln\delta\right)^2}$ | $\left( \dfrac{p_1(1-p_1)}{\kappa} + p_2(1-p_2) \right)$ $\times \left( \dfrac{z_\alpha + z_\beta}{p_2^* - p_1^* - \delta} \right)^2$ | $\dfrac{1}{\pi_1 \pi_2 p_t} \dfrac{\left(z_\alpha + z_\beta\right)^2}{\left(\ln\theta^* - \ln\delta\right)^2}$ |
| Margin $(\delta)$ | 0.82 | 0.29 | 0.61 |
| Other parameters | $p_o = 0.14;\ \theta^* = 1$ | $p_1 = 0.26; p_2 = 0.55; p_2^* - p_1^* = 0$ | $p_t = 0.4;\ \theta^* = 1$ |
| Sample size | 6213 | 45 | 351 |

*Note:* 1. For one-sided test of non-inferiority with significance level $\alpha = 0.05$ and expected power $1 - \beta = 0.9$. $H_a$ denotes the alternative hypothesis, $\delta$ is the non-inferiority margin, $\theta = h_1/h_2$ is the hazard ratio, $p_i$ is the response rate for sample $i$, $\pi_1$ and $\pi_2$ are the proportions of the sample size allocated to the two groups, $p_o$ is the overall probability of death occurring within the study period, $p_t$ is the overall probability of disease progression occurring within the study period, $\ln\theta$ is the natural logarithm of the hazard ratio, $\kappa = n_1/n_2$ is the sample size ratio, and $z_\alpha = \Phi^{-1}(1-\alpha)$ is $100(1-\alpha)\%$ quantile of the standard normal distribution. 2. We assume a balanced design, i.e., $\kappa = 1$ and $\pi_1 = \pi_1 = \dfrac{1}{2}$. The sample size calculation formulas are obtained in Chow et al. (2018). The non-inferiority margin and other parameters are based on the descriptive statistics for the 560 patients with PD-L1-positive tumors in Motzer et al. (2019), where the margin is selected as the improvement of clinically meaningful difference.

been accepted for use in phase III clinical trials but lack sufficient evidence to be used for full approval. Validated surrogates are biomarkers for which there are data to show that the biomarker does predict the clinical outcomes. For example, hemoglobin A1C is a validated surrogate endpoint that has been used as the basis for full/regular approval of therapeutic agents to treat diabetes. The accelerated and conditional approval pathways offer the potential to use surrogate endpoint(s) or intermediate clinical endpoints that are determined to be reasonably likely to predict the clinical benefit. The agencies may grant accelerated/conditional approval based on these surrogate endpoints, with full approval being granted based on subsequent confirmatory studies using the well-established and well-defined clinical outcomes.

There is no validated surrogate endpoint in NASH. Resolution of NASH (i.e., absence of ballooning with no or minimal inflammation by histology) is considered a surrogate endpoint reasonable likely to be associated with outcomes. Therefore, it is suggested that the approval process must involve a two-step approach, i.e., an initial accelerated or conditional approval followed by a final approval after confirming clinical benefit in preventing

progression to cirrhosis and liver outcomes, including cirrhosis decompensa-
tion events, overall death, and liver transplantation (Filozof et al., 2017). The
NAFLD activity score (NAS) is a histology-based validated scoring system
(an unweighted composite of steatosis, inflammation, and cellular ballooning scores) that can be used to assess pathologic changes in clinical trials
(Kleiner et al, 2005). The histologic endpoint of improvement in activity as
assessed by a reduction in at least two points in NAS (with at least 1-point
reduction in hepatocellular ballooning, a key feature required for the diagnosis of NASH) can be an acceptable marker of improvement if it is associated with no progression in fibrosis. However, there is no evidence to support
that NAS is reasonably likely associated with outcomes; thus, resolution of
NASH with no worsening of fibrosis (assessed as an increase in the fibrosis
stage by histology) and/or improvement in fibrosis with no progression of
steatohepatitis, i.e., no increase in activity as assessed by NAS, is required for
trials to support a marketing application. Filozof et al. (2017) indicated that
both resolution of NASH and improvement in fibrosis have been accepted
as surrogate endpoints and considered *reasonably likely to predict clinical
benefit* and have the potential to lead to an accelerated/conditional approval.
To provide a better understanding, Table 4.3 provides a list of endpoints and
populations in NASH clinical trials.

As it can have been seen from Table 4.3, there are no validated and widely
accepted endpoints for NASH clinical trials. Filozof et al. (2017) proposed
the idea of using a utility function to develop a therapeutic index for the
assessment of treatment effect for NASH. The use of a scalar or vector utility
function allows all endpoints to be linked for the overall analysis, providing

**TABLE 4.3**

Endpoints and Populations in Clinical Trial in NASH

| Phase | Primary Endpoint | Target Population |
|---|---|---|
| Early-phase trials/ proof-of-concept trials | Endpoint should be based on the mechanism of drug<br>Reduction in liver fat with a sustained improvement in transaminases<br>Improvement in biomarkers of liver inflammation, apoptosis, and/or fibrosis<br>Consider using improvement in NAS (ballooning and inflammation) and/or fibrosis | Ideal to enroll patients with biopsy-proven NASH but acceptable to enroll patients at high risk for NASH (i.e., evidence of fatty liver, two components of the metabolic syndrome, evidences of liver stiffness by imaging) |
| Dose ranging/ phase 2 | Resolution of NASH without worsening of fibrosis alternatively, improvement in disease activity (NAS)/improvement in ballooning/ information without worsening of fibrosis | Biopsy-proven NASH and NAS ≥ 4, include patients with NASH and liver fibrosis, include a sufficient number of patients with NAS and fibrosis stage 2/3 to inform phase 3 |

*(Continued)*

**TABLE 4.3 (*Continued*)**

Endpoints and Populations in Clinical Trial in NASH

| Phase | Primary Endpoint | Target Population |
|---|---|---|
| Trials to support a marketing application: phase 3 | Resolution of steatohepatitis and no worsening of fibrosis<br>Improvement in fibrosis with no worsening of steatohepatitis<br>A co-primary endpoint of the above depending on the mode of action, either one or the other can be used | Patients with biopsy-confirmed NASH with moderate/advanced fibrosis (F2/F3) |
| Trials to support a marketing application phase 4 (post-marketing part) | Clinical outcome trial underway by the time of submission<br>Composite endpoint: histopathologic progression to cirrhosis; MELD score change by >2 points or MELD increase to >15 in population enrolled with MELD < 13; death; transplant; cirrhosis decompensation events | Patients with biopsy-confirmed NASH with moderate/advanced fibrosis (F2/F3) |

*Abbreviation*: MELD, model for end-stage liver disease.

a reliable method to assess the treatment effect (see also Chow and Huang, 2019b). The therapeutic index function is briefly described in the next section of this chapter.

More details regarding design and analysis of NASH clinical trials are given in Chapter 14.

## 4.3 Therapeutic Index Function

Following the proposal of Filozof et al. (2017), Chow and Huang (2019b) studied the development of therapeutic index function for multiple endpoints in clinical trials, which is briefly described below.

### 4.3.1 Utility Function

In a clinical trial, suppose there are $J$ study endpoints, denoted by $e_i, i = 1, \ldots, J$. Let

$$e = \left(e_1, e_2, \ldots, e_J\right)'$$

be the $J$ clinical endpoints at baseline. The therapeutic index is defined as:

$$I_i = f_i(\omega_i, e), \quad i = 1, \ldots, K, \tag{4.1}$$

where $\omega_i = \left(\omega_{i1}, \omega_{i2}, ..., \omega_{ij}\right)'$ is a vector of weights with $\omega_{ij}$ being the weight for $e_j$ with respect to index $I_i$, and $f_i(\bullet)$ is a utility function, which is referred to as a therapeutic index function, to construct therapeutic index $I_i$ based on $\omega_i$ and $e$. Generally, $e_j$ can be of different data types (e.g., continuous, binary, time-to-event endpoints), and $\omega_{ij}$ is pre-specified (or calculated by pre-specified criteria) and can be different for different therapeutic index $I_i$. Moreover, the therapeutic index function typically generates a vector of index $\left(I_1, I_2, ..., I_K\right)'$; if $K = 1$, it reduces to a single (composite) index. As an example, consider

$$I_i = \sum_{j=1}^{J} \omega_{ij} e_j,$$

then $I_i$ is simply a linear combination of the endpoints; moreover, if

$$\omega_i = \left(\frac{1}{J}, \frac{1}{J}, ..., \frac{1}{J}\right)',$$

then $I_i$ is the average over all the endpoints.

### 4.3.2 Selection of $\omega_i$

For the therapeutic index function given in (3.1), one of the important concerns is how to select the weights $\omega_i$. There might be various ways of specifying the weights (e.g., based on variabilities associated with the individual study endpoints), while we would like to propose selecting weights based on the $p$-values observed from the individual study endpoints. In clinical trials, the $p$-values are indicators of the levels of substantial evidences regarding the safety and effectiveness of the test treatment under investigation provided by the individual endpoints. Specifically, denote

$$\theta_j, \ j = 1, ..., J$$

as the treatment effect assessed by the endpoint $e_j$. Without the loss of generality, $\theta_j$ is tested by the following hypotheses:

$$H_{0j}: \ \theta_j \leq \delta_j \text{ versus } H_{aj}: \ \theta_j > \delta_j, \tag{4.2}$$

where $\delta_j, j = 1, ..., J$ are the pre-specified margins. Under some appropriate assumptions, we can calculate the $p$-value $p_j$ for each $H_{0j}$ based on the sample of $e_j$ and the weights $\omega_i$ can be constructed based on

$$\boldsymbol{p} = \left(p_1,\, p_2,\, ...,\, p_J\right)',$$

That is,

$$\omega_{ij} = \omega_{ij}\left(\boldsymbol{p}\right), \tag{4.3}$$

which is reasonable since each $p$-value indicates the significance of the treatment effect based on its responding endpoint; thus, it is possible to use all the information available to construct effective therapeutic indexes. Note that $\omega_{ij}(\cdot)$ should be constructed such that high value of $\omega_{ij}$ is associated with low value of $p_j$. For example,

$$\omega_{ij} = \frac{1}{p_j} \Big/ \sum_{j=1}^{J} \frac{1}{p_j}.$$

### 4.3.3 Determination of $f_i(\cdot)$

Another important issue is how to select the utility function $f_i(\cdot)$ for the construction of the therapeutic index. In practice, $f_i(\cdot)$ could be linear or non-linear, or with more complicated forms. For simplicity, we consider $f_i(\cdot)$ as a linear function here. Thus, (4.1) reduces to

$$I_i = \sum_{j=1}^{J} \omega_{ij} e_j = \sum_{j=1}^{J} \omega_{ij}(\boldsymbol{p}) e_j, \quad i = 1,..., K. \tag{4.4}$$

### 4.3.4 Assumption for the Distribution of $e$

In order to study the statistical properties of the developed therapeutic index described in (4.1), we need to specify the distribution of $e$. For simplicity, we assume $e$ follows the multidimensional normal distribution $N(\theta, \Sigma)$, where

$$\theta = \left(\theta_1,..., \ \theta_J\right)'$$

and

$$\Sigma = \left(\sigma_{jj'}^2\right)_{J \times J},$$

with

$$\sigma_{jj'}^2 = \sigma_j^2, \ j' = j \text{ and } \sigma_{jj'}^2 = \rho_{jj'}\sigma_j\sigma_{j'}, \ j' \neq j.$$

## 4.4 Statistical Evaluation of the Therapeutic Index

### 4.4.1 Evaluation Criteria

Although $e_j$ can be of different data types, without the loss of generality and for illustration purpose, we assume they are of the same type at this step. On the one hand, we would like to investigate the predictability of $I_i$ given that $e_j$ can inform the disease (drug) status; on the other hand, we are also interested in the predictability of $e_j$ given that $I_i$ is informative. Particularly, we are interested in the following two conditional probabilities:

$$\text{(i) } p_{1ij} = \Pr\left(I_i \mid e_j\right), \ i = 1,\dots, K \ \& \ j = 1,\dots, J \tag{4.5}$$

and

$$\text{(ii) } p_{2ij} = \Pr\left(e_j \mid I_i\right), \ i = 1,\dots, K \ \& \ j = 1,\dots, J. \tag{4.6}$$

Intuitively, we would expect that $p_{1ij}$ to be relatively large given that $e_j$ is informative since $I_i$ is a function of $e_j$, especially when relatively high weight is assigned to $e_j$; on the other hand, $p_{2ij}$ could be small even if $I_i$ is predictive since the information contained in $I_i$ may be attributed to another endpoint $e_{j'}$ rather than other $e_j$.

In what follows, equations (4.5) and (4.6) will be derived under the assumption that (i) the specified weights $\omega_i$ based on $p$-values, and (ii) the distribution of $e$ and the functions $f_i(\cdot)$ are described in the previous section.

### 4.4.2 Derivation of $\Pr\left(I_i \mid e_j\right)$ and $\Pr\left(e_j \mid I_i\right)$

Suppose $n$ subjects are independently and randomly selected from the population for the clinical trial. For each baseline endpoint $e_j$ and hypothesis $H_{0j}$, a test statistic $\hat{e}_j$ is constructed based on the observations of the $n$ subjects and the corresponding $p$-value $p_j$ is calculated. $e_j$ is informative and is equivalent to $\hat{e}_j > c_j$ for some pre-specified critical value $c_j$, based on $\delta_j$, significance level $\alpha$, and the variance of $\hat{e}_j$. The estimate of the therapeutic index $I_i$ in (4.4) can be accordingly constructed as

$$\hat{I}_i = \omega_i' \hat{e} = \sum_{j=1}^{J} \omega_{ij}\hat{e}_j, \ i = 1,\dots, K, \tag{4.7}$$

where

$$\omega_i = \left(\omega_{i1}, \omega_{i2}, \dots, \omega_{ij}\right)'$$

and

$$\omega_{ij} = \omega_{ij}(p)$$

is calculated based on the $p$-values on

$$p = (p_1, p_2, \ldots, p_J)',$$

and

$$\hat{e} = (\hat{e}_1, \hat{e}_2, \ldots, \hat{e}_J)'.$$

$\hat{I}_i$ is informative if $\hat{I}_i > d_i$ for some pre-specified threshold $d_i$. Thus, (4.5) and (4.6) become

$$\text{(i)} \quad p_{1ij} = \Pr\left(\hat{I}_i > d_i \middle| \hat{e}_j\rangle c_j\right), \quad i = 1, \ldots, K \ \& \ j = 1, \ldots, J \qquad (4.8)$$

and

$$\text{(ii)} \quad p_{2ij} = \Pr\left(\hat{e}_j > c_j \middle| \hat{I}_i\rangle d_i\right), \quad i = 1, \ldots, K \ \& \ j = 1, \ldots, J. \qquad (4.9)$$

Without the loss of generality, suppose $\hat{e}$ is the vector of sample means, then $\hat{e}$ follows the multidimensional normal distribution $N(\theta, \Sigma/n)$ based on the normality assumption of $e$. Moreover, $\hat{I}_i$ follows the normal distribution $N(\varphi_i, \eta_i^2/n)$, where

$$\varphi_i = \omega_i'\theta = \sum_{j=1}^{J} \omega_{ij}\theta_j$$

and

$$\eta_i^2 = \omega_i'\Sigma\omega_i.$$

Further, $\left(\hat{e}_j, \hat{I}_i\right)'$ jointly follows a binormal distribution $N(\mu, \Gamma/n)$, where

$$\mu = (\theta_j, \varphi_i)'$$

and

$$\Gamma = \begin{pmatrix} \sigma_j^2 & 1_j'\Sigma\omega_i \\ 1_j'\Sigma\omega_i & \eta_i^2 \end{pmatrix} = \begin{pmatrix} \sigma_j^2 & \rho_{ji}^*\sigma_j\eta_i \\ \rho_{ji}^*\sigma_j\eta_i & \eta_i^2 \end{pmatrix},$$

where $\mathbf{1}_j$ is a $J$ dimensional vector of 0 except the $j$th item, which equals 1, and thus

$$\rho_{ji}^* = \mathbf{1}_j' \Sigma \omega_i / (\sigma_j \eta_i) = \sum_{j'=1}^{J} \omega_{ij'} \sigma_{jj'}^2 / (\sigma_j \eta_i) = \sum_{j'=1}^{J} \omega_{ij'} \rho_{jj'} \sigma_j \sigma_{j'} / (\sigma_j \eta_i)$$

$$= \frac{1}{\eta_i} \sum_{j'=1}^{J} \omega_{ij'} \rho_{jj'} \sigma_{j'}.$$

Thus, the conditional probabilities (4.8) and (4.9) become

$$\text{(i)} \; p_{1ij} = \frac{\Pr\left(\hat{I}_i > d_i, \hat{e}_j > c_j\right)}{\Pr\left(\hat{e}_j > c_j\right)} = \frac{1 - \Phi\left(\dfrac{\sqrt{n}\left(c_j - \theta_j\right)}{\sigma_j}\right) - \Phi\left(\dfrac{\sqrt{n}\left(d_i - \varphi_i\right)}{\eta_i}\right) + \Psi\left(\dfrac{\sqrt{n}\left(c_j - \theta_j\right)}{\sigma_j}, \dfrac{\sqrt{n}\left(d_i - \varphi_i\right)}{\eta_i}, \rho_{ji}^*\right)}{1 - \Phi\left(\dfrac{\sqrt{n}\left(c_j - \theta_j\right)}{\sigma_j}\right)} \quad (4.10)$$

and

$$\text{(ii)} \; p_{2ij} = \frac{\Pr\left(\hat{I}_i > d_i, \hat{e}_j > c_j\right)}{\Pr\left(\hat{I}_i > d_i\right)} = \frac{1 - \Phi\left(\dfrac{\sqrt{n}\left(c_j - \theta_j\right)}{\sigma_j}\right) - \Phi\left(\dfrac{\sqrt{n}\left(d_i - \varphi_i\right)}{\eta_i}\right) + \Psi\left(\dfrac{\sqrt{n}\left(c_j - \theta_j\right)}{\sigma_j}, \dfrac{\sqrt{n}\left(d_i - \varphi_i\right)}{\eta_i}, \rho_{ji}^*\right)}{1 - \Phi\left(\dfrac{\sqrt{n}\left(d_i - \varphi_i\right)}{\eta_i}\right)} \quad (4.11)$$

Moreover

$$\frac{p_{2ij}}{p_{1ij}} = \frac{\Pr\left(\hat{e}_j > c_j\right)}{\Pr\left(\hat{I}_i > d_i\right)} = \frac{1 - \Phi\left(\dfrac{\sqrt{n}\left(c_j - \theta_j\right)}{\sigma_j}\right)}{1 - \Phi\left(\dfrac{\sqrt{n}\left(d_i - \varphi_i\right)}{\eta_i}\right)}, \quad (4.12)$$

where $\Phi(x)$ and $\Psi(x, y, \rho)$ denote the cumulative distribution functions for standard single variate normal and bivariate normal distributions, respectively. Note that both conditional probabilities (4.10) and (4.11) depend on the parameters $\theta$, $\Sigma$; the sample size $n$; the number of baseline endpoints $J$; the pre-specified weights $\omega_i$; and the pre-specified thresholds $c_j$, $d_i$, which further depend on the hypothesis testing margins $\delta_j$ and pre-specified type I error rate(s), among others. Intuitively, there are not simple formulas for (4.10) and (4.11) that can be derived directly. Although methods like Taylor expansion may be employed to approximate (4.10) and (4.11), it is still non-trivial and could be quite complicated. However, note that $\Phi(x)$ is monotonic increasing, based on (4.12) we have

$$\frac{p_{2ij}}{p_{1ij}} < 1 \Leftrightarrow \frac{c_j - \theta_j}{\sigma_j} > \frac{d_i - \varphi_i}{\eta_i}. \tag{4.13}$$

Moreover, we assume

$$c_j = \delta_j + z_\alpha \frac{\sigma_j}{\sqrt{n}}$$

conventionally, and $d_i$ is a linear combination of $c_j$s, i.e.,

$$d_i = \sum_{j=1}^{J} \omega_{ij} c_j = \sum_{j=1}^{J} \omega_{ij}\delta_j + \frac{z_\alpha}{\sqrt{n}} \sum_{j=1}^{J} \omega_{ij}\sigma_j, \quad i = 1, \dots, K, \tag{4.14}$$

Then (4.13) is further expressed as

$$\frac{p_{2ij}}{p_{1ij}} < 1 \Leftrightarrow \frac{c_j - \theta_j}{\sigma_j} > \frac{d_i - \varphi_i}{\eta_i}$$

$$\tag{4.15}$$

$$\Leftrightarrow \left(1 - \frac{\sigma_j}{\eta_i}\omega_{ij}\right)\left(\Delta\theta - \frac{z_\alpha}{\sqrt{n}}\sigma_j\right) < \frac{\sigma_j}{\eta_i}\left(\Delta\theta_i^{(-j)} - \frac{z_\alpha}{\sqrt{n}}\sigma_i^{(-j)}\right),$$

where

$$\Delta\theta_j = \theta_j - \delta_j,$$

$$\Delta\theta = \left(\Delta\theta_1, \Delta\theta_2, \dots, \Delta\theta_J\right)',$$

$$\Delta\theta_i^{(-j)} = \omega_i^{(-j)'}\Delta\theta = \sum_{j'\neq j}^{J} \omega_{ij'}\Delta\theta_{j'},$$

$$\sigma_i^{(-j)} = \sum_{j' \neq j}^{J} \omega_{ij'} \sigma_{j'},$$

$\omega_i^{(-j)}$ equals $\omega_i$ except the $i$th item equals 0. To obtain more insights of (4.15), we assume $J = 2$, $K = 1$, and focus on $j = 1$ without the loss of generality. Then, the last inequality in (4.15) can be simplified as

$$\left(1 - \frac{\sigma_1}{\eta}\omega_1\right)\left(\Delta\theta_1 - \frac{z_\alpha}{\sqrt{n}}\sigma_1\right) < \frac{\sigma_1}{\eta}\omega_2\left(\Delta\theta_2 - \frac{z_\alpha}{\sqrt{n}}\sigma_2\right), \tag{4.16}$$

where $\omega_1$ and $\omega_2$ are the weights for the two endpoints, respectively, with $\omega_1 + \omega_2 = 1$, and

$$\eta = \sqrt{\omega_1^2\sigma_1^2 + 2\rho\omega_1\omega_2\sigma_1\sigma_2 + \omega_2^2\sigma_2^2}$$

and $\rho$ is the correlation coefficient of the two endpoints. Obviously, (4.16) depends on the variabilities of the endpoints and their correlation, the underlined effect sizes of both endpoints, the weights, and the sample size. We illustrate several special scenarios of (4.16) in Table 4.4.

From Table 4.3, we can see a remarkable situation that when $\rho = 1$, $\sigma_1 = \frac{1}{\tau}\sigma_2$, whether $p_{1ij}$ is greater than $p_{2ij}$ depends on whether the underlined effect size $\Delta\theta_1$ is less than $\frac{1}{\tau}\Delta\theta_2$ only, regardless of the weights. For other situations, the relation between $p_{1ij}$ and $p_{2ij}$ varies for different combinations of weights, variabilities and correlations, underlined effect sizes, and sample sizes.

### 4.4.3 Remarks

In this section, for simplicity, we assume different single endpoints are of the same data type, i.e., continuous endpoint. In practice, however, different endpoints of different data types may be considered in a clinical trial like a cancer clinical trial. In this case, the development of therapeutic index is much more difficult, if it is not impossible, because it is not easy to figure out the joint distribution of endpoints with different data types (e.g., a continuous endpoint, a binary endpoint, and/or a time-to-event endpoint) without some plausible assumptions.

For illustration of the idea of the use of the developed therapeutic index in clinical trials, we consider data transformations such that all the endpoints are continuous in order to overcome these difficulties. In this case, the multinormal approximation can be assumed for the joint distribution of the endpoints for the evaluation of the developed therapeutic index. Specifically, for continuous endpoint, we considered standardization of the data, i.e., subtract the mean and then divide it by the standard deviation.

**TABLE 4.4**

Illustrated Inequalities (4.16) with Respect to Different Parameter Settings

| Parameters | Inequality (4.16) |
|---|---|
| | $$\left(1-\frac{\sigma_1}{\sqrt{\omega_1^2\sigma_1^2+2\rho\omega_1\omega_2\sigma_1\sigma_2+\omega_2^2\sigma_2^2}}\omega_1\right)$$ $$\left(\Delta\theta_1-\frac{z_\alpha}{\sqrt{n}}\sigma_1\right)<\frac{\sigma_1}{\sqrt{\omega_1^2\sigma_1^2+2\rho\omega_1\omega_2\sigma_1\sigma_2+\omega_2^2\sigma_2^2}}\omega_2\left(\Delta\theta_2-\frac{z_\alpha}{\sqrt{n}}\sigma_2\right)$$ |
| $\omega_1=\omega_2=1/2$ | $$\left(1-\frac{\sigma_1}{\sqrt{\sigma_1^2+2\rho\sigma_1\sigma_2+\sigma_2^2}}\right)\left(\Delta\theta_1-\frac{z_\alpha}{\sqrt{n}}\sigma_1\right)<\frac{\sigma_1}{\sqrt{\sigma_1^2+2\rho\sigma_1\sigma_2+\sigma_2^2}}\left(\Delta\theta_2-\frac{z_\alpha}{\sqrt{n}}\sigma_2\right)$$ |
| $\rho=0$ | $$\left(1-\frac{\omega_1\sigma_1}{\sqrt{\omega_1^2\sigma_1^2+\omega_2^2\sigma_2^2}}\right)\left(\Delta\theta_1-\frac{z_\alpha}{\sqrt{n}}\sigma_1\right)<\frac{\omega_2\sigma_1}{\sqrt{\omega_1^2\sigma_1^2+\omega_2^2\sigma_2^2}}\left(\Delta\theta_2-\frac{z_\alpha}{\sqrt{n}}\sigma_2\right)$$ |
| $\rho=1$ | $$\left(1-\frac{\omega_1\sigma_1}{\omega_1\sigma_1+\omega_2\sigma_2}\right)\left(\Delta\theta_1-\frac{z_\alpha}{\sqrt{n}}\sigma_1\right)<\frac{\omega_2\sigma_1}{\omega_1\sigma_1+\omega_2\sigma_2}\left(\Delta\theta_2-\frac{z_\alpha}{\sqrt{n}}\sigma_2\right)$$ |
| $\sigma_1=\sigma_2$ | $$\left(1-\frac{\omega_1}{\sqrt{\omega_1^2+2\rho\omega_1\omega_2+\omega_2^2}}\right)\left(\Delta\theta_1-\frac{z_\alpha}{\sqrt{n}}\sigma_1\right)<\frac{\omega_2}{\sqrt{\omega_1^2+2\rho\omega_1\omega_2+\omega_2^2}}\left(\Delta\theta_2-\frac{z_\alpha}{\sqrt{n}}\sigma_1\right)$$ |
| $\omega_1=\omega_2=1/2$ <br> $\rho=0$ | $$\left(1-\frac{\sigma_1}{\sqrt{\sigma_1^2+\sigma_2^2}}\right)\left(\Delta\theta_1-\frac{z_\alpha}{\sqrt{n}}\sigma_1\right)<\frac{\sigma_1}{\sqrt{\sigma_1^2+\sigma_2^2}}\left(\Delta\theta_2-\frac{z_\alpha}{\sqrt{n}}\sigma_2\right)$$ |
| $\omega_1=\omega_2=1/2$ <br> $\rho=1$ | $$\left(1-\frac{\sigma_1}{\sigma_1+\sigma_2}\right)\left(\Delta\theta_1-\frac{z_\alpha}{\sqrt{n}}\sigma_1\right)<\frac{\sigma_1}{\sigma_1+\sigma_2}\left(\Delta\theta_2-\frac{z_\alpha}{\sqrt{n}}\sigma_2\right)$$ |
| $\omega_1=\omega_2=1/2$ <br> $\sigma_1=\sigma_2$ | $$\left(1-\frac{1}{\sqrt{2+2\rho}}\right)\Delta\theta_1-\frac{1}{\sqrt{2+2\rho}}\Delta\theta_2<\left(1-\frac{2}{\sqrt{2+2\rho}}\right)\frac{z_\alpha}{\sqrt{n}}\sigma_1$$ |
| $\rho=0,\sigma_1=\sigma_2$ | $$\Delta\theta_1<\frac{\sqrt{\omega_1^2+\omega_2^2}+\omega_1}{\omega_2}\Delta\theta_2-\frac{\sqrt{\omega_1^2+\omega_2^2}+\omega_1-\omega_2}{\omega_2}\frac{z_\alpha}{\sqrt{n}}\sigma_1$$ |
| $\omega_1=\omega_2=1/2$ <br> $\rho=0,\ \sigma_1=\sigma_2$ | $$\Delta\theta_1<\left(\sqrt{2}+1\right)\Delta\theta_2-\frac{\sqrt{2}z_\alpha}{\sqrt{n}}\sigma_1$$ |
| $\omega_1=1/3,$ <br> $\omega_2=2/3$ <br> $\rho=0,\sigma_1=\sigma_2$ | $$\Delta\theta_1<\frac{\sqrt{5}+1}{2}\Delta\theta_2-\frac{\sqrt{5}-1}{2}\frac{z_\alpha}{\sqrt{n}}\sigma_1$$ |
| $\rho=1,\ \sigma_1=\frac{1}{\tau}\sigma_2$ | $$\Delta\theta_1<\frac{1}{\tau}\Delta\theta_2$$ |

For binary endpoint, treat the binary endpoint as the response variable and fit a regression model (e.g., logistic model, probit model, or log-log model, among others) with some reasonable independent variable(s) (with input from clinician). Then, standardize the estimated success probabilities. For time-to-event endpoint, we may either subtract the median and then divide it by the standard deviation of the median, or subtract the mean and then divide it by the standard deviation of the mean. The transformed endpoints are then used as the surrogate endpoints to construct the therapeutic indices.

In what follows, the above idea is illustrated using a numerical example based on the real practices.

## 4.5 A Numerical Example

In this section, we will demonstrate how to develop a therapeutic index that combines a binary endpoint and a survival endpoint based on the simulated patient data, which mimic a recent cancer trial.

### 4.5.1 Simulate Patient Data

For illustration purpose, we simulated patient data which contain a binary endpoint and a survival endpoint using parameters with respect to patients with PD-L1-positive tumors reported in a recent cancer trial (Motzer et al., 2019).

Without the loss of generality and for simplicity, we consider a single-arm study with $n = 300$ patients. Assume that the underlying success probability for the selected binary endpoint is about $p_b = 0.55$ and the median PFS for the selected time-to-event endpoint is about $t_s = 13.8$ month. Denote the binary endpoint as $e_1$ and the survival endpoint as $e_2$. Thus, $e = (e_1, e_2)'$. The procedure for the generation of the patient data is summarized below.

Step 1. Generate a two-dimensional random vector $X = (X_1, X_2)'$ from a binormal distribution

$$N\left(\mu_0, \left(\sigma_{jj'}^2\right)_{2\times 2}\right),$$

where

$$\mu_0 = \left(\log\frac{p_b}{1-p_b}, \log(t_s)\right)',$$

and

$$\sigma_{jj}^2 = 1, \quad \sigma_{12}^2 = 0.5, \quad j = 1, 2.$$

Step 2. Generate a random error vector $\varepsilon = (\varepsilon_1, \varepsilon_2)'$ from a binormal distribution

$$N\left( \begin{pmatrix} 0 \\ 0 \end{pmatrix}, \; 0.2 \begin{pmatrix} 1 & 0 \\ 0 & 1 \end{pmatrix} \right).$$

Step 3. Calculate

$$e_1^* = \frac{\exp(X_1 + \varepsilon_1)}{1 + \exp(X_1 + \varepsilon_1)},$$

and generate the binary endpoint $e_1$ from the Bernoulli distribution with the success probability of $e_1^*$.

Step 4. Calculate $e_2^* = \exp(X_2 + \varepsilon_2)$ and generate a censoring variable $e_C$ with

$$\log(e_C) \sim N(0.7*\log(t_s), 1).$$

Let

$$T^* = \min(e_2^*, e_C),$$

$$T = \max(T^* + t_s - \text{median}(T^*), 0),$$

and

$$C = 1\{e_2^* < e_C\}.$$

The survival endpoint is $e_2 = \{T, C\}$.

For the generated patient data, the average of $e_1$ is about 0.53, which is close to $p_b$; the median of $e_2$ is equal to $t_s$ with a censor rate about 29.3%. Moreover, the correlation between the underlined success probability $e_1^*$ of the binary endpoint and the survival endpoint $e_2$ is about 0.14, which can be adjusted by modifying $\sigma_{12}^2$ in the above procedure. Assume the random variable $X_1$ is also obtained such that it can be used as the explanatory variable to fit the regression model for $e_1$. Thus, we generated a data set which is comprised of a binary endpoint $e_1$, a survival endpoint $e_2$, and a continuous variable $X_1$ to use as an explanatory variable for fitting a logistic regression model.

### 4.5.2 Data Transformation, Normal Approximation, and Illustration

Fitting the logistic regression model with the explanatory variable $X_1$, we have

$$\log \frac{p_b}{1 - p_b} = a + bX_1, \tag{4.17}$$

where $p_b$ is the underlying success probability of $e_1$. The fitted value of success probability has a sample mean of 0.53 and a sample standard deviation of 0.50, and its standardized form is denoted as $\tilde{e}_1$.

For survival data, median is more commonly used than mean. However, the variance of the median is intractable (Rider, 1960) such that median may not be a good option for data standardization. Thus, the mean is adopted for data standardization of the survival endpoint $e_2$. The same mean and standard deviation of $e_2$ are 15.39 and 5.43, respectively, and the standardized form of $e_2$ is denoted as $\tilde{e}_2$.

The two standardized endpoints $\tilde{e}_1$ and $\tilde{e}_2$ are then assumed to approximately follow the binormal distribution with mean zero and variance 1, and their correlation is estimated from the standardized observations (about 0.17 in the simulated patient data).

Note that the two conditional probabilities (4.8) and (4.9), and $p_{1ij} > p_{2ij}$ is equivalent to that probability (4.16), hold when there are two endpoints with joint normal assumption (or approximation). To illustrate an instance of (4.16), assume $\delta_1 = 0.39$ and $\delta_2 = 13.8$. Note that $\theta_1 = 0.53$ and $\theta_2 = 15.39$; thus, $\Delta\theta_1 = \theta_1 - \delta_1 = 0.14$ and $\Delta\theta_2 = \theta_2 - \delta_2 = 1.59$. Moreover, $\sigma_1 = 0.50$, $\sigma_2 = 5.43$, $n = 300$, $\rho = 0.17$, and $z_\alpha = 1.96$ given $\alpha = 0.025$. Note that (4.16) can be expressed as

$$\frac{\sigma_1 \omega_1 + \varepsilon \sigma_2 \omega_2}{\sqrt{\sigma_1^2 \omega_1^2 + 2\rho \sigma_1 \sigma_2 \omega_1 \omega_2 + \sigma_2^2 \omega_2^2}} > 1, \tag{4.18}$$

where $\varepsilon = \left( \Delta\theta_2 - \frac{z_\alpha}{\sqrt{n}} \sigma_2 \right) \Big/ \left( \Delta\theta_1 - \frac{z_\alpha}{\sqrt{n}} \sigma_1 \right)$, given that $\Delta\theta_1 - \frac{z_\alpha}{\sqrt{n}} \sigma_1 > 0$. On the one hand, if we assume equal weights, i.e., $\omega_1 = \omega_2 = 1/2$, the left-hand side of (4.18) is 10.78 and (4.18) holds. On the other hand, recall the hypothesis testing in (4.4), the ratio of p-values of the two alternative hypotheses $H_{a1}$: $\theta_1 > \delta_1$ and $H_{a2}$: $\theta_2 > \delta_2$ is $p_1/p_2 = 3.38/2.03$; if we let $\omega_j = \frac{1}{p_j} \Big/ \sum_{j=1}^{J} \frac{1}{p_j}$, then $\omega_1 = 0.37$ and $\omega_2 = 0.63$, the left-hand side of (4.18) is 10.96 and (4.18) holds again. Thus, the hypothetical example supports the intuition of $p_{1ij} > p_{2ij}$, i.e., the conditional probability that the therapeutic index $I_i$ is informative given that one original endpoint $e_j$ is informative is larger than the probability that one original endpoint $e_j$ is informative given that the therapeutic index $I_i$ is informative.

## 4.6 Concluding Remarks

In clinical trials, endpoint selection is always a concern, especially when multiple endpoints are available. Based on the available endpoints, a number of derived endpoints (including individual endpoints, co-primary endpoints, and composite endpoints) may be obtained. In practice, it is often a concern which (derived), which can best inform the disease status and/or treatment effect, should be used for the evaluation of the safety and effectiveness of the test treatment under investigation as some endpoints may be achieved and some may not be. Alternatively, the concept of therapeutic index that is able to combine and fully utilize the information collected from the individual and derived endpoints is developed. In this chapter, for simplicity and illustration purpose, a therapeutic index was developed under the assumption that (i) different endpoints are of the same data type, (ii) the utility function is linear, and (iii) the selection of weights is based on the observed p-values of individual endpoints. The results indicate that the performance of the developed therapeutic index is better than those of the individual endpoints and/or co-primary endpoints in terms of false-positive and false-negative rates.

For the case where different endpoints are of different data types (e.g., continuous, binary, or time-to-event endpoint), as discussed in Section 2, to calculate the conditional distributions of (4.5) and (4.6), it is necessary to obtain the joint distribution of the baseline endpoints and the therapeutic indices. This is, however, difficult if not impossible to obtain. To overcome these difficulties, it is suggested appropriate data transformation be considered for obtaining normal approximation in order to use the theoretical results derived in this chapter. The idea of data transformation for an application in cancer trial is illustrated in Section 4.4.

It should be noted that for studying the joint distribution of different endpoints, there are results available in the literature. For example, for studying the joint distribution of two binary endpoints $e_{b1}$, $e_{b2}$ with success probabilities

$$p_{bk} = \Pr(e_{bk} = 1), \quad k = 1,2$$

and

$$p_{b12} = \Pr(e_{b1} = 1, \ e_{b2} = 1),$$

their correlation coefficient is (Leisch et al., 1998)

$$\gamma_{b12} = \frac{p_{b12} - p_{b1}p_{b2}}{\sqrt{p_{b1}(1 - p_{b1})p_{b2}(1 - p_{b2})}}.$$

When there are more than two binary endpoints, their joint distribution and pairwise correlations can be similarly extended. Moreover, the marginal and conditional distributions of a multivariate Bernoulli distribution are still Bernoulli (Dai et al., 2013). However, the distributions of therapeutic indices constructed based on the binary endpoints can be very complicated. One simple exception is that the index follows a binomial distribution given that it is the sum of mutually independent binary endpoints with the same success probability. On the other hand, for time-to-event endpoints, the joint distribution and correlation structure can be more complicated than those of binary endpoints. Although there are a few methods to estimate the correlation of bivariate failure times under censoring, such as the normal copula approach (NCE) and the iterative multiple imputation (IMI) method (Schemper et al., 2013), it is still very difficult to derive the joint distribution of survival endpoints, which often require strong model assumptions, in addition to the complexity of the joint and conditional distributions of the survival endpoints and the indices.

# 5

## Strategy for Margin Selection

### 5.1 Introduction

For the approval of a proposed biosimilar product, the sponsor is required to provide totality-of-the-evidence for the demonstration of high similarity and show that there is no clinically meaningful difference in safety, purity, and potency between the proposed biosimilar (test) product and its innovative (reference) product (FDA, 2015). The United States (US) Food and Drug Administration (FDA) recommends a stepwise approach for obtaining the totality-of-the-evidence for demonstrating biosimilarity between the test product and the reference product. The stepwise approach includes analytical similarity assessment for functional and structural characterizations of critical quality attributes that are relevant to clinical outcomes during the manufacturing process, pharmacokinetic and pharmacodynamic (PK/PD) studies for pharmacological activities, and immunogenicity and clinical studies for safety and potency of the proposed biosimilar product. For the assessment of similarity in analytical, PK/PD, and clinical studies, including immunogenicity studies, the two one-sided tests (TOST) procedure proposed by Schuirmann (1987) was recommended. Schuirmann's TOST procedure is the first to test non-inferiority (one side) at the $\alpha$-level of significance and then to test non-superiority (the other side) at the $\alpha$-level of significance after the non-inferiority has been established. Schuirmann's TOST procedure is a size-$\alpha$-test (Chow and Shao, 2002). Note that in many cases, Schuirmann's TOST procedure at the $\alpha$-level of significance is operationally equivalent to a $(1 - 2\alpha) \times 100\%$ confidence interval approach for testing similarity.

In testing of similarity between a test product and a reference product, the selection of similarity margin is critical. For comparative clinical studies, a scientific justification regarding clinical judgment and statistical rationale based on previous studies and/or historical data is essential for selecting an appropriate similarity margin in support of a demonstration that there are no clinically meaningful differences between the proposed biosimilar and the reference products with respect to the chosen study endpoints. In 2000, the International Conference on Harmonization (ICH) published a guideline on *Choice of Control Group and Related Design and Conduct Issues in Clinical Trials* to assist the sponsors for the selection of an appropriate non-inferiority margin (ICH, 2000). As indicated in the ICH E10 guideline, non-inferiority margins may be selected based on past experience in placebo-controlled trials under similar conditions to the new trial. ICH E10 also pointed out that the selection of a non-inferiority should be suitably conservative and reflect uncertainties in the evidence on which the choice is based on. In 2016, United States (US) Food and Drug Administration (FDA) also published Draft Guidance on Non-Inferiority Clinical Trials (FDA, 2016). In practice, following the FDA 2016 guidance, non-inferiority margin is often obtained based on a meta-analysis that usually combines clinical data from a number of previous studies. In practice, sponsor's proposed margin can be wider to be accepted by the FDA. In this case, the sponsors are usually asked to provide scientific justification, including clinical judgment and statistical rationale for the wider margin.

The remaining of this chapter is organized as follows. In the next section, non-inferiority (similarity) margin obtained from a meta-analysis is briefly described. In Section 5.3, a strategy based on a risk-and-benefit analysis to facilitate the communication between the sponsor and the FDA in making final decision for margin selection is proposed. Risk assessment with continuous endpoints is illustrated in Section 5.4. The performance of the proposed strategy is evaluated both theoretically and via numerical studies for various scenarios in Section 5.5. An example is given in Section 5.6 to illustrate the proposed clinical strategy. Concluding remarks are given in the last section of this chapter.

## 5.2 Similarity Margin Selection

The goal of a comparative clinical study for a biosimilar development program is to assess clinically meaningful differences between the products through demonstrating that the difference between two products is small enough, which determines the margin. With no requirement to establish the efficacy of the biosimilar products, biosimilar comparative studies choose the most sensible endpoint to assess clinically meaningful difference. For the selection of similarity margin, a meta-analysis that combines data from a number of previous studies is commonly considered, which is briefly described below. In addition, FDA's current thinking for determining similarity margin based on historical data is outlined.

It should be noted that the determination of a similarity margin in comparative clinical biosimilar studies can be different from the determination of non-inferiority margins, although they have a lot in common. In this chapter, without the loss of generality, we will focus on the determination of one-sided similarity margin (corresponds to non-inferiority margin in non-inferiority trials) under the assumption of symmetric similarity margin.

One goal for a non-inferiority trial is to demonstrate that the experimental treatment effect $T$ is superior to the placebo effect $P$ by rejecting the null hypothesis $H_0$: $\mu_T - \mu_P \leq 0$, where $\mu_T$ and $\mu_P$ are the mean responses under the experimental treatment and placebo, respectively. In a fixed margin approach, a margin $\delta$ is pre-defined based on the effect size (ES) of the active control, which is estimated considering historical trials in which the active control product was previously compared to a placebo. Under the fixed margin approach, we declare non-inferiority through implicitly rejecting the null hypothesis $H_0$: $\mu_T - \mu_P \leq 0$ and also by rejecting the hypothesis $K_0$: $\mu_T - \mu_C \leq \delta$, where $\mu_C$ is the mean response under the control of the non-inferiority trial.

The margin $\delta$ must be chosen so that rejection of $K_0$: $\mu_T - \mu_C \leq \delta$ implies the rejection of $H_0$: $\mu_T - \mu_P \leq 0$. For example, under the constancy assumption, $\delta$ may be chosen as $\hat{\mu}_{CP} - z_\alpha \sigma_{CP}$, the lower bound of the $100(1 - 2\alpha)\%$ confidence interval of the treatment effect of the active control versus placebo. Here, $\hat{\mu}_{CP} = \hat{\mu}_C - \hat{\mu}_P$, $\sigma_{CP}$ is the standard error, and $z_\alpha$ is the $100(1 - \alpha)\%$ quantile of a standard normal distribution.

A popular way of determining the non-inferiority margin is to estimate the treatment effect using historical information available for the control therapy. When there are many historical trials available, meta-analysis is typically used to assess the treatment effect of the control versus placebo. Once that effect is estimated, the margin is set to ensure the difference between the experimental treatment and the control treatment, if greater than $\delta$ will rule out the possibility of the experimental product being less effective than placebo by some preselected amount.

Note that when the endpoint is a clinical endpoint, we may start with the non-inferiority. However, this margin may or may not be appropriate to determine no clinically meaningful difference. When the clinical endpoint is a surrogate endpoint like a PD marker, we may start with the non-inferiority in this endpoint. Whether this margin may be appropriate to be a similarity margin depends on its relationship with the clinical endpoint. The discrepancy between the sponsor's proposal and FDA's recommendation could be due to scientific and feasibility consideration.

## 5.3 Proposed Strategy for Similarity Margin Determination

In practice, similarity margins proposed by sponsors and the FDA are likely to be different. Similarity margins are often derived based on previous studies or historical data without universally accepted statistical, clinical, and regulatory justification. While some sponsors may accept the FDA's proposal, some other sponsors may argue for a wider margin to reduce the sample size required for achieving the study objective with a desired power. In this section, we will focus a strategy proposed by Nie et al. (2020) for the selection of an appropriate similarity margin based on a risk-and-benefit assessment.

### 5.3.1 Five Steps in Determining Similarity Margin

Nie et al. (2020)'s proposed strategy for similarity margin selection can be summarized in the following steps:

Step 1. The sponsors are to identify historical studies available that are accepted by the FDA to determine similarity margin.

Step 2. Based on a meta-analysis that combines these identified historical studies, the similarity margin is determined; hence, the corresponding sample size is required for testing similarity. Power calculation for the required sample size is obtained based on the sponsor's proposed margin.

Step 3. At the same time, the FDA will propose a similarity margin by taking clinical judgment, statistical rationale, and regulatory feasibility into considerations.

Step 4. A risk assessment is then conducted for the evaluation of the sponsor's proposed margin assuming that the FDA's proposal is true.

Step 5. The risk assessment is then reviewed by the FDA review team and communicated with the sponsor in order to reach agreement on the final similarity margin.

### 5.3.2 Criteria for Risk Assessment

In this section, we focus on step 4 from of Nie et al. (2020)'s strategy, which quantifies the risk of different margins based on several criteria. This will assist sponsors to adjust their margins according to the maximum risk which is allowed by FDA. Four criteria are considered regarding different aspects of the similarity test. Numerical derivations are given in the next section based on continuous endpoints (e.g., normality assumption). Let $\epsilon$ be the true difference between the proposed biosimilar product and its reference product, i.e., $\epsilon = \mu_B - \mu_R$, where $\mu_B$ and $\mu_R$ are the treatment effect of the biosimilar product and the reference product, respectively. It is also assumed that a positive value of $\epsilon$ means that the biosimilar product is more efficacious than the reference product in the selected efficacy endpoint. Let $\delta_{Sponsor}$ and $\delta_{FDA}$ be the sponsor's proposed margin and the FDA's recommended margin, respectively. In this chapter, we assume

$$0 < |\epsilon| < \delta_{FDA} < \delta_{Sponsor}.$$

**Criterion 1: Sample Size Ratio (SSR)** – When fixing the power of the similarity test, sample size is a decreasing function of similarity margin; i.e., the smaller the similarity margin is, the bigger the sample size is required. In clinical trial studies, large sample size corresponds to more costs on sponsors. The goal is to move sponsor-proposed margin toward the FDA-recommended margin, with a moderate increase in sample size, but maintain the power at the desired level. Let $n_{FDA}$ be the sample size required to maintain $1 - \beta$ power under $\delta_{FDA}$, similarly for $n_{Sponsor}$.

Define

$$SSR = \frac{n_{FDA}}{n_{Sponsor}}$$

by the sample size ratio. Then, sample size difference (SSD)

$$SSD = n_{FDA} - n_{Sponsor} = (SSR - 1) \cdot n_{Sponsor}$$

can be viewed as the amount of the information lost for the use of a wider margin (i.e., $\delta_{Sponsor}$) assuming that $\delta_{FDA}$ is the true margin. By plotting the curve of SSR, we can choose a threshold $SSR_M$, which serves as a guideline for margin determination, say 105%, 110%, 115%, and 120%, which is corresponding to 5%, 10%, 15%, and 20% loss based on $n_{Sponsor}$.

**Criterion 2: Relative Difference in Power (RED)** – When fixing sample size for the similarity test, power is an increasing function of similarity margin; i.e., the larger the margin is, the bigger the power is. Let $Power_{Sponsor}$ and $Power_{FDA}$ be the power under $\delta_{Sponsor}$ and $\delta_{FDA}$, respectively.

Since $\delta_{FDA} < \delta_{Sponsor}$, $Power_{FDA} < Power_{Sponsor}$. This is due to the wider region of the alternative hypothesis under wider margin $\delta_{Sponsor}$ and because wider margins have smaller type II errors. Although we gain some power (smaller type II error rate) by using a wider margin, we also weaken the result (or say accuracy) when rejecting null hypothesis. Define

$$RED = Power_{Sponsor} - Power_{FDA}.$$

RED is the gain in power by sacrificing accuracy, i.e., by using wider margin. In order to close up the gap between $\delta_{FDA}$ and $\delta_{Sponsor}$, we need to minimize RED. So we can set a threshold distance $RED_M$, say 0.05, 0.10, 0.15, and 0.20, between $Power_{FDA}$ and $Power_{Sponsor}$ to be the maximum power gain by using a wider margin.

**Criterion 3: Relative Ratio in Power/Relative Risk (RR)** – The power described in the last section is the probability of concluding biosimilarity. Given the same sample size under both FDA's and sponsor's margins, $Power_{FDA} < Power_{Sponsor}$; i.e., the probability of concluding biosimilarity is bigger under $\delta_{Sponsor}$ than its under $\delta_{FDA}$. It means that among all the biosimilar products which conclude biosimilarity under the sponsor-proposed margin, only a portion of them will be biosimilar under the FDA-recommended margin. The rest is wrongly claimed as biosimilar by sponsors according to FDA's margin. This is regarded as a risk factor for using wider margins. Define RR as the probability in which a product is not concluded as biosimilar under $\delta_{FDA}$ given that it is concluded as biosimilar under $\delta_{Sponsor}$,

$$RR = 1 - \frac{Power_{FDA}}{Power_{Sponsor}} = \frac{Power_{Sponsor} - Power_{FDA}}{Power_{Sponsor}}.$$

Under the FDA-recommended margin, RR is the risk of wrongly concluding biosimilarity of a biosimilar drug using sponsor's margin. Furthermore, among all biosimilar drugs concluded using sponsor's margin, $100 \cdot RR$ of them would have been failed under the FDA-recommended margin. Thus, RR is the risk of using sponsor-proposed margin. Wider margins lead to larger risks. Thus, we may choose an appropriate margin by assuring that the risk is smaller than a maximum risk $RR_M$ that is considered acceptable by the FDA (say 0.15). Let $\delta_M$ be the margin, which is corresponding to $RR_M$. We will derive an asymptotically analytical form of $\delta_M$ in the next section based on the continuous endpoint.

**Criterion 4: Type I Error Inflation (TERI)** –TERI is the probability, assuming the smaller margin is the true difference, by rejecting a null hypothesis based on the wider margin in a study powered to rule out the wider margin (i.e., the type I error rate of the test under the "FDA" null). This will be greater than 5%, and the degree of its inflation is probably relevant information. The inflation is also an increasing function of similarity margins. Bigger margins lead to larger inflations, i.e., larger type I error rates. We can set up a threshold value of TERI for choosing the largest margin that is allowed.

---

## 5.4 Risk Assessment with Continuous Endpoints

In this section, both analytic and asymptotic forms of the four criteria proposed in the last section are derived. Without loss of generality, we only consider biosimilar products that have continuous endpoints. All calculations below can be derived in a similar fashion for biosimilar products with categorical endpoints.

Let $\delta > 0$ be the similarity margin, and the null hypothesis of the similarity test is

$$H_0 : |\epsilon| \geq \delta.$$

Rejection of the null hypothesis implies similarity between the biosimilar product and the reference product. For simplicity reason, we assume that samples from both biosimilar group and reference group follow normal distributions with mean $\mu_B$ and $\mu_R$, respectively, and same unknown variance $\sigma^2$, which means the within-subject variances of both biosimilar and reference products are the same.

$$x_1^B, x_2^B, \ldots, x_{n_B}^B \sim N\left(\mu_B, \sigma_B^2\right), x_1^R, x_2^R, \ldots, x_{n_R}^R \sim N\left(\mu_R, \sigma_R^2\right),$$

where $n_B$ and $n_R$ are the sample sizes for the biosimilar group and the reference group. Let

$$\hat{\mu}_{BR} = \hat{\mu}_B - \hat{\mu}_R$$

be the estimated treatment effect of the biosimilar product relative to the reference product with standard error of

$$\hat{\sigma}_{BR} = \hat{\sigma}\sqrt{1/n_B + 1/n_R},$$

where

$$\hat{\mu}_B = \frac{1}{n_B}\sum_{i=1}^{n_B} x_i^B,$$

$$\hat{\mu}_R = \frac{1}{n_R}\sum_{i=1}^{n_R} x_i^R,$$

and

$$\hat{\sigma}^2 = \frac{1}{n_B + n_R - 2}\left\{\sum_{i=1}^{n_B}\left(x_i^B - \hat{\mu}_B\right)^2 + \sum_{i=1}^{n_R}\left(x_i^R - \hat{\mu}_R\right)^2\right\}.$$

Note that $\hat{\sigma}_{BR}$ depends on sample sizes (and treatment effects in some scenarios).

The rejection region for testing $H_0$ with statistical significance level $\alpha$ is

$$R = \left\{\frac{\hat{\mu}_{BR} + \delta}{\hat{\sigma}_{BR}} > t_{\alpha,\, n_B + n_R - 2}\right\} \cap \left\{\frac{\hat{\mu}_{BR} - \delta}{\hat{\sigma}_{BR}} < -t_{\alpha,\, n_B + n_R - 2}\right\}.$$

Thus, the power of the study is

$$\text{Power} = P\left(\frac{\hat{\mu}_{BR} + \delta}{\hat{\sigma}_{BR}} > t_{\alpha,\, n_B + n_R - 2} \text{ and } \frac{\hat{\mu}_{BR} - \delta}{\hat{\sigma}_{BR}} < -t_{\alpha,\, n_B + n_R - 2}\right)$$

$$\approx 1 - T_{n_B + n_R - 2}\left(t_{\alpha,\, n_B + n_R - 2}\Big|\frac{\delta - \epsilon}{\sigma\sqrt{\dfrac{1}{n_B} + \dfrac{1}{n_R}}}\right)$$

$$- T_{n_B + n_R - 2}\left(t_{\alpha,\, n_B + n_R - 2}\Big|\frac{\delta + \epsilon}{\sigma\sqrt{1/n_B + 1/n_R}}\right),$$

where $T_k(\cdot\,|\,\theta)$ is the cumulative distribution function of a noncentral $t$-distribution with $k$ degrees of freedom and the noncentrality parameter $\theta$ and $-\delta < \epsilon < \delta$ under $H_a$.

**Sample Size Ratio (SSR)** – Assume that $n_B = \kappa n_R$, the sample size $n_R$ needed to achieve power $1 - \beta$ can be obtained by setting the power to $1 - \beta$. Since the power is larger than

$$1 - 2T_{n_B + n_R - 2}\left(t_{\alpha,\, n_B + n_R - 2}\Big|\frac{\delta - |\epsilon|}{\sigma\sqrt{1/n_B + 1/n_R}}\right),$$

a conservative approximation to the sample size $n_R$ can be obtained by solving

$$T_{(1+\kappa)n_R - 2}\left(t_{\alpha,\, (1+\kappa)n_R - 2}\Big|\frac{\sqrt{n_R}\,(\delta - |\epsilon|)}{\sigma\sqrt{1 + 1/\kappa}}\right) = \frac{\beta}{2}.$$

If sample size $n_R$ is sufficiently large,

$$t_{\alpha,\, (1+\kappa)n_R - 2} \approx z_\alpha, \text{ and } T_{(1+\kappa)n_R - 2}(t\,|\,\theta) \approx \Phi(t - \theta),$$

then

$$\frac{\beta}{2} = T_{(1+\kappa)n_R - 2}\left(t_{\alpha,\, (1+\kappa)n_R - 2}\Big|\frac{\sqrt{n_R}\,(\delta - |\epsilon|)}{\sigma\sqrt{1 + 1/\kappa}}\right) \approx \Phi\left(z_\alpha - \frac{\sqrt{n_R}\,(\delta - |\epsilon|)}{\sigma\sqrt{1 + 1/\kappa}}\right).$$

As a result, the sample size needed to achieve power $1 - \beta$ can be obtained by solving the following equation:

$$z_\alpha - \frac{\sqrt{n_R}\left(\delta - |\epsilon|\right)}{\sigma\sqrt{1 + 1/\kappa}} = z_{1-\beta/2} = -z_{\beta/2}.$$

This leads to

$$n_R = \frac{\left(z_\alpha + z_{\beta/2}\right)^2 \sigma^2 \left(1 + 1/\kappa\right)}{\left(\delta - |\epsilon|\right)^2}.$$

Thus,

$$n_R^{\text{FDA}} = \frac{\left(z_\alpha + z_{\beta/2}\right)^2 \sigma^2 \left(1 + 1/\kappa\right)}{\left(\delta_{\text{FDA}} - |\epsilon|\right)^2},$$

$$n_R^{\text{Sponsor}} = \frac{\left(z_\alpha + z_{\beta/2}\right)^2 \sigma^2 \left(1 + 1/\kappa\right)}{\left(\delta_{\text{Sponsor}} - |\epsilon|\right)^2},$$

and with

$$\delta_{\text{Sponsor}} = \lambda \delta_{\text{FDA}},$$

we have

$$\text{SSR} = \frac{n_{\text{FDA}}}{n_{\text{Sponsor}}} = \frac{(1+\kappa)n_R^{\text{FDA}}}{(1+\kappa)n_R^{\text{Sponsor}}} = \left(\frac{\lambda \delta_{\text{FDA}} - |\epsilon|}{\delta_{\text{FDA}} - |\epsilon|}\right)^2$$

and

$$\lambda_M = \sqrt{\text{SSR}_M} + \frac{|\epsilon|}{\delta}\left(1 - \sqrt{\text{SSR}_M}\right).$$

**Relative Difference in Power (RED)** – Let

$$B(\epsilon,\ \delta,\ n_R,\ \sigma,\ \kappa) = T_{(1+\kappa)n_R - 2}\left(t_{\alpha,\ (1+\kappa)n_R - 2} \bigg| \frac{\sqrt{n_R}\left(\delta + \epsilon\right)}{\sigma\sqrt{1 + 1/\kappa}}\right),$$

based on the calculation above, we immediately have

$$\text{RED} = \text{Power}_{\text{Sponsor}} - \text{Power}_{\text{FDA}}$$

$$\approx B(\epsilon,\ \delta_{\text{FDA}},\ n_R,\ \sigma,\ \kappa) - B(\epsilon,\ \lambda\delta_{\text{FDA}},\ n_R,\ \sigma,\ \kappa)$$

$$+ B(-\epsilon,\ \delta_{\text{FDA}},\ n_R,\ \sigma,\ \kappa) - B(-\epsilon,\ \lambda\delta_{\text{FDA}},\ n_R,\ \sigma,\ \kappa).$$

When $n_R$ is sufficiently large, let

$$\tilde{\Phi}\left(\epsilon, \delta, n_R, \sigma, \kappa\right) = \Phi\left(z_\alpha - \frac{\sqrt{n_R}\left(\delta_{\text{FDA}} + \epsilon\right)}{\sigma\sqrt{1 + 1/\kappa}}\right),$$

by using the same approximation in last section,

$$\text{RED} \approx \left[\Phi\left(z_\alpha - \frac{\sqrt{n_R}\left(\delta_{\text{FDA}} + \epsilon\right)}{\sigma\sqrt{1 + 1/\kappa}}\right) - \Phi\left(z_\alpha - \frac{\sqrt{n_R}\left(\lambda\delta_{\text{FDA}} + \epsilon\right)}{\sigma\sqrt{1 + 1/\kappa}}\right)\right]$$

$$+ \left[\Phi\left(z_\alpha - \frac{\sqrt{n_R}\left(\delta_{\text{FDA}} - \epsilon\right)}{\sigma\sqrt{1 + 1/\kappa}}\right) - \Phi\left(z_\alpha - \frac{\sqrt{n_R}\left(\lambda\delta_{\text{FDA}} - \epsilon\right)}{\sigma\sqrt{1 + 1/\kappa}}\right)\right]$$

$$\approx \tilde{\Phi}\left(\epsilon, \delta_{\text{FDA}}, n_R, \sigma, \kappa\right) - \tilde{\Phi}\left(\epsilon, \lambda\delta_{\text{FDA}}, n_R, \sigma, \kappa\right)$$

$$+ \tilde{\Phi}\left(-\epsilon, \delta_{\text{FDA}}, n_R, \sigma, \kappa\right)\tilde{\Phi}\left(-\epsilon, \lambda\delta_{\text{FDA}}, n_R, \sigma, \kappa\right).$$

**Relative Ratio in Power/Relative Risk (RR)** – Let $S_{\text{FDA}} = \{\text{reject } H_0 \text{ when } \delta = \delta_{\text{FDA}}\}$ and $S_{\text{Sponsor}} = \{\text{reject } H_0 \text{ when } \delta = \delta_{\text{Sponsor}}\}$. Since $\delta_{\text{FDA}} < \delta_{\text{Sponsor}}$, $|\epsilon| \leq \delta_{\text{FDA}}$ leads to $|\epsilon| \leq \delta_{\text{Sponsor}}$. Therefore, rejecting $H_0$ under $\delta_{\text{FDA}}$ leads to the rejection of $H_0$ under $\delta_{\text{Sponsor}}$, which means $S_{\text{FDA}} \subseteq S_{\text{Sponsor}}$ and $S_{\text{FDA}} \cap S_{\text{Sponsor}} = S_{\text{FDA}}$. Let $p_s$ be the probability of concluding biosimilarity under $\delta_{\text{FDA}}$ given the concluding biosimilarity under $\delta_{\text{Sponsor}}$. Then, based on the relationship between $S_{\text{FDA}}$ and $S_{\text{Sponsor}}$, we have

$$p_s = \text{Pr}\left(\text{conclude similarity under } \delta_{\text{FDA}} \mid \text{conclude similarity under } \delta_{\text{Sponsor}}\right)$$

$$= \frac{\text{Pr}\left(\text{reject } H_0 \text{ when } \delta = \delta_{\text{FDA}}\right)}{\text{Pr}\left(\text{reject } H_0 \text{ when } \delta = \delta_{\text{Sponsor}}\right)} = \frac{\text{Power}_{\text{FDA}}}{\text{Power}_{\text{Sponsor}}}$$

$$\approx \frac{1 - B\left(\epsilon, \delta_{\text{FDA}}, n_R, \sigma, \kappa\right) - B\left(-\epsilon, \delta_{\text{FDA}}, n_R, \sigma, \kappa\right)}{1 - B\left(\epsilon, \delta_{\text{Sponsor}}, n_R, \sigma, \kappa\right) - B\left(-\epsilon, \delta_{\text{Sponsor}}, n_R, \sigma, \kappa\right)}$$

Thus, based on the definition of RR in Criterion 3, we have

$$\text{RR} = 1 - p_s \approx \frac{\text{RED}}{1 - B\left(\epsilon, \lambda\delta_{\text{FDA}}, n_R, \sigma, \kappa\right) - B\left(-\epsilon, \lambda\delta_{\text{FDA}}, n_R, \sigma, \kappa\right)}.$$

For large $n_R$, we have

$$\text{RR} \approx \frac{\text{RED}}{1 - \tilde{\Phi}\left(\epsilon, \lambda\delta_{\text{FDA}}, n_R, \sigma, \kappa\right) - \tilde{\Phi}\left(-\epsilon, \lambda\delta_{\text{FDA}}, n_R, \sigma, \kappa\right)}.$$

**Type I Error Inflation (TERI)** – When assuming the smaller margin is the true difference, i.e., $\epsilon = \pm\delta_{FDA}$ and $\delta_{FDA} < \delta_{Sponsor}$, TERI is calculated as follows:

Type I Error $| \epsilon = \pm\delta_{FDA}$

$$= P\left( \frac{\hat{\mu}_{BR} + \delta_{Sponsor}}{\hat{\sigma}_{BR}} > t_{\alpha, n_B + n_R - 2} \ and \right.$$

$$\left. \frac{\hat{\mu}_{BR} - \delta_{Sponsor}}{\hat{\sigma}_{BR}} < -t_{\alpha, n_B + n_R - 2} \Big| \epsilon = \pm\delta_{FDA} \right)$$

$$= 1 - T_{n_B + n_R - 2}\left( t_{\alpha, n_B + n_R - 2} \Big| \frac{\delta_{Sponsor} + \delta_{FDA}}{\sigma\sqrt{1/n_B + 1/n_R}} \right)$$

$$- T_{n_B + n_R - 2}\left( t_{\alpha, n_B + n_R - 2} \Big| \frac{\delta_{Sponsor} - \delta_{FDA}}{\sigma\sqrt{1/n_B + 1/n_R}} \right)$$

$$= 1 - B(\delta_{FDA}, \lambda\delta_{FDA}, n_R, \sigma, \kappa) - B(-\delta_{FDA}, \lambda\delta_{FDA}, n_R, \sigma, \kappa)$$

For large sample, we have

$$\text{Inflation} \approx 1 - \alpha - \Phi\left( z_\alpha - \frac{\sqrt{n_R}\,(\lambda + 1)}{\sigma\sqrt{1 + 1/\kappa}} \cdot \delta_{FDA} \right) - \Phi\left( z_\alpha - \frac{\sqrt{n_R}\,(\lambda - 1)}{\sigma\sqrt{1 + 1/\kappa}} \cdot \delta_{FDA} \right).$$

## 5.5 Numerical Studies

In this section, simulation studies for all four criteria are conducted and risk curves are plotted. Based on the results, suggestions on choosing a reasonable threshold are discussed for different scenarios. The validity of large sample approximation is investigated for small sample sizes. During this section, type I error rate and type II error rate are fixed to be 0.05 and 0.2, respectively.

**Sample Size Ratio (SSR)** – For SSR, it can be expressed as follows:

$$\sqrt{SSR} = \frac{\delta_{FDA}}{\delta_{FDA} - |\epsilon|}\lambda - \frac{|\epsilon|}{\delta_{FDA} - |\epsilon|}.$$

To further investigate the relationship between SSR and $\lambda$, we consider $\sqrt{SSR}$ instead of SSR since $\sqrt{SSR}$ is a linear function of $\lambda$ with $\dfrac{\delta_{FDA}}{\delta_{FDA} - |\epsilon|}$ as slope and $-\dfrac{|\epsilon|}{\delta_{FDA} - |\epsilon|}$ as intercept. $\sqrt{SSR}$ increases by a portion of $\dfrac{\delta_{FDA}}{\delta_{FDA} - |\epsilon|}$. For example, if $\delta_{\text{Sponsor}}$ is 10% wider than $\delta_{\text{FDA}}$, then $\sqrt{\text{SSR}}$ is increased by $\dfrac{0.1\delta_{\text{FDA}}}{\delta_{\text{FDA}} - |\epsilon|}$.

So smaller values of $\delta_{\text{FDA}} - |\epsilon|$ lead to steeper lines. In other words, if $\delta_{\text{FDA}}$ is set to be closer to $|\epsilon|$, sample size of sponsor increases more rapidly when $\delta_{\text{Sponsor}}$ moves toward $\delta_{\text{FDA}}$. We can observe this from the plot (Figure 5.1).

Let $\text{SSR}_{\text{cur}}$ be the sample according to the current $\delta_{\text{Sponsor}}$. For a safe choice, we propose to use $\delta_{\text{new}}$, which is corresponding to $\text{SSR}_{\text{cur}} - \Delta$, where $\Delta$ can be ranging from 0.2 to 0.3. This will bring the gap between $\delta_{\text{FDA}}$ and $\delta_{\text{Sponsor}}$ smaller. But this is not a universal choice. Different thresholds should be chosen based on a case-by-case basis. We also plot the SSD curve (Figure 5.2). It follows the same pattern as SSR curve.

**Relative Difference in Power (RED)** – Since the large sample approximation is used in deriving the asymptotic form of *RED*, we first investigate the validity of this approximation when sample size is small. As we can see from the four plots (Figure 5.3), when sample size of a single arm is 15, the approximation is still close to the original. For the sample size of 30, two

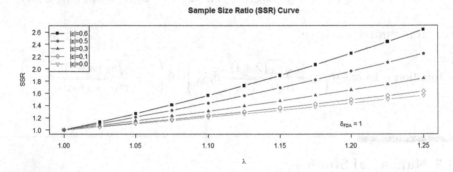

**FIGURE 5.1**
Plot of SSR versus $\lambda$.

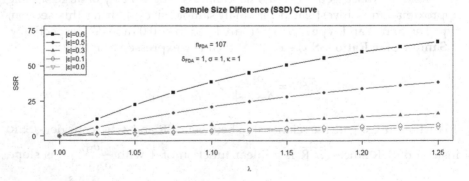

**FIGURE 5.2**
Plot of SSD versus $\lambda$.

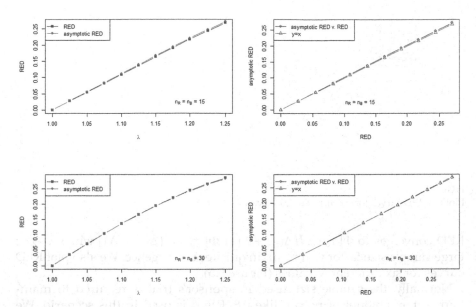

**FIGURE 5.3**
Plots of RED versus $\lambda$ and plots of asymptotic RED versus RED ($\epsilon < 0, \delta_{\text{FDA}} = 1.0, \sigma = 1, \kappa = 1$).

curves look identical to each other. The condition of the normal approximation of $t$-distribution is that the degree of freedom is greater than 30. In this case, $n_B + n_R - 2 > 30$, which is not difficult to satisfy in practice. For simplicity reason, we will use the asymptotic form instead of the original one in the following discussion. The rest of the parameters used in sample size comparison plot are set as follows: $\epsilon = -0.5$, $\delta_{\text{FDA}} = 1.0$, $\sigma = 1$, and $\kappa = 1$. Since RED is symmetric about $\epsilon$, we only plot when $\epsilon < 0$ (Figure 5.3).

Next, to eliminate some of the parameters in RED, we rewrite it in terms of ES $\Delta = -\epsilon/\sigma$ and $\Delta_{\text{FDA}} = \delta_{\text{FDA}}/\sigma$, and let $N = \sqrt{\dfrac{n_R}{1+1/\kappa}}$ be the sample size factor, then

$$\text{RED} \approx \Phi\left[z_\alpha - N\left(\Delta_{\text{FDA}} + \Delta\right)\right] - \Phi\left[z_\alpha - N\left(\lambda\Delta_{\text{FDA}} + \Delta\right)\right]$$

$$+ \Phi\left[z_\alpha - N\left(\Delta_{\text{FDA}} - \Delta\right)\right] - \Phi\left[z_\alpha - N\left(\lambda\Delta_{\text{FDA}} - \Delta\right)\right].$$

We plot RED curves for six different $\text{ES}_{\text{FDA}}$ values, which are corresponding to 0%, 5%, 10%, 15%, 20%, and 25% increase from ES = 0.5 (Figure 5.4). Large $\text{ES}_{\text{FDA}}$ leads to steeper curve, i.e., the drastic increase in RED for smaller values of $\lambda$. For large $\text{ES}_{\text{FDA}}$, narrowing the same portion of $\delta_{\text{FDA}}$ will yield more decrement in RED value. Therefore, under current parameter setting, for large $\text{ES}_{\text{FDA}}$, we recommend choosing the margin (value of $\lambda$) such that RED is in the range of (0.20, 0.40); for small $\text{ES}_{\text{FDA}}$, less than 0.20 is preferred.

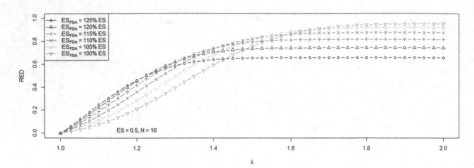

**FIGURE 5.4**
Plots of RED curves for six different $ES_{FDA}$ values.

RED converges to $\Phi\left[z_\alpha - N\left(\Delta_{FDA} + \Delta\right)\right] + \Phi\left[z_\alpha - N\left(\Delta_{FDA} - \Delta\right)\right]$ when $\lambda \to \infty$; large sample size factor will result in quicker convergence. We also plot RED curves for six different $N$ values (Figure 5.5).

Normally, the sample size used in sponsor's trial is required to maintain certain amount of power, like 0.8. $RED^\beta$ is used in this scenario. We plot five different values of $\delta_{FDA}$, which are gradually increasing from $\epsilon$ (see Figure 5.6). The sample sizes used for each value of $\lambda$ here maintain $1 - \beta$ power for sponsor's text. $RED^\beta$ is different from RED; i.e., larger value of $\delta_{FDA}$ leads to slow growth of difference in power. For a large value of $\delta_{FDA}$, $\lambda$ that leads to $RED^\beta$ in the range of $(0.1, 0.2)$ is recommended; for a small value, less than 0.3 is preferred.

**Relative Ratio in Power/Relative Risk (RR)** – The definition of RR in the last section is also based on multiple steps of large sample approximation of noncentral $t$-distribution and its quantile. We first check the validity of the large sample approximation when sample size is small. As we can see from the following four plots (Figure 5.7), even when sample size is as small as 15 (single arm), the original RR and the asymptotic one look identical. Therefore, we will only use the asymptotic expression in our following

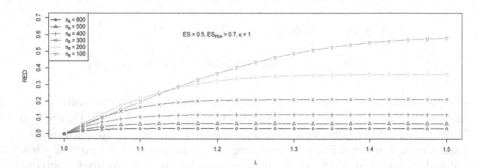

**FIGURE 5.5**
Plots of RED curves for six different $N$ values.

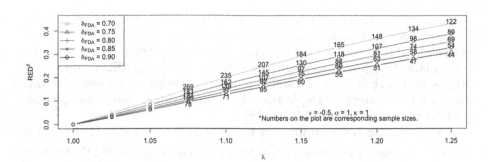

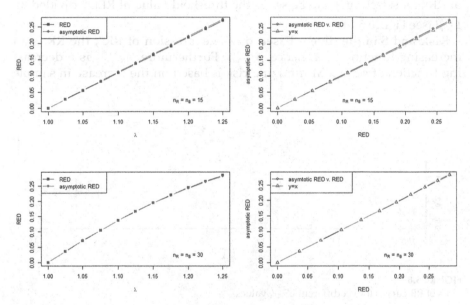

**FIGURE 5.6**
Plots of $RED^\beta$ curves with five different values of $\delta_{FDA}$.

**FIGURE 5.7**
Plots of RED versus $\lambda$ and plots of asymptotic RED versus RED ($\epsilon = -0.5, \delta_{FDA} = 1.0, \sigma = 1, \kappa = 1$).

decision-making. The rest of the parameters used in sample size comparison plot are set as follows: $\epsilon = -0.5$, $\delta_{FDA} = 1.0$, $\sigma = 1$, and $\kappa = 1$.

The relationship between RR and RED can be described as follows:

$$RR \approx \frac{RED}{1 - \Phi\left[z_\alpha - N\left(\lambda\Delta_{FDA} + \Delta\right)\right] - \Phi\left[z_\alpha - N\left(\lambda\Delta_{FDA} - \Delta\right)\right]}.$$

Based on the expression of RR, we can see RR is the regularized version of RED. But different fromRED, RR has a clear definition in terms of risk, which is the probability of wrongly concluding biosimilarity of a biosimilar drug using sponsor's margin. So smaller value of RR is preferred. We rewrite RR based on the expression of RED and plot six curves according to six different values of $ES_{FDA}$, which are the same as in the RED plot. From the figure below, large $ES_{FDA}$ leads to smaller risk. RR converges to RED when $\lambda \to \infty$; large sample size factor will result in quicker convergence. We also plot RED curves for six different sample sizes $n_R$. As we can see, larger sample size leads to lower risk (Figures 5.8 and 5.9).

When plugging in the sample size which retains power as $1-\beta$, the shape of the following five curves is almost identical to those in the $RED^\beta$ plot. This is because $RR^\beta$ is approximately proportional to $\dfrac{RED^\beta}{1-\beta}$. The threshold value of $RR^\beta$ for selecting $\lambda$ can be set as the threshold value of $RED^\beta$ divided by $1-\beta$ (see Figure 5.10).

**Risk and Sample Size** – Based on the expression of $RR^\beta$, the RR is an increasing function of $\lambda$, hence, $\delta_{Sponsor}$. Furthermore, $n_R^{Sponsor}$ is a decreasing function of $\delta_{Sponsor}$. Minimizing risk is based on the increase in sample

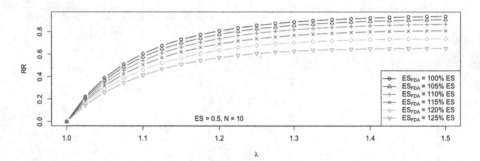

**FIGURE 5.8**
Plots of RR curves for six different $ES_{FDA}$ values.

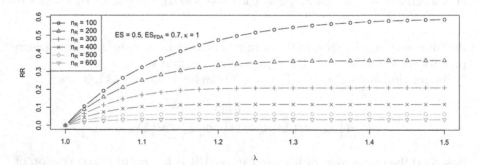

**FIGURE 5.9**
Plots of RR curves for six different $N$ values.

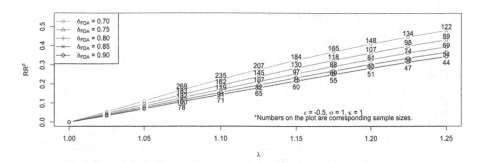

**FIGURE 5.10**
Plots of $RR^\beta$ curves with five different values of $\delta_{FDA}$.

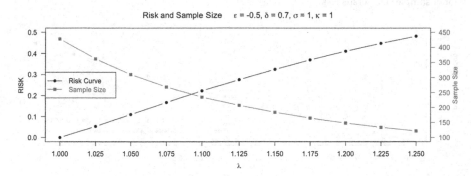

**FIGURE 5.11**
Plots of risk curve and sample size curve against $\lambda$.

size, which leads to the increase in the cost of clinical trial for sponsor. When minimizing risk, sample size should be considered at the same time. A compromise should be made between the risk and sample size when shrinking the gap between $\delta_{FDA}$ and $\delta_{Sponsor}$. To provide a better understanding, both risk curve and sample size curve were plotted on the sample plot (Figure 5.11). The values of parameters are $\epsilon = -0.5$, $\delta_{FDA} = 0.7$, $\sigma = 1$, $\kappa = 1$, $\alpha = 0.05$, and $\beta = 0.2$.

As it can be seen from Figure 5.11, under the condition that sample size should be chosen while maintaining 0.8 as power, risk is increasing as $\delta_{Sponsor}$ moves away from $\delta_{FDA}$ (can be seen as $\lambda$ increases) and at the meantime, sample size is needed to maintain 0.8 as power decreases. It is reasonable that large sample size is needed to keep the risk low. For this case, we can choose $\lambda = 1.075$. It leads to about 30% risk, but only requires half of the sample size as for the FDA-recommended margin. Next, we directly plot the sample size against risk (Figure 5.12). The relationship is almost linear with a negative slope.

**Type I Error Inflation (TERI)** – Here, we plot the TERI when $\delta_{Sponsor}$ moves away from $\delta_{FDA}$. The values of parameters are $\epsilon = -0.5$, $\delta_{FDA} = 0.7$, $\sigma = 1$, $\kappa = 1$, $\alpha = 0.05$, and $\beta = 0.2$. The sample size used in the plot is $n_R^{Sponsor}$ (see Figure 5.13),

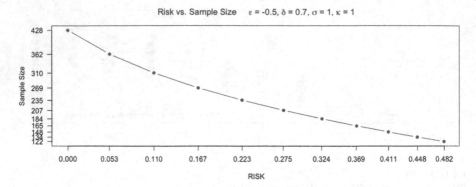

**FIGURE 5.12**
Plot of sample size versus risk.

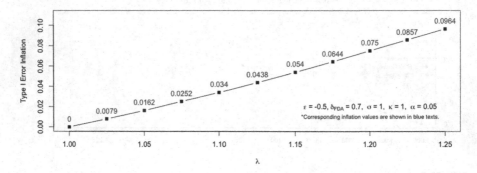

**FIGURE 5.13**
Plot of TERI against $\lambda$.

i.e., maintains 0.8 power. Only the asymptotic expression of TERI is used here. Type I error rate can also be seen as a risk factor. Minimizing the inflation caused by wider margin is the goal here. In this case, about 50% of the inflation is acceptable since type I error is 0.05 here. Thus, we can choose $\lambda = 1.15$ and $\delta_{Sponsor} = 0.805$.

## 5.6 An Example

In this section, we present a synthetic example to demonstrate the strategy proposed in this paper and four criteria in margin selection. Assume after the clinical trial, we observe the following settings: $\hat{\mu}_B = 2.55$, $\hat{\mu}_R = 2.75$, $\hat{\sigma} = 1.35$, $n_B = 140$, $n_R = 200$, $\delta_{FDA} = 0.25$, and $\delta_{Sponsor} = 0.35$. Based on the FDA-recommended margin, the sample size required to maintain 0.8 power is around 274 for reference group (assuming true difference is zero in sample

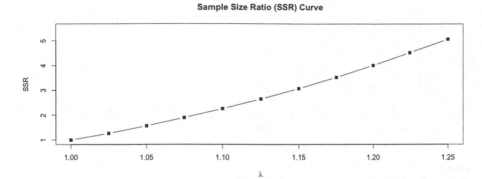

**FIGURE 5.14**
Plot of SSR against $\lambda$.

size calculation), which is more than the sample size used in the clinical trial. This may cost too much for sponsor to adjust. Based on the sponsor-proposed margin, 200 samples for reference group are more than enough to retain the 0.8 power. Obviously, some compromises are needed here to benefit both parties.

**Sample Size Ratio (SSR)** – SSR here is 9, which is too big for sponsor to accommodate. Based on the SSR plot (Figure 5.14), we can choose RSS to be between 3 and 4. Thus, $\delta_{\text{Sponsor}} = 1.15 * 0.25 = 0.2875$.

**Relative Difference in Power (RED)** – Based on the RED plot, we see that power difference is not big in this case even for large values of $\lambda$. So $\delta_{\text{Sponsor}}$ does not need to move too much toward $\delta_{\text{FDA}}$. Therefore, any value of $\lambda$ between 1.15 and 1.20 is acceptable (Figure 5.15).

**Relative Ratio in Power/Relative Risk (RR)** – After regularization, the risk is more understandable than the previous two criteria. Consider that it has a clear meaning in terms of the probability of wrongly concluding similarity. Figure 5.16 plots $RR^{\beta}$ against $\lambda$. As it can be seen from Figure 5.16, anything larger than 40% may be too risky. So 40% can be our maximum risk, and $\lambda = 1.125$ and $\delta_{\text{Sponsor}} = 0.281$.

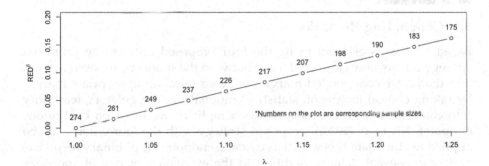

**FIGURE 5.15**
Plot of $RED^{\beta}$ against $\lambda$.

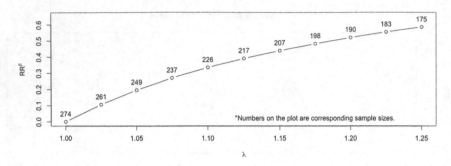

**FIGURE 5.16**
Plot of $RR^\beta$ against $\lambda$.

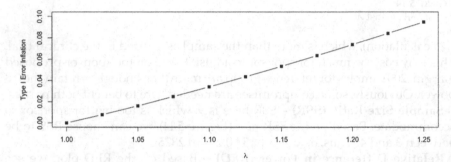

**FIGURE 5.17**
Plot of TERI against $\lambda$.

**Type I Error Inflation (TERI)** – Figure 5.17 plots TERI against $\lambda$. Since the significance level here is 0.05, after the inflation we want the significance level not to go beyond 0.1. Thus, the maximum inflation allowed is 0.05, and $\lambda = 1.15$ and $\delta_{Sponsor} = 0.2875$.

## 5.7 Concluding Remarks

Based on risk assessment using the four proposed criteria, the proposed strategy can not only close up the gap between the sponsor-proposed margin and the FDA-recommended margin, but also select an appropriate margin by taking clinical judgment, statistical rationale, and regulatory feasibility into consideration. In this chapter, for simplicity, we focus on continuous endpoint. Nie et al. (2020)'s proposed strategy with the four criteria can be applied to other data types such as discrete endpoints (e.g., binary response) and time-to-event data. In addition to the evaluation of risk of sponsor's proposed margin, we can also assess the risk of FDA-recommended margin assuming that the margin proposed by the sponsor is the true margin.

# 6

## Probability of Inconclusiveness

### 6.1 Introduction

For the approval of a new drug product, the sponsor is required to provide substantial evidence regarding safety and efficacy of the drug product under investigation. In practice, a typical approach is to conduct adequate and placebo-controlled clinical studies and test the following point hypotheses:

$$H_0: \text{ineffectiveness versus } H_a: \text{effectiveness.} \tag{6.1}$$

The rejection of the null hypothesis of *in*effectiveness is in favor of the alternative hypothesis of effectiveness. Most researchers interpret that the rejection of the null hypothesis is the demonstration of the alternative hypothesis of effectiveness. It, however, should be noted that "in favor of effectiveness" does not imply "the demonstration of effectiveness." Alternatively, Chow and Huang (2019a) indicated that hypotheses (6.1) should be

$$H_0: \text{ineffectiveness versus } H_a: \text{not ineffectiveness.} \tag{6.2}$$

As it can be seen from $H_a$ in (6.1) and (6.2), the concept of *effectiveness* and the concept of *not-ineffectiveness* are not the same. Not-ineffectiveness does not imply effectiveness. Thus, the traditional approach for clinical evaluation of the drug product under investigation can only demonstrate *not-ineffectiveness* but not *effectiveness*. In practice, we typically test the null hypothesis at the $\alpha = 5\%$ level of significance. However, many researchers prefer testing the null hypothesis at the $\alpha = 1\%$ level of significance. If the observed $p$-value falls between 1% and 5%, we claim the test result is *inconclusive*. In placebo-controlled studies, conceptually, *not-ineffectiveness* includes the portion of *inconclusiveness* and *effectiveness* (see also Figure 6.1).

As illustrated in Figure 6.1, let $\theta$ be the true treatment effect (which is unknown) and $(\theta_L, \theta_U)$ be the corresponding $(1-\alpha) \times 100\%$ confidence interval of $\theta$. Then, hypotheses (6.1) can be rewritten as follows:

$$H_0: \theta \leq \theta_L \text{ versus } H_a: \theta > \theta_U. \tag{6.3}$$

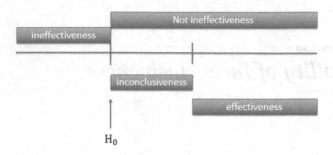

**FIGURE 6.1**
The relationship between "effectiveness" and "not-ineffectiveness" in placebo-controlled studies.

Similarly, hypotheses (6.2) can be rewritten as

$$H_0: \; \theta \leq \theta_L \text{ versus } H_a: \; \theta > \theta_L. \tag{6.4}$$

The set of hypotheses (6.3) is similar to the hypotheses set in Simon's two-stage optimal design for cancer research (Simon, 1989). At the first stage, Simon suggested testing whether the response rate has exceeded a pre-specified undesirable response rate. If yes, then proceed to test whether the response rate has achieved a pre-specified desirable response rate. Note that Simon's hypotheses testing actually is an interval hypotheses testing. On the other hand, hypotheses (6.4) is a typical one-sided test for non-inferiority of the test treatment as compared to the placebo. Thus, the rejection of inferiority leads to the conclusion of non-inferiority, which consists of equivalence (the area of inconclusiveness, i.e., $\theta_L < \theta \leq \theta_U$) and superiority (i.e., effectiveness). For a given sample size, the traditional approach for clinical evaluation of the drug product under investigation can only demonstrate that the drug product is *not ineffective* when the null hypothesis is rejected. For demonstrating the drug product is truly effective, we need to perform another test to rule out the possibility of inconclusiveness (i.e., to reduce the probability of inconclusiveness).

In practice, however, we typically test point hypotheses of equality at the $\alpha = 5\%$ level of significance. The rejection of the null hypothesis leads to the conclusion that there is a treatment effect. An adequate sample size is then selected to have a desired power (say 80%) to determine whether the observed treatment effect is clinically meaningful and hence claim that the effectiveness is demonstrated. For testing point hypotheses of no treatment effect, many researchers prefer testing the null hypothesis at the $\alpha = 1\%$ rather than $\alpha = 5\%$ level of significance in order to account for the possibility of inconclusiveness. In other words, if the observed $p$-value falls between 1% and 5%, we claim the test result is *inconclusive*. It should be noted that the concept of point hypotheses testing for no treatment effect is very different from interval hypotheses testing (6.1) and one-sided hypotheses testing for non-inferiority (6.2). In practice, however, point hypotheses testing, interval hypotheses testing, and one-sided hypotheses for testing non-inferiority have been mixed up and used in pharmaceutical research and development.

In this chapter, we intend to study (i) the sampling distribution of $p$-value under a two-group parallel design and (ii) the probability of inconclusiveness for clarifying the concept between effectiveness and not-ineffectiveness for clinical evaluation of a drug product under investigation. In addition, statistical test for effectiveness and sample size required for ruling out the possibility of inconclusiveness are derived under a proposed two-stage adaptive trial design.

## 6.2 Sampling Distribution of $p$-Value

Suppose we have observed an independent, identically distributed sample $X_n = \{X_1, X_2, \ldots, X_n\}$. The purpose is to test hypotheses (6.3) that $H_0 : \theta \le \theta_L$ versus $H_a: \theta > \theta_U$, where $\theta$ is the treatment effect and $\theta_L < \theta_U$. In this case, a test statistic $T$ can then be obtained based on the sample $X_n$, i.e.,

$$T = T(X_n).$$

Denote the cumulative function of $T$ under $H_0$ as $F_0$, and the $p$-value as $P$. Thus, $P = 1 - F_0(T)$. It can be verified that under $H_0$, $P$ follows the uniform distribution between 0 and 1 (Wasserstein and Lazar, 2016).

Suppose the true cumulative distribution function of $T$ is $F$, the distribution of $P$ is

$$F(p) = \Pr(P \le p) = \Pr(F_0(T) \ge 1 - p) = 1 - F(F_0^{-1}(1 - p)). \qquad (6.5)$$

However, although the null distribution $F_0$ might be obtained, the true distribution $F$ is difficult to obtain. Instead, we may consider the asymptotical distribution of $P$. Suppose the test statistic $T$ is the sample mean/proportion

$$\bar{X} = \frac{1}{n} \sum_{i=1}^{n} X_i,$$

then

$$\sqrt{n}[T - \theta] \xrightarrow{d} N(0, \sigma^2),$$

where $\sigma^2$ is the population variance of $X_i$ and $\xrightarrow{d}$ means the convergence in distribution. Note that $\theta = \theta_L$ under $H_0$. Thus, the cumulative distribution function of $P$ (5.5) equals

$$F(p) = 1 - \Phi\left( z_p - \frac{\theta - \theta_L}{\sigma/\sqrt{n}} \right), \tag{6.6}$$

where $\Phi(\cdot)$ is the cumulative distribution function of the standard normal distribution and

$$z_p = \Phi^{-1}(1-p).$$

Yet, it is not trivial to figure out the mean, variance, and density function of $P$ based on (6.6).

In practice, we may consider to approximate the distribution of $P$ by the delta method. Denote the density function of $T$ under $H_0$ as $f_0$, and assume it is continuous. In this case, we have

$$\sqrt{n}\left[ P - (1 - F_0(\theta)) \right] \xrightarrow{d} N\left( 0, \ \sigma^2 f_0^2(\theta) \right), \tag{6.7}$$

That is,

$$P \xrightarrow{d} N\left( 1 - F_0(\theta), \ \frac{\sigma^2 f_0^2(\theta)}{n} \right)$$

based on the delta method. Note that by the central limit theorem, $F_0(\theta)$ and $f_0(\theta)$ can be approximated by

$$\Phi\left( \frac{\theta - \theta_L}{\sigma/\sqrt{n}} \right)$$

and

$$\frac{1}{\sigma/\sqrt{n}} \phi\left( \frac{\theta - \theta_L}{\sigma/\sqrt{n}} \right),$$

respectively, where $\phi(\cdot)$ is the density distribution function of the standard normal distribution. Thus, $P$ asymptotically follows the normal distribution $N\left( \theta_P, \sigma_P^2 \right)$ with mean

$$\theta_P = 1 - \Phi\left( \frac{\theta - \theta_L}{\sigma/\sqrt{n}} \right),$$

and variance

$$\sigma_P^2 = \phi^2 \left( \frac{\theta - \theta_L}{\sigma / \sqrt{n}} \right).$$

To provide a better understanding, Figure 6.2 illustrates some relationships between the asymptotical mean and variance of the p-value, and $n$, $\theta - \theta_L$, and $\sigma$.

Note that $\theta_P$ and $\sigma_P^2$ converge to 0 as $n$ goes to infinity given $\theta - \theta_L > 0$. In other words, one can always reject the null hypothesis with large enough sample size, given $\theta - \theta_L > 0$. Besides, larger $\theta - \theta_L$ and/or smaller $\sigma$ result in smaller $\theta_P$ and $\sigma_P^2$, given $\theta - \theta_L > 0$. From Figure 6.2, we can also see that in some situations, the asymptotical mean of the $p$-value could be between 1% and 5% or larger for some small sample size $n$ even though the true effect size $\theta$ is greater than $\theta_U$, which is in turn greater than $\theta_L$. Thus, the $p$-value based on a random sample with small size (which is common in rare diseases) could be between 1% and 5% or even larger although the underlined effect size is in fact greater than $\theta_U$. Fortunately, this could be overcome by increasing the sample size. For example, suppose $\theta_U \in (\theta_L, \theta_L + 0.5)$ and consider the upper left panel of Figure 6.2, where $\sigma = 1$ and $\theta = \theta_L + 0.5$. The asymptotical mean of the $p$-value decreases from 0.110 to 0.006 as the sample size increases from 6 to 25 although $\theta > \theta_U$. More specifically, the asymptotical mean of the $p$-value varies between 0.048 and 0.010 when the sample size varies between 11 and 22. Thus, if we are able to set the sample size to be greater than 11, the possibility of insignificant conclusion can be reduced. Further, if we can increase the sample size greater than 22, the possibility of inconclusiveness could be ruled out.

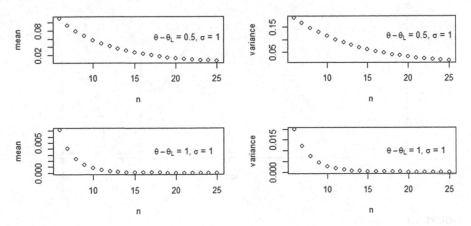

**FIGURE 6.2**
Illustrations of the relation between the asymptotical mean and variance of the p-value, and $n$, $\theta - \theta_L$, and $\sigma$.

In practice, however, we may not be able to draw sufficient random samples within a relatively short period of time, especially for rare diseases clinical trials. To overcome this problem, we propose testing the hypotheses (6.3) under a two-stage adaptive trial design. At the first stage, we test $\theta > \theta_L$ by relaxing the conventional significance level $\alpha_1$ (e.g., we may choose $\alpha_1 = 10\%$) with a limited sample size $n_1$. In the second stage, we then obtain another sample with size $n_2$ and test $\theta > \theta_U$ by ruling out (or controlling) the probability of inconclusiveness at a significance level of $\alpha_2$. The proposed two-stage adaptive trial design will be able to control the overall type I error rate at the $\alpha$-level of significance (see, e.g., Chang, 2007).

## 6.3 Evaluation of the Probability of Inconclusiveness

As mentioned above, the area where $\theta_L < \theta \le \theta_U$ is considered as the area of inconclusiveness. Suppose we have rejected the null hypothesis that $\theta \le \theta_L$ and decided that $\theta > \theta_L$. As mentioned above, to account for the possibility of inconclusiveness, many researchers tend to test the null hypothesis $\theta \le \theta_L$ at the $\alpha = 1\%$ rather than $\alpha = 5\%$ level of significance. Although intuitively meaningful, this may be inaccurate in testing hypotheses (6.3) since whether it is inconclusive also depends on the value of $\theta_U$. Different values of $\theta_U$ can lead to different results. We demonstrate this in Figure 6.3.

Assume the population mean and variance equal $\theta_L$ and $\sigma^2$, respectively. Given a sample $\mathcal{X}_n$ with test statistics $T = \bar{X}$, under the normal assumption or approximation and the null of hypotheses (6.4), the range of the critical value is $\left(\theta_\alpha, \theta_{\alpha*}\right)$ for $p$-value between $\alpha = 5\%$ and $\alpha_* = 1\%$, where

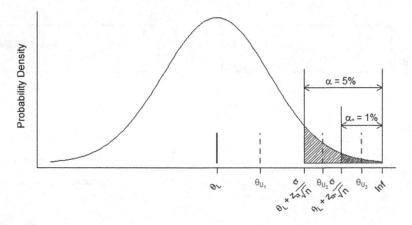

**FIGURE 6.3**
Demonstration of the relations between the $p$-values $\alpha$, $\alpha_*$ and their corresponding critical values given the sample size $n$, true population mean $\theta_L$, and variance equal $\sigma^2$. And illustrate the possible values of $\theta_U$, say $\theta_{U_1}$, $\theta_{U_2}$, and $\theta_{U_3}$, and their relations with the $p$-values.

$$\theta_\alpha = \theta_L + z_\alpha \frac{\sigma}{\sqrt{n}}$$

and

$$\theta_{\alpha^*} = \theta_L + z_{\alpha^*} \frac{\sigma}{\sqrt{n}}.$$

Here, we consider three possible values of $\theta_U$, say $\{\theta_{U_1}, \theta_{U_2}, \theta_{U_3}\}$, where

$$\theta_L < \theta_{U_1} < \theta_\alpha < \theta_{U_2} < \theta_{\alpha^*} < \theta_{U_3},$$

as shown in Figure 6.3. If $\theta_U = \theta_{U_1}$, the $p$-value of 5% may be small enough to rule out the inconclusiveness and demonstrate effectiveness since $\theta_{U_1} < \theta_\alpha$; if $\theta_U = \theta_{U_2} \in (\theta_\alpha, \theta_{\alpha^*})$, the $p$-value of $\alpha = 5\%$ results in inconclusiveness, while the $p$-value of $\alpha^* = 1\%$ may lead to inconclusiveness or effectiveness; if $\theta_U = \theta_{U_3}$, the $p$-value of $\alpha^* = 1\%$ still cannot demonstrate effectiveness since $\theta_{\alpha^*} < \theta_{U_3}$. Moreover, think about the upper boundary is

$$\theta_L + \frac{1}{2}(\theta_U - \theta_L)$$

instead of $\theta_U$, the conclusion of the above discussion may be totally different. Thus, we need to rely on $\theta_U$ to determine the probability of inconclusiveness instead of intuitively using of 1% to 5% significance level.

For a significance level of $\alpha_1$, given a sample

$$X_{n_1} = \{X_{11}, X_{12}, \ldots, X_{1n_1}\}$$

and a test statistic $T_1 = T(X_{n_1})$, we can conclude *not-ineffectiveness* if

$$T_1 > \theta_L + z_{\alpha_1} \frac{\sigma}{\sqrt{n_1}}$$

and *effectiveness* if

$$T_1 > \theta_U + z_{\alpha_1} \frac{\sigma}{\sqrt{n_1}}.$$

Then, the area where

$$T_1 \in \left( \theta_L + z_{\alpha_1} \frac{\sigma}{\sqrt{n_1}}, \theta_U + z_{\alpha_1} \frac{\sigma}{\sqrt{n_1}} \right)$$

is considered as the area of inconclusiveness. We denote the probability of inconclusiveness as $P_I$ and define it as the conditional probability that the test statistic $T_1$ falls within

$$\left( \theta_L + z_{\alpha_1} \frac{\sigma}{\sqrt{n_1}}, \; \theta_U + z_{\alpha_1} \frac{\sigma}{\sqrt{n_1}} \right)$$

given that

$$T_1 > \theta_L + z_{\alpha_1} \frac{\sigma}{\sqrt{n_1}}.$$

That is,

$$P_I = Pr\left( T_1 \in \left( \theta_L + z_{\alpha_1} \frac{\sigma}{\sqrt{n_1}}, \; \theta_U + z_{\alpha_1} \frac{\sigma}{\sqrt{n_1}} \right) \middle| T_1 \right) \theta_L + z_{\alpha_1} \frac{\sigma}{\sqrt{n_1}} \right)$$

$$= \left( \Phi\left( \frac{\theta_U - \theta}{\sigma/\sqrt{n_1}} + z_{\alpha_1} \right) - \Phi\left( \frac{\theta_L - \theta}{\sigma/\sqrt{n_1}} + z_{\alpha_1} \right) \right) \middle/ \left( 1 - \Phi\left( \frac{\theta_L - \theta}{\sigma/\sqrt{n_1}} + z_{\alpha_1} \right) \right). \quad (6.8)$$

At the lower margin of the null hypothesis of (6.3), where $\theta = \theta_L$, (6.8) becomes

$$P_I = \left( \Phi\left( \frac{\theta_U - \theta_L}{\sigma/\sqrt{n_1}} + z_{\alpha_1} \right) - (1 - \alpha_1) \right) / \alpha_1. \quad (6.9)$$

At the upper margin of the null hypothesis of (6.3), where $\theta = \theta_U$, (6.8) becomes

$$P_I = \left( (1 - \alpha_1) - \Phi\left( -\frac{\theta_U - \theta_L}{\sigma/\sqrt{n_1}} + z_{\alpha_1} \right) \right) \middle/ \left( 1 - \Phi\left( -\frac{\theta_U - \theta_L}{\sigma/\sqrt{n_1}} + z_{\alpha_1} \right) \right). \quad (6.10)$$

See Table 6.1 and Figure 6.4 for some examples of the probability of inconclusiveness. As shown in Table 6.1 and Figure 6.4, if the true $\theta$ belongs to $(\theta_L, \theta_U)$, $P_I$ is large; if the true $\theta > \theta_U$, $P_I$ monotonically decreases as $\theta$ increases. Besides, given $\theta > \theta_U$, $P_I$ decreases as the sample size $n_1$ increases. In addition, given $\theta > \theta_U$, $P_I$ increases as $\theta_U$ approaches $\theta$. Moreover, given $\theta > \theta_U$, $P_I$ increases as the significance level of $\alpha_1$ decreases, which means that it will be hard to separate inconclusiveness and effectiveness if we choose a level $\alpha_1$ that is highly significant to test not-ineffectiveness (6.4). In other words, we should begin with a mild to moderate significance level $\alpha_1$ for the first stage in the two-stage design.

To estimate $P_I$, we obtain another sample

$$X_{n_2} = \{ X_{21}, X_{22}, \ldots, X_{2n_2} \}$$

and replace $\theta$ in (6.8) by

$$T_2 = T(X_{n_2}),$$

**TABLE 6.1**

Demonstration of the Relations between the Probability of Inconclusiveness with the True Value of $\theta$, the Sample Size $n$, Possible values of $\theta_U$, and the significance level $\alpha_1$

| | | | | | | | | | | |
|---|---|---|---|---|---|---|---|---|---|---|
| | $\theta_L = 0$, $\theta_U = 0.5$, $\sigma = 1$, $\alpha_1 = 0.05$, $n_1 = 20$ | | | | | | | | | |
| $\theta$ | −0.1 | 0.1 | 0.3 | 0.5 | 0.7 | 0.9 | 1.1 | 1.3 | 1.5 | 1.7 |
| $P_I$ | 1.000 | 0.997 | 0.985 | 0.931 | 0.757 | 0.438 | 0.149 | 0.026 | 0.002 | 0 |
| | $\theta_L = 0$, $\theta_U = 0.5$, $\theta = 1$, $\sigma = 1$, $\alpha_1 = 0.05$ | | | | | | | | | |
| $n_1$ | 5 | 10 | 15 | 20 | 25 | 30 | 35 | 40 | 45 | 50 |
| $P_I$ | 0.586 | 0.493 | 0.377 | 0.275 | 0.196 | 0.137 | 0.095 | 0.065 | 0.044 | 0.029 |
| | $\theta_L = 0$, $\theta = 1$, $\sigma = 1$, $\alpha_1 = 0.05$, $n_1 = 20$ | | | | | | | | | |
| $\theta_U$ | 0.05 | 0.15 | 0.25 | 0.35 | 0.45 | 0.55 | 0.65 | 0.75 | 0.85 | 0.95 |
| $P_I$ | 0.002 | 0.013 | 0.041 | 0.101 | 0.206 | 0.355 | 0.531 | 0.700 | 0.835 | 0.922 |
| | $\theta_L = 0$, $\theta_U = 0.5$, $\theta = 1$, $\sigma = 1$, $n_1 = 20$ | | | | | | | | | |
| $\alpha_1$ | 0.005 | 0.020 | 0.035 | 0.050 | 0.065 | 0.080 | 0.095 | 0.110 | 0.125 | 0.140 |
| $P_I$ | 0.622 | 0.423 | 0.333 | 0.275 | 0.234 | 0.202 | 0.177 | 0.156 | 0.138 | 0.124 |

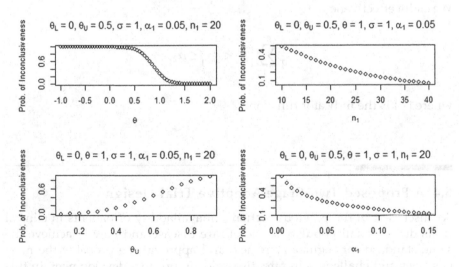

**FIGURE 6.4**

Demonstration of the relations between the probability of inconclusiveness with the true value of $\theta$, the sample size $n$, possible values of $\theta_U$, and the significance level $\alpha_1$.

and estimate $P_I$ as

$$\hat{P}_I = \left( \Phi\left( \frac{\theta_U - T_2}{\sigma/\sqrt{n_1}} + z_{\alpha_1} \right) - \Phi\left( \frac{\theta_L - T_2}{\sigma/\sqrt{n_1}} + z_{\alpha_1} \right) \right) \Bigg/ \left( 1 - \Phi\left( \frac{\theta_L - T_2}{\sigma/\sqrt{n_1}} + z_{\alpha_1} \right) \right). \quad (6.11)$$

If $\hat{P}_I$ is small enough, we would reject the null hypothesis of (6.3) and claim effectiveness.

Note that the determination of the threshold in the hypotheses testing is critical and may be rather complicated. In this case, we may consider the bootstrap method for the hypothesis test (see, e.g., Efron and Tibshirani, 1994). Specifically, we create $B$ pseudo samples with replacement

$$\left\{X_{n_2}^1, X_{n_2}^2, ..., X_{n_2}^B\right\},$$

where

$$\mathbf{X}_{n_2}^b = \left\{X_{21}^b, X_{22}^b, ..., X_{2n_2}^b\right\}, b = 1,..., B.$$

Then, we calculate a set of bootstrap estimates of $T_2$ and $P_I$, say

$$\left\{T_2^1, T_2^2, ..., T_2^B\right\} \text{ and } \left\{\hat{P}_I^1, \hat{P}_I^2, ..., \hat{P}_I^B\right\}.$$

We claim effectiveness if

$$\frac{1}{B}\sum_{b=1}^B \mathbb{I}\left\{\hat{P}_I^b < \hat{P}_I\right\} \leq \alpha_2,$$

where $\mathbb{I}\{\cdot\}$ is the indicator function.

## 6.4 A Proposed Two-Stage Adaptive Trial Design

As discussed in the previous sections, unavailability of patients in clinical trials due to small population size of rare disease and how to achieve the same standard for regulatory review and approval are probably the most obstacles and challenges in rare disease drug product development. In this section, to address these dilemmas, we propose a two-stage adaptive trial design for demonstrating "not-ineffectiveness" at the first stage and then demonstrating "effectiveness" at the second stage of rare disease drug product. The proposed two-stage adaptive trial design is briefly outline below.

*Stage 1.* Constructing a $(1-\alpha)\times 100\%$ confidence interval for $\theta$ based on previous/pilot studies or literature review. Then, based on $n_1$ subjects available at stage 1, test hypotheses for non-inferiority (i.e., test for not-ineffectiveness) at the $\alpha_1$ level, a pre-specified level of significance. Specifically, we construct a test statistic

$$T_1 = T(\mathcal{X}_{n_1}) = \frac{1}{n_1}\sum_{i=1}^{n_1} X_{1i}$$

based on the observation of the $n_1$ subjects

$$X_{n_1} = \{X_{11}, X_{12}, ..., X_{1n_1}\}.$$

Denote the corresponding $p$-value as $P_1$, which is equal to or less than

$$1 - \Phi\left(\frac{T_1 - \theta_L}{\sigma/\sqrt{n_1}}\right)$$

in the null parameter space, and reject the null hypothesis of ineffective-ness if

$$P_1 \le \alpha_1 \left(\text{or } T_1 \ge \theta_L + z_{\alpha_1}\frac{\sigma}{\sqrt{n_1}}\right).$$

If fails to reject the null hypothesis of ineffectiveness, then stop the trial due to futility. Otherwise, proceed to the next stage.

*Stage 2.* Recruit additional $n_2$ subjects at the second stage. At this stage, sample size re-estimation may be performed for achieving the desirable sta-tistical assurance (say 80%) for the establishment of effectiveness of the test treatment under investigation. At the second stage, a statistical test is per-formed to assure that the probability of the area of inclusiveness is within an acceptable range at the $\alpha_2$ level, a pre-specified level of significance. Specifically, denote the additional sample as

$$X_{n_2} = \{X_{21}, X_{22}, ..., X_{2n_2}\}.$$

the test statistic as

$$T_2 = T(X_{n_2}) = \frac{1}{n_2}\sum_{i=1}^{n_2} X_{2i}$$

and the estimated probability of inconclusiveness as $\hat{P}_I$ based on (6.11). Denote the $B$ bootstrap pseudo samples, test statistics, and probabilities of inconclu-siveness as

$$\{X_{n_2}^1, X_{n_2}^2, ..., X_{n_2}^B\},$$

$$\{T_2^1, T_2^2, ..., T_2^B\},$$

and

$$\left\{ \hat{P}_I^1, \hat{P}_I^2, ..., \hat{P}_I^B \right\},$$

respectively.

Denote

$$P_2 = \frac{1}{B} \sum_{b=1}^{B} \mathbb{I}\left\{ \hat{P}_I^b < \hat{P}_I \right\}.$$

We claim effectiveness if $P_2 \leq \alpha_2$.

Under the proposed two-stage adaptive trial design, it can be showed that the overall type I error rate is a function of $\alpha_1$ and $\alpha_2$. Thus, with appropriate choice of $\alpha_1$, we may reduce the sample size required for demonstrating "not-ineffectiveness." However, it is suggested that the selection of $\alpha_1$ and $\alpha_2$ be specified in the study protocol. The post-study adjustment is not encouraged.

For review and approval of rare diseases drug products, we propose to first demonstrate *not-ineffectiveness* with limited information available at a pre-specified level of significance $\alpha_1$, and then collect additional information to rule out *inconclusiveness* for the demonstration of effectiveness at a pre-specified level of significance $\alpha_2$ under the proposed two-stage adaptive trial design.

Note that the idea of the proposed two-stage trial design is to first demonstrate not-ineffectiveness at a pre-specified level of significance $\alpha_1$ at the first stage, and then rule out inconclusiveness and conclude the demonstration of effectiveness at a pre-specified level of significance $\alpha_2$. Under a two-stage adaptive trial design, the overall type I error rate is a function of $\alpha_1$ and $\alpha_2$ if a test statistic based on $p$-values such as method of individual $p$-values (MIP), method of sum of $p$-values (MSP), or method of product of $p$-values (MPP) is used (Chang, 2007).

For illustration purpose, consider MIP described in Section 2.4 of Chapter 2. That is, the test statistic at the $k$th stage is given by

$$T_k = P_k, \ k = 1, ..., K.$$

Under the two-stage adaptive trial design (i.e., $K=2$), we have

$$\alpha = \alpha_1 + \alpha_2 (\beta_1 - \alpha_1), \tag{6.12}$$

where

$$\begin{cases} \text{Stop for efficacy} & \text{if } T_k \leq \alpha_k, \\ \text{Stop for futility} & \text{if } T_k > \beta_k, \\ \text{Continue with adaptations} & \text{if } \alpha_k < T_k \leq \beta_k. \end{cases}$$

**TABLE 6.2**

Stopping Boundaries with MIP for One-Sided $\alpha = 0.05$

| | $\alpha_1$ | 0.000 | 0.005 | 0.010 | 0.015 | 0.020 |
|---|---|---|---|---|---|---|
| $\beta_1$ | | | | | | |
| 0.10 | $\alpha_2$ | 0.5000 | 0.4737 | 0.4444 | 0.4118 | 0.3750 |
| 0.15 | | 0.3333 | 0.3103 | 0.2857 | 0.2593 | 0.2308 |
| 0.20 | | 0.2500 | 0.2308 | 0.2105 | 0.1892 | 0.1667 |
| 0.30 | | 0.1667 | 0.1525 | 0.1379 | 0.1228 | 0.1071 |
| 0.40 | | 0.1250 | 0.1139 | 0.1026 | 0.0909 | 0.0789 |
| 0.50 | | 0.1000 | 0.0909 | 0.0816 | 0.0722 | 0.0625 |
| 0.75 | | 0.0667 | 0.0604 | 0.0541 | 0.0476 | 0.0411 |
| 1.00 | | 0.0500 | 0.0452 | 0.0404 | 0.0355 | 0.0306 |

Thus, with an appropriate choice of $\alpha_1$ and $\alpha_2$, the overall type I error rate can be controlled at the $\alpha$-level of significance. We list some possible combinations of $\alpha_1$, $\beta_1$, and $\alpha_2$ in Table 6.2 according to (6.12) based on MIP (Chow and Chang, 2011).

As it can be seen from Table 6.2, if $\alpha_1$ and $\beta_1$ are chosen to be 0.05 and 0.30, respectively, then $\alpha_2$ given by 0.1525 will control the overall type I error at the 5% level of significance.

## 6.5 Application of Rare Diseases Drug Development

In rare disease drug development, one of the major challenges is probably smaller patient population (unavailability of patients for clinical trials), which is required for providing substantial evidence regarding the safety and effectiveness of the test treatment under investigation (FDA, 2015b). With the smaller patient population available, one of the major obstacles is how to achieve the same standards for regulatory approval as compared to new drug development. To overcome these dilemmas, the proposed two-stage adaptive trial design demonstrating *not-ineffectiveness* at the first stage and then controlling the probability of *inconclusiveness* at the second stage may be useful.

For illustration of the use of proposed two-stage adaptive trial design for rare disease drug development, we consider the example concerning the cellular therapy for Lyell disease, of which the incidence is estimated at 2 per 1 million inhabitants in Europe (Miller et al., 2018). It is an acute disease with approximately 22% mortality in Europe. The primary endpoint for the efficacy evaluation is complete healing at day 7, and the response rate of the current treatment is $\theta_c = 0.5$. Since the treatment costs are high, a single-arm phase I/II trial is planned to investigate the effect of the new cellular therapy. Assume the response rate of the new treatment is $\theta_t = 0.6$ and consider a significance level of $\alpha = 5\%$ and a power of $1 - \beta = 80\%$. The sample size needed for the right-sided test using traditional approach is

$$N_0 = \bar{\theta}\left(1 - \bar{\theta}\right)\left(z_\alpha + z_\beta\right)^2 / \left(\theta_t - \theta_c\right)^2 = 153$$

roughly, where

$$\bar{\theta} = \left(\theta_c + \theta_t\right)/2 = 0.55.$$

And the $(1 - \alpha) \times 100\%$ confidence interval of $\theta_t$ given $N$ is

$$(\theta_L, \theta_U) = (0.53, 0.67).$$

Consider the proposed two-stage adaptive design. In the first stage, we test

$$H_0: \theta_t \le \theta_L = 0.53 \text{ versus } H_a: \theta_t > \theta_L.$$

At the second stage, the alternative hypothesis is

$$H_a: \theta_t > \theta_U = 0.67.$$

Given a combination of $(\alpha_1, \beta_1, \alpha_2)$, we want to calculate the sample sizes $n_1$ in the first stage and $n_2$ in the second stage to maintain the total type I error rate $\alpha$ and the pre-specified power $(1 - \beta)$. Similar to the idea of Simon's two-stage optimal design, we define the probability of rejecting the alternative hypothesis with response rate $\theta$ as

$$\Pr\left(T_1 \le c_1 \mid \theta, n_1\right) + \int_{T_1 > c_1}^{\infty} f_{T_1}\left(t \mid \theta, n_1\right) \Pr\left(T_2 \in (c_{21}, c_{22}) \mid T_1 = t, \theta, n_2\right) dt \tag{6.13}$$

$$= \Pr\left(T_1 \le c_1 \mid \theta, n_1\right) + E_{T_1 > c_1}\left[\Pr\left(T_2 \in (c_{21}, c_{22}) \mid T_1, \theta, n_2\right)\right]$$

Suppose there is not early stop for efficacy in the first stage, then $\alpha_1 = 0$. Consider

$$c_1 = \theta_L + z_{\beta_1}\sqrt{\frac{\theta_L(1 - \theta_L)}{n_1}},$$

$$c_{21} = -\infty,$$

$$c_{22} = \theta_L + z_{\alpha_2}\sqrt{\frac{\theta_L(1 - \theta_L)}{n_2}}.$$

Then (6.13) becomes

$$\Pr(T_1 \le c_1 \mid \theta, n_1) + \Pr(T_1 > c_1 \mid \theta, n_1)\Pr(T_2 \le c_{22} \mid \theta, n_2). \qquad (6.14)$$

Under the margin of the null where $\theta = \theta_L$, (6.14) equals

$$1 - \beta_1 + \beta_1(1 - \alpha_2) = (1 - \alpha),$$

which maintains the type I error rate. Under the margin of the alternative, where $\theta = \theta_L$, (6.14) becomes

$$\Phi\left(\frac{\theta_L - \theta_U}{\sigma_U/\sqrt{n_1}} + z_{\beta_1}\frac{\sigma_L}{\sigma_U}\right) + \left(1 - \Phi\left(\frac{\theta_L - \theta_U}{\sigma_U/\sqrt{n_1}} + z_{\beta_1}\frac{\sigma_L}{\sigma_U}\right)\right)\Phi\left(\frac{\theta_L - \theta_U}{\sigma_U/\sqrt{n_2}} + z_{\alpha_2}\frac{\sigma_L}{\sigma_U}\right),$$

$$(6.15)$$

where

$$\sigma_L = \sqrt{\theta_L(1 - \theta_L)}, \quad \sigma_U = \sqrt{\theta_U(1 - \theta_U)}.$$

Without the loss of generality, assume $n_2 = \lambda n_1$; thus, the total sample size is

$$N = (1 + \lambda)n_1.$$

We want to figure out the minimum value of $n_1$ such that (6.15) is not greater than $\beta$ under the alternative hypothesis. Table 6.3 lists some possible combinations of $(\lambda, n_1, N)$ and the expectation of total sample size $EN$ under the null hypothesis given several combinations of $(\alpha_1, \beta_1, \alpha_2)$ that satisfy this constraint.

As it can be seen from Table 6.3, for different combinations of $(\alpha_1, \beta_1, \alpha_2)$, we can select the value of $\lambda$ from a reasonable range such that the proposed two-stage adaptive design requires smaller total sample size than the traditional design, sufficiently smaller first-stage sample size, and smaller expected total sample size under the null hypothesis.

From Table 6.3, we can see choosing the value of $\lambda$ around 1 will result in smaller sample size. Besides, it is of interest to study the relationship of required sample size with the combination of $(\alpha_1, \beta_1, \alpha_2)$. Figure 6.5 plots the required sample sizes $n_1$ and $N$ to maintain the pre-specified type I error rate and power for different values of $\beta_1$ and four possible values of $\lambda$, and compares with $N_0$. For all the scenarios, the two-stage adaptive design requires less sample size than the traditional design.

**TABLE 6.3**

Some Possible Combinations of $(\lambda, n_1, N)$ to Maintain Power of 80%

| $\lambda$ | 0.2 | 0.25 | 0.33 | 0.5 | 1 | 2 | 3 | 4 | 5 |
|---|---|---|---|---|---|---|---|---|---|
| | | | $\alpha_1 = 0, \beta_1 = 0.15, \alpha_2 = 0.33$ | | | | | | |
| $n_1$ | 119 | 103 | 88 | 72 | 56 | 49 | 48 | 48 | 48 |
| $N$ | 143 | 129 | 117 | 108 | 112 | 147 | 192 | 240 | 288 |
| $EN$ | 139 | 125 | 113 | 103 | 104 | 132 | 170 | 211 | 252 |
| | | | $\alpha_1 = 0, \beta_1 = 0.2, \alpha_2 = 0.25$ | | | | | | |
| $n_1$ | 150 | 128 | 103 | 77 | 54 | 42 | 39 | 38 | 38 |
| $N$ | 180 | 160 | 137 | 116 | 108 | 126 | 156 | 190 | 228 |
| $EN$ | 174 | 154 | 130 | 108 | 97 | 109 | 133 | 160 | 190 |
| | | | $\alpha_1 = 0, \beta_1 = 0.25, \alpha_2 = 0.2$ | | | | | | |
| $n_1$ | 150 | 150 | 119 | 85 | 54 | 39 | 34 | 32 | 31 |
| $N$ | 180 | 188 | 158 | 128 | 108 | 117 | 136 | 160 | 186 |
| $EN$ | 172 | 178 | 148 | 117 | 94 | 98 | 110 | 128 | 147 |

$\theta_L = 0.53$, $\theta_U = 0.67$, $\alpha = 0.05$, $\beta = 0.2$. EN is the expectation of total sample size under the null hypothesis. Traditional design requires a sample size of $N_0 = 153$.

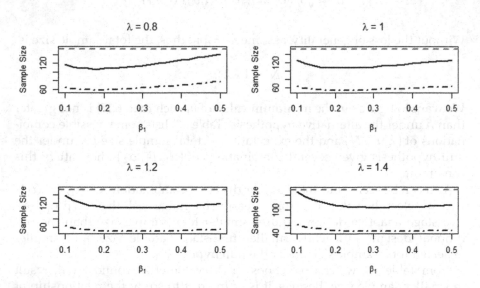

**FIGURE 6.5**

The required sample sizes $n_1$ and $N$ to maintain the pre-specified type I error rate and power for different values of $\beta_1$. The solid line is the total sample size $N$, the dash-dot line is the first-stage sample size $n_1$, and the dash horizontal line is the traditional sample size $N_0$, which is 153. $\theta_L = 0.53$, $\theta_U = 0.67$, $\alpha = 0.05$, $\beta = 0.2$, $\alpha_1 = 0$. Four possible values of $\lambda$ are considered, e.g., {0.8, 1, 1.2, 1.4}.

## 6.6 Concluding Remarks

In this chapter, we introduce the concept of demonstrating "not-ineffectiveness" rather than demonstrating "effectiveness" for rare disease drug development. The concept of the traditional way for demonstrating effectiveness for new drug development is very different from the concept of demonstrating "not-ineffectiveness." The concept of demonstrating "not-ineffectiveness" is based on testing for non-inferiority of the test treatment under investigation; i.e., we test the null hypothesis of inferiority (i.e., ineffectiveness) against the alternative hypothesis of non-inferiority (i.e., not-ineffectiveness), which consists of the concept of equivalence (i.e., inclusiveness) and superiority (i.e., effectiveness).

As indicated in the FDA draft guidance for rare disease drug development, one of the major obstacles and challenges is how to achieve the same standards for regulatory review/approval based on a limited number of patients available in a relatively smaller patient population with rare disease under investigation. The FDA encourages innovative thinking, trial design, and statistical method for data analysis/interpretation for rare disease drug development. For this purpose, in this chapter, a two-stage adaptive trial design was proposed. Under the two-stage adaptive trial design, we first demonstrate "not-ineffectiveness" of the test drug under investigation at a pre-specified level of significance ($\alpha_1$) in the first stage. Once the "not-ineffectiveness" has been demonstrated, we then test whether the probability of inconclusiveness is controlled at a pre-specified level of significance ($\alpha_2$) at the second stage. If the probability of inconclusiveness is within a pre-specified range of acceptance (i.e., it is controlled), we then claim that the effectiveness of the test drug under investigation has been demonstrated. The proposed two-stage adaptive trial design for rare disease drug development is useful by appropriately choosing $\alpha_1$ and $\alpha_2$ at the first and second stage, respectively. In this case, we will be able to control the overall type I error rate at the $\alpha$-level of significance.

# 7

## *Probability Monitoring Procedure for Sample Size*

### 7.1 Introduction

In clinical research, a typical approach for sample size calculation is based on power analysis, which controls type II error rate at a pre-specified level. Power analysis is to select an appropriate sample size for achieving a desired probability of correctly detecting a clinically meaningful difference at a pre-specified level of significance if such a difference truly exists. In practice, however, power analysis may not be feasible when the intended clinical trial is with extremely low incidence rate. This is because power analysis may require a huge sample size for detecting a relatively small difference. For example, if the incidence rate is 3 per 10,000, a sample size of 100,000 may be required for detecting a clinically meaningful difference of one per 10,000 at a pre-specified level of significance. In addition, power analysis for sample size calculation may not be feasible for rare disease clinical trials. In rare disease clinical trials, there are usually limited patients available due to the small patient population size. For example, assume a two-arm parallel rare disease clinical trial where the control arm has a response rate of 5% and the new treatment has 2% higher response rate. Consider a significance level of 5% and a targeted power of 90%. Assume equal sample sizes for both arms, the total sample size needed for the right-sided test with zero margin based on the power analysis is around 4,822, which is unrealistic in practice for rare disease drug development. Thus, in these cases, other procedures for sample size determination with certain statistical assurance are needed.

As indicated in Chow, Shao et al. (2017), alternatively, we may consider the precision analysis which controls type I error rate at a pre-specified significance level (or confidence level). In the above hypothetical example, the width of the confidence interval of the response rate difference is around 1.12% based on the power analysis. We may increase the width of the confidence interval to reduce the sample size requirement. For example, if we double the width, the required sample size is reduced to around 1,208, which is a quarter of the sample size based on the power analysis. However, the reduced sample size is still not feasible in rare disease clinical trials.

In addition, one may consider reproducibility analysis for sample size calculation. In practice, under certain circumstances, it may be of interest to conduct a second clinical trial to evaluate whether the clinical results from the first trial are reproducible (Shao and Chow, 2002). The sample size required in the second trial is calculated based on the estimates of parameters obtained from the first trial. It should be noted that the method of reproducibility analysis for sample size controls both the effect size and the variability associated with the observed effect size.

In this chapter, an innovative method based on a probability monitoring procedure is proposed for sample size determination. The concept is to select an appropriate sample size for controlling the probability of crossing safety and/or efficacy boundaries. For rare disease clinical development, an adaptive probability monitoring procedure may be applied if a multiple-stage adaptive trial design is used.

In the next section, traditional power analysis, precision analysis, and reproducibility analysis for sample size calculation are briefly described and compared. The proposed probability monitoring procedure for the sample size calculation is outlined in Section 7.3. An example concerning a rare disease clinical trial is given in Section 7.4. Some concluding remarks are provided in Section 7.5.

## 7.2 Traditional Methods for Sample Size Calculation

Without loss of generality, for simplicity and illustration purpose, consider non-inferiority hypotheses comparing a test treatment and an active control based on continuous endpoints:

$$H_0: \mu_1 - \mu_2 \leq \delta \text{ versus } H_a: \mu_1 - \mu_2 > \delta, \qquad (7.1)$$

where $\mu_1$ and $\mu_2$ are the means of the study endpoint of the two parallel groups of the test treatment and the active control agent, and $\delta$ is the non-inferiority margin, which is assumed to be negative. Denote the sample sizes of the two groups as $n_1$ and $n_2$, and their ratio as $\kappa = \dfrac{n_1}{n_2}$.

### 7.2.1 Procedures for Sample Size Calculation

In clinical trials, a pre-study power analysis is often performed for sample size calculation under the hypotheses (7.1). In practice, power analysis is not the only way for sample size calculation (Chow, Shao et al., 2017). In addition to power analysis, precision analysis and reproducibility analysis are probably the most commonly considered alternatives for sample size calculation in clinical trials. These methods are briefly described below.

**Power Analysis** – Under hypotheses (7.1), based on the power analysis, the sample sizes are given by (Chow, Shao et al., 2017):

$$n_1 = \kappa n_2 \text{ and } n_2 = \left( \frac{\sigma_1^2}{\kappa} + \sigma_2^2 \right) \left( \frac{z_\alpha + z_\beta}{\mu_1 - \mu_2 - \delta} \right)^2, \tag{7.2}$$

where $\sigma_1^2$ and $\sigma_2^2$ are the variances of the two groups, respectively; $\alpha$ and $\beta$ are the type I and II error rates; $z_\alpha = \Phi^{-1}(1-\alpha)$; and $\Phi(\cdot)$ is the cumulative distribution functions of the standard normal distribution.

**Precision Analysis** – For the precision analysis, the width of the $(1-\alpha) \times 100\%$ confidence interval for $\mu_1 - \mu_2$ can be obtained as

$$\omega = z_\alpha \sqrt{\frac{\sigma_1^2}{n_1} + \frac{\sigma_2^2}{n_2}}. \tag{7.3}$$

Thus, the method of precision analysis for the sample size calculation yields:

$$n_1 = \kappa n_2 \text{ and } n_2 = \left( \frac{\sigma_1^2}{\kappa} + \sigma_2^2 \right) \left( \frac{z_\alpha}{\omega} \right)^2. \tag{7.4}$$

**Reproducibility Analysis** – Assume the estimated means are $\hat{\mu}_1$ and $\hat{\mu}_2$ and the estimated standard deviations are $\hat{\sigma}_1$ and $\hat{\sigma}_2$ based on sample sizes $n_1$ and $n_2$ in a previous or pilot trial. The power of the intended trial is:

$$p(\theta) = 1 - \Phi \left( \frac{\mu_1 - \mu_2 - \delta}{\sigma} + z_\alpha \right), \tag{7.5}$$

where $\theta = (\mu_1, \mu_2, \delta, \sigma)$ and $\sigma' = \sqrt{\frac{\sigma_1^2}{\kappa n_2} + \frac{\sigma_2^2}{n_2}}$. Replace $\mu_1$, $\mu_2$, $\sigma'$, and $\delta$ by their estimates $\hat{\mu}_1$, $\hat{\mu}_2$, $\hat{\sigma}'$, and $\hat{\delta}$, we obtain the following probability of reproducibility, i.e., the empirical power (Shao and Chow, 2002):

$$\hat{P} = p(\hat{\theta}) = 1 - \Phi \left( \frac{\hat{\mu}_1 - \hat{\mu}_2 - \hat{\delta}}{\hat{\sigma}'} + z_\alpha \right). \tag{7.6}$$

Suppose $P_0$ is the desired reproducibility probability. Thus, the sample size required for achieving $P_0$ can be obtained as:

$$n_1^* = \kappa n_2^* \text{ and } n_2^* = \left(\frac{\hat{\sigma}_1^2}{\kappa} + \hat{\sigma}_2^2\right)\left(\frac{z_\alpha + z_{1-P_0}}{\hat{\mu}_1 - \hat{\mu}_2 - \hat{\delta}}\right)^2. \tag{7.7}$$

It should be noted that the method of reproducibility analysis for sample size controls both $\mu_1 - \mu_2$ and the variability associated with the observed difference $\hat{\mu}_1 - \hat{\mu}_2$.

## 7.2.2 A Comparison

For an easy comparison and without the loss of generality, we assume $\sigma_1 = \sigma_2 = \sigma$, $\kappa = 1$. Table 7.1 lists the formulas for the sample size calculation of the three methods based on a non-inferiority test.

It can be seen from Table 7.1 that the sample sizes calculated by power analysis and precision analysis share the same factor of $2\sigma^2$, while they differ in the factors of

$$\left(\frac{z_\alpha + z_\beta}{\mu_1 - \mu_2 - \delta}\right)^2 \text{ or } \left(\frac{z_\alpha}{\omega}\right)^2.$$

Thus, the difference of power analysis and precision analysis depends on the type I error rate $\alpha$, the desired power $1 - \beta$ (or type II error rate $\beta$), the difference of $\mu_1 - \mu_2$, the margin $\delta$, and the specified confidence interval width $\omega$. Assuming $\mu_1 - \mu_2 - \delta > 0$, power analysis and precision analysis will yield the same sample size if

$$\omega = \frac{z_\alpha}{z_\alpha + z_\beta}(\mu_1 - \mu_2 - \delta).$$

**TABLE 7.1**

Sample Size Calculation Formula for Non-Inferiority Test of Continuous Endpoints

| Method | Sample Size Calculation Formula ($n_2$) | Comment |
|---|---|---|
| Power analysis | $\dfrac{2(z_\alpha + z_\beta)^2 \sigma^2}{(\mu_1 - \mu_2 - \delta)^2}$ | Control type II error rate ($\beta$) |
| Precision analysis | $\dfrac{2z_\alpha^2 \sigma^2}{\omega^2}$ | Control type I error rate ($\alpha$) |
| Reproducibility analysis | $\dfrac{2(z_\alpha + z_{1-P_0})^2 \hat{\sigma}^2}{(\hat{\mu}_1 - \hat{\mu}_2 - \hat{\delta})^2}$ | Control both $\mu_1 - \mu_2$ and $\sigma$ |

*Note:* We assume $\sigma_1 = \sigma_2 = \sigma, \kappa = 1$. $n_1 = \kappa n_2 = n_2$.

Power analysis will give a larger sample size than precision analysis if

$$\omega > \frac{z_\alpha}{z_\alpha + z_\beta}(\mu_1 - \mu_2 - \delta)$$

and a smaller sample size if

$$\omega < \frac{z_\alpha}{z_\alpha + z_\beta}(\mu_1 - \mu_2 - \delta).$$

On the other hand, reproducibility analysis exhibits the same pattern with that of power analysis; yet, $u_1$, $\mu_2$, $\sigma$, $\delta$, and $z_\beta$ for power analysis are replaced by $\hat{\mu}_1$, $\hat{\mu}_2$, $\hat{\sigma}$, $\hat{\delta}$, and $z_{1-P_0}$, respectively, for reproducibility analysis. Thus, the difference in power analysis and reproducibility analysis depends on the pre-specified means $(u_1, \mu_2)$, variance $(\sigma^2)$, non-inferiority margin $(\delta)$, and their estimates $\hat{\mu}_1$, $\hat{\mu}_2$, $\hat{\sigma}$, $\hat{\delta}$ from a pilot trial, and the pre-specified power $1 - \beta$ and the desired reproducibility probability $P_0$. Assuming

$$P_0 = 1 - \beta,\ \mu_1 - \mu_2 - \delta > 0$$

and

$$\hat{\mu}_1 - \hat{\mu}_2 - \hat{\delta} > 0,$$

power analysis and reproducibility analysis will yield the same sample size if

$$\frac{\sigma}{\mu_1 - \mu_2 - \delta} = \frac{\hat{\sigma}}{\hat{\mu}_1 - \hat{\mu}_2 - \hat{\delta}}.$$

Power analysis will give a larger sample size than reproducibility analysis if

$$\frac{\sigma}{\mu_1 - \mu_2 - \delta} > \frac{\hat{\sigma}}{\hat{\mu}_1 - \hat{\mu}_2 - \hat{\delta}}$$

and a smaller sample size if

$$\frac{\sigma}{\mu_1 - \mu_2 - \delta} < \frac{\hat{\sigma}}{\hat{\mu}_1 - \hat{\mu}_2 - \hat{\delta}}.$$

A following concern is the probability of

$$\frac{\sigma}{\mu_1 - \mu_2 - \delta} > \frac{\hat{\sigma}}{\hat{\mu}_1 - \hat{\mu}_2 - \hat{\delta}},$$

which depends on $u_1$, $\mu_2$, $\sigma$, $\delta$ and the distribution of

$$\frac{\hat{\sigma}}{\hat{\mu}_1 - \hat{\mu}_2 - \hat{\delta}}$$

that further depends on the estimates $\hat{\mu}_1$, $\hat{\mu}_2$, $\hat{\sigma}$, $\hat{\delta}$ and the underlining distribution of the data. One may derive this probability analytically or approximate it numerically through simulation, yet we will not discuss further here since it is beyond the scope of this chapter.

Thus, the three methods result in the same sample size if $\omega$ equals the product of

$$\frac{z_\alpha}{z_\alpha + z_\beta} \text{ and } (\mu_1 - \mu_2 - \delta),$$

and the estimate

$$\frac{\hat{\sigma}}{\hat{\mu}_1 - \hat{\mu}_2 - \hat{\delta}}$$

based on the pilot trial equals

$$\frac{\sigma}{\mu_1 - \mu_2 - \delta},$$

given $\mu_1 - \mu_2 - \delta$ and its estimate $\hat{\mu}_1 - \hat{\mu}_2 - \hat{\delta}$ positive and the desired reproducibility probability equaling the desired power. This situation rarely happens; however, the three methods still share some common properties. For example, the required sample size decreases as the variance $\sigma^2$ (and thus its estimate $\hat{\sigma}^2$) decreases and/or the pre-specified type I error rate $\alpha$ increases. The required sample size will also decrease if the desired power level $1 - \beta$ and/or reproducibility probability $P_0$ decrease based on power analysis and reproducibility analysis, respectively. Moreover, the required sample size also decreases as $\mu_1 - \mu_2 - \delta$, $\omega$, and $\hat{\mu}_1 - \hat{\mu}_2 - \hat{\delta}$ increase with respect to the three methods, respectively. See Figure 7.1 for the comparison of sample size calculations based on the three methods under some arbitrary settings.

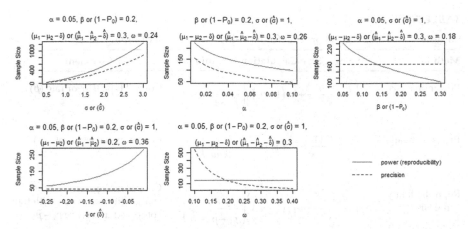

**FIGURE 7.1**
Comparison of sample size calculation based on the three methods.

In addition, for power analysis (and reproducibility analysis), it is of interest to investigate the relation between $\sigma$ and $\delta$ to guarantee that the required sample size is unchanged given the other parameters fixed. By simple algebraic operation, we have:

$$\sigma = \frac{\sqrt{n_2}}{\sqrt{2}\left(z_\alpha + z_\beta\right)}\left(\mu_1 - \mu_2 - \delta\right)$$

and/or

$$\hat{\sigma} = \frac{\sqrt{n_2}}{\sqrt{2}\left(z_\alpha + z_{1-P_0}\right)}\left(\hat{\mu}_1 - \hat{\mu}_2 - \hat{\delta}\right). \tag{7.8}$$

Thus, given the other quantities fixed, $\sigma$ $(\hat{\sigma})$ is negatively linear with $\delta$ $(\hat{\delta})$.

### 7.2.3 Remarks

In addition to the continuous endpoint, we provide the formulas to calculate the sample size for binary endpoints in Table 7.2 with respect to non-inferiority hypotheses comparing a test treatment and an active control:

$$H_0: p_1 - p_2 \le \delta \text{ versus } H_a: p_1 - p_2 > \delta, \tag{7.9}$$

where $p_1$ and $p_2$ are the incidence rates of the two groups of the test treatment and the active control agent. Note that for binary responses, the variability

**TABLE 7.2**

Sample Size Calculation Formula for Non-Inferiority Test of Binary Endpoints

| Method | Sample Size Calculation Formula ($n_2$) | Comment |
|---|---|---|
| Power analysis | $\left( \dfrac{p_1(1-p_1)}{\kappa} + p_2(1-p_2) \right)\left( \dfrac{z_\alpha + z_\beta}{p_1 - p_2 - \delta} \right)^2$ | Control type II error rate ($\beta$) |
| Precision analysis | $\left( \dfrac{p_1(1-p_1)}{\kappa} + p_2(1-p_2) \right)\left( \dfrac{z_\alpha}{\omega} \right)^2$ | Control type I error rate ($\alpha$) |
| Reproducibility analysis | $\left( \dfrac{\hat{p}_1(1-\hat{p}_1)}{\kappa} + \hat{p}_2(1-\hat{p}_2) \right)\left( \dfrac{z_\alpha + z_{1-P_0}}{\hat{p}_1 - \hat{p}_2 - \hat{\delta}} \right)^2$ | Control both $p_1 - p_2$ and the variability associated with the observed difference $\hat{p}_1 - \hat{p}_2$ |

*Note:*   1. $n_1 = \kappa n_2$. 2. $\omega$ is the width of the $(1-\alpha) \times 100\%$ confidence interval for $p_1 - p_2$. $\hat{p}_1$, $\hat{p}_2$ and $\hat{\delta}$ are the estimated incidence rates and estimated margin in the first trial. 3. $P_0$ is the desired reproducibility probability.

associated with the observed difference is determined by $p_1$ and $p_2$, and reaches its maximum of

$$0.25\left(1 + \frac{1}{\kappa}\right)$$

when $p_1 = p_2 = 0.5$. The variability becomes smaller as $p_1$ and $p_2$ move far away from 0.5.

Moreover, we compare the sample sizes calculated by power analysis, precision analysis, and reproducibility analysis by listing some possible (and reasonable) combinations of $\left(\alpha, \beta, p_1, p_2, \delta, \omega, \hat{p}_1, \hat{p}_2, \hat{\delta}, P_0\right)$ in Table 7.3. For demonstration purposes, we set $\hat{p}_1 = p_1$, $\hat{p}_2 = p_2$, $\hat{\delta} = \delta$, $P_0 = 1 - \beta$; then,

the sample sizes calculated by reproducibility analysis are the same as those calculated by power analysis; thus, for simplification and without the loss of generalization, we don't distinguish the results of power analysis and reproducibility analysis.

Note that it can be seen from Table 7.3, the two groups are balanced; i.e., $\kappa = 1$. $n_{pow}$, $n_{pre}$, and $n_{rep}$ are the sample sizes calculated by the power analysis, the precision analysis, and the reproducibility analysis, respectively. In this case, we don't distinguish the power analysis and reproducibility analysis for simplification. A blank item means the item value is the same as that in the previous column.

From Table 7.3, we can see that all the scenarios except one require sample sizes more than 1,000, which is typically infeasible in rare disease clinical trials.

**TABLE 7.3**

Comparison of Sample Size Calculation by Power Analysis, Precision Analysis and Reproducibility Analysis by Varying $\alpha$, $\beta$, $p_1$, $\hat{p}_1$, $\delta$, $\omega$, $\hat{p}_1$, $\hat{p}_2$, $\hat{\delta}$, $P_0$

| $\alpha$ | $\beta$ $(1 - P_0)$ | $p_1$ $(\hat{p}_1)$ | $p_2$ $(\hat{p}_2)$ | $\delta$ $(\hat{\delta})$ | $\omega$ | $n_{\text{pow}}$ $(n_{\text{rep}})$ | $n_{\text{pre}}$ |
|---|---|---|---|---|---|---|---|
| 0.05 | 0.10 | 0.07 | 0.05 | 0 | 1% | 4,822 | 6,094 |
| | | | | | 2% | | 1,524 |
| | | | | | 2.5% | | 976 |
| | | | | −0.5% | | 3,086 | |
| | | | | −1.5% | | 1,574 | |
| | 0.20 | | | 0 | | 3,482 | |
| | | | | −0.5% | | 2,228 | |
| | | | | −1.5% | | 1,138 | |
| 0.025 | 0.10 | | | 0 | 1% | 5,916 | 8,652 |
| | | | | | 2% | | 2,164 |
| | | | | | 2.5% | | 1,384 |
| | | | | −0.5% | | 3,786 | |
| | | | | −1.5% | | 1,932 | |
| | 0.20 | | | 0 | | 4,420 | |
| | | | | −0.5% | | 2,828 | |
| | | | | −1.5% | | 1,444 | |
| 0.05 | 0.10 | 0.08 | | 0 | 1% | 2,306 | 6,554 |
| | | | | | 2% | | 1,638 |
| | | | | | 2.5% | | 1,048 |
| | | | | −0.5% | | 1,694 | |
| | | | | −1.5% | | 1,024 | |
| | | 0.07 | 0.06 | 0 | 1% | 20,810 | 6,574 |
| | | | | | 2% | | 1,644 |
| | | | | | 2.5% | | 1,052 |
| | | | | −0.25% | | 13,318 | |
| | | | | −0.75% | | 6,796 | |

## 7.3 Probability Monitoring Procedure

As it can be seen from Table 7.3 that neither power analysis, precision analysis, nor reproducibility analysis is feasible for clinical trials with extremely low incidence rates or rare disease drug development due to the small patient population available. Thus, it is desirable to propose innovative methods like methods based on probability statement (Chow et al., 2017) for sample size calculation while maintaining some statistical assurance. For this purpose, Huang and Chow (2019) proposed a probability monitoring procedure.

The concept of *probability monitoring procedure* is to pre-specify the total sample size throughout the entire clinical study in advance based on clinical, budget, and other considerations, and then carry out continual probability monitoring on a sequence of accumulated sub-samples to maintain the statistical assurance such as the probability of crossing safety and/or efficacy boundaries.

Moreover, there are two versions of the probability monitoring procedure: One is non-adaptive and the other one is adaptive. For better understanding, we describe the two versions of *probability monitoring procedure* and the selection of the safety/efficacy boundaries through the following hypotheses testing.

Consider sample size calculation for (i) a phase II clinical trial with an extremely low incidence rate or (ii) a rare disease clinical. Consider the following non-inferiority hypotheses:

$$H_0: p \geq p_0 \text{ versus } H_a: p < p_0, \tag{7.10}$$

where $p_0$ is a pre-specified clinically meaningful threshold of incidence rate. As discussed earlier, sample size calculation based on either one of the three methods (i.e., power analysis, precision analysis, and reproducibility analysis) may not be feasible. Thus, Huang and Chow (2019) proposed a probability monitoring procedure for a pre-selected sample size to maintain some statistical assurance. We describe the non-adaptive and adaptive versions of monitoring procedures in the following two sections separately.

### 7.3.1 Non-Adaptive Probability Monitoring Procedure

The non-adaptive probability monitoring procedure contains three steps:

*Step 1*. Select an appropriate total sample size $n$ based on clinical, economical, and other considerations, and specify the probability monitoring procedure. For example, assume the total sample size is $n = 800$, and the monitoring procedure consists of a sequence of $Q = 4$ accumulated sub-samples $\{s_1, s_2, ..., s_Q\}$ with sizes $n_1, n_2, ..., n_Q$, where $s_1 \subset s_2 \subset ... \subset s_Q$ and $n_1 = 200$, $n_2 = 400$, $n_3 = 600$, and $n_Q = n = 800$.

*Step 2*. Specify the probability (efficacy and/or safety) boundary $P_b$, e.g., $P_b = 0.05$.

*Step 3*. For a given sequential sub-sample $s_q$, denote the number of incidents as $r_q$. Calculate the probability $P_q = B(r_q; n_q, p_0)$, where $B$ denote the cumulative binomial distribution. If $P_q$ is less than the boundary $P_b$ (or $r_q$ is less than the pre-specified threshold determined by $n_q$, $p_0$, and $P_b$), the study is continued with high confidence of success; otherwise, the boundary is crossed and it is highly possible that the study will fail. Repeat the monitoring procedure until the study is completed.

Huang and Chow (2019) demonstrated the non-adaptive probability monitoring procedure in Table 7.4 using a hypothetical example. In the hypothetical example, for simplicity, we assume that the total sample size is $n = 800$, and the monitoring procedure consists of a sequence of $Q = 8$ accumulated

**TABLE 7.4**

Non-Adaptive Probability Monitoring Procedure

| | | $p_0 = 0.05$ | | | | $p_0 = 0.05$ | | | | $p_0 = 0.01$ | | | |
| | | $P_b = 0.05$ | | | | $P_b = 0.05$ | | | | $P_b = 0.05$ | | | |
|---|---|---|---|---|---|---|---|---|---|---|---|---|---|
| $q$ | $n_q$ | $P_q^*$ | $r_q^*$ | $P_q^{**}$ | $r_q^{**}$ | $P_q^*$ | $r_q^*$ | $P_q^{**}$ | $r_q^{**}$ | $P_q^*$ | $r_q^*$ | $P_q^{**}$ | $r_q^{**}$ |
| 1 | 100 | 0.037 | 1 | 0.118 | 2 | | | | | | | | |
| 2 | 200 | 0.026 | 4 | 0.062 | 5 | 0.018 | 0 | 0.089 | 1 | | | | |
| 3 | 300 | 0.034 | 8 | 0.065 | 9 | 0.017 | 1 | 0.060 | 2 | 0.049 | 0 | 0.198 | 1 |
| 4 | 400 | 0.036 | 12 | 0.061 | 13 | 0.041 | 3 | 0.097 | 4 | 0.018 | 0 | 0.090 | 1 |
| 5 | 500 | 0.034 | 16 | 0.056 | 17 | 0.028 | 4 | 0.065 | 5 | 0.040 | 1 | 0.123 | 2 |
| 6 | 600 | 0.032 | 20 | 0.050 | 21 | 0.044 | 6 | 0.087 | 7 | 0.017 | 1 | 0.061 | 2 |
| 7 | 700 | 0.045 | 25 | 0.066 | 26 | 0.030 | 7 | 0.060 | 8 | 0.029 | 2 | 0.081 | 3 |
| 8 | 800 | 0.039 | 29 | 0.057 | 30 | 0.042 | 9 | 0.075 | 10 | 0.042 | 3 | 0.098 | 4 |

sub-samples. For the given true $p_0$ and pre-specified boundary $P_b$, we figure out the maximum possible probability that hasn't crossed the boundary and its corresponding number of incidences, and the minimum possible probability that crosses the boundary and its corresponding number of incidences.

Table 7.4 illustrates the non-adaptive probability monitoring procedure. Assume that the total sample size is $n = 800$, and the monitoring procedure consists of a sequence of $Q = 8$ accumulated sub-samples. $P_q^*$ is the maximum of the possible cumulative probabilities that haven't crossed the boundary $P_b$ for sub-sample $s_q$, and $r_q^*$ is the corresponding number of incidences. $P_q^{**}$ is the minimum of the possible cumulative probabilities that cross the boundary $P_b$ for sub-sample $s_q$, and $r_q^{**}$ is the corresponding number of incidences. The blank items mean inapplicable.

The best situation is that $P_q$ never crosses $P_b$ under each individual monitoring $q$. However, it should be pointed out that $P_q$ crosses $P_b$ for some individual monitoring $q$, which doesn't mean the study fails but highly implies the possibility of failure only, since it is possible that there is no additional incident for the subsequent sub-samples until the end of the study. For example, in Table 7.4 with $p_0 = 0.02$, $P_b = 0.05$, assume the number of incidences is $r_q = 6$, $q = 4$. Then, $P_q = 0.311$, which substantially crosses the boundary $P_b$ and highly implies the possibility of failure. Yet, it is still possible that no more incident happens subsequently and ends up with $r_8 = 6$ and $P_8 = B(r_8; n, p_0) = 0.004$, which denies the conclusion of failure.

However, it is highly risky to continue the study given a high observation of $r_q$ and the corresponding high value of $P_q$ for an individual monitoring $q$, since the study is highly possible to fail, which leads to waste of time and resource. Yet, the non-adaptive version of probability monitoring procedure cannot stop the study early due to futility; otherwise, it may cause the loss of power. Thus, it is necessary to propose an adaptive version of probability monitoring procedure.

### 7.3.2 Adaptive Probability Monitoring Procedure

The adaptive probability monitoring procedure differ from the non-adaptive version in that (i) instead of fixing the probability $p_0$, we update it for each sub-sample $s_q$ based on the estimated incidence rate from the previous sub-sample $s_{q-1}$ and (ii) it is possible to stop the study early under some stopping criteria (i.e., $P_q$ crossed the futility/safety and/or efficacy boundaries). Note that it may be necessary to update the boundaries after each monitoring to maintain the overall type I error rate and avoid losing of power due to potential early stops. On the one hand, to maintain the overall type I error rate, one may select the efficacy boundaries based on the idea of test statistic $p$-values such as method of individual $p$-values (MIP), method of sum of $p$-values (MSP), or method of product of $p$-values (MPP) (Chow and Chang, 2011). On the other hand, to avoid potentially losing of power, one should avoid too stringent futility boundaries in the early stages of the monitoring.

Thus, the adaptive probability monitoring procedure shares the same first step to the non-adaptive probability monitoring procedure, and the remaining two steps are as follows:

*Step 2a.* Determine the futility/safety probability boundaries $P_{fq}$ and/or efficacy boundaries $P_{eq}$, $q = 1,\ldots,Q$.

*Step 3a.* For a given sequential sub-sample $s_q$, denote the number of incidents as $r_q$. Calculate the probability $P_q = B(r_q; n_q, p_q)$, where $p_q$ is estimated based on the observed incidence rate from $s_q$. One of the reasonable estimates is

$$p_q = \min\left( p_0, \hat{p}_{q-1} + z_\varphi \sqrt{\frac{\hat{p}_{q-1}\left(1-\hat{p}_{q-1}\right)}{n_{q-1}}} \right),$$

where

$$\hat{p}_{q-1} = \frac{r_{q-1}}{n_{q-1}},\ p_1 = p_0$$

and $\varphi$ is a pre-specified value namely $\varphi = 0.025$. If $P_q < P_{eq}$, the study is stopped due to efficacy; if $P_q \geq P_{fq}$, the study is stopped due to futility; otherwise, the study is continued. Repeat Step 3a until the study is stopped early or the monitoring procedure is completed.

Huang and Chow (2019) demonstrated the adaptive probability monitoring procedure in Table 7.5 using a hypothetical example. For simplicity, we only consider the situation of early stopping due to futility/safety; i.e., we only specify the futility/safety boundaries and set the same futility/safety boundaries $P_{fq} = P_f$ for each $q$ without the loss of generality. Note that the overall type I error rate will not be inflated since there is no early stop due to efficacy. As the same for the non-adaptive probability monitoring, we assume the total sample size is $n = 800$,

**TABLE 7.5**

Adaptive Probability Monitoring Procedure

| | | $p_0 = 0.05$ | | | $p_0 = 0.05$ | | | $p_0 = 0.10$ | | |
| | | $P_f = 0.15$ | | | $P_f = 0.10$ | | | $P_f = 0.10$ | | |
| $q$ | $n_q$ | $P_q^m$ | $r_q^m$ | $p_q$ | $P_q^m$ | $r_q^m$ | $p_q$ | $P_q^m$ | $r_q^m$ | $p_q$ |
|---|---|---|---|---|---|---|---|---|---|---|
| 1 | 100 | 0.118 | 2 | 0.05 | 0.037 | 1 | 0.05 | 0.058 | 5 | 0.10 |
| 2 | 200 | 0.084 | 5 | 0.047 | 0.064 | 2 | 0.030 | 0.064 | 12 | 0.093 |
| 3 | 300 | 0.170 | 10 | 0.047 | 0.073 | 3 | 0.024 | 0.099 | 21 | 0.093 |
| 4 | 400 | 0.150 | 15 | 0.05 | 0.072 | 4 | 0.021 | 0.085 | 31 | 0.099 |
| 5 | 500 | 0.179 | 21 | 0.05 | 0.070 | 5 | 0.020 | 0.075 | 40 | 0.10 |
| 6 | 600 | 0.151 | 24 | 0.05 | 0.068 | 6 | 0.019 | 0.096 | 50 | 0.10 |
| 7 | 700 | 0.170 | 29 | 0.05 | 0.066 | 7 | 0.018 | 0.091 | 59 | 0.10 |
| 8 | 800 | 0.187 | 34 | 0.05 | 0.064 | 8 | 0.017 | 0.085 | 68 | 0.10 |

and the monitoring procedure contains a sequence of $Q = 8$ accumulated sub-samples in the hypothetical example. Besides, we assume $\varphi = 0.025$ and the observed number of incidents $r_q$ for each sub-sample $s_q$ is the maximum value that allows the study to continue.

Table 7.5 illustrates the adaptive probability monitoring procedure. Assume the total sample size is $n = 800$, and the monitoring procedure consists of a sequence of $Q = 8$ accumulated sub-samples. $P_q^m$ is the maximum of the possible cumulative probabilities that are less the futility boundary $P_f$ for sub-sample $s_q$, and $r_q^m$ is the corresponding maximum number of incidences. Assume the observed number of incidents $r_q$ for each sub-sample $s_q$ is $r_q^m$, such that

$$p_q = \min\left( p_0, \hat{p}_{q-1} + z_\varphi \sqrt{\frac{\hat{p}_{q-1}\left(1 - \hat{p}_{q-1}\right)}{n_{q-1}}} \right), \ \hat{p}_{q-1} = \frac{r_{q-1}^m}{n_{q-1}}, \ \varphi = 0.025$$

and

$$p_1 = p_0.$$

## 7.4 An Example

To demonstrate the use of the proposed non-adaptive and adaptive probability monitoring procedures in rare disease clinical research, consider the example of evaluation of treatments applicable to congenitally abnormal fetuses, which has an incidence rate of one in 10,000 pregnancies (Lilford et al., 1995). A category of patient with fairly advanced primary biliary cirrhosis for whom a new drug is proposed as an alternative to conventional ursodeoxycholic treatment is considered. It is assumed five-year mortality among patients with this stage of primary ciliary cirrhosis is $p_0 = 30\%$ with current treatment. Since the prevalence is so low, a single-arm trial is planned to investigate the effect of the drug, with the expectation of improvement in mortality from 30% to 26%, i.e., $\Delta p = 4\%$. Consider a significance level of $\alpha = 5\%$ and a power of $1 - \beta = 90\%$, and the width of the $(1-\alpha)\%$ confidence interval of $\omega = 2\%$; the sample size needed for test (7.8) is

$$n_{\text{pow}} = p_0(1-p_0)(z_\alpha + z_\beta)^2 \big/ \Delta p^2 = 1,124$$

and

$$n_{\text{pre}} = p_0(1-p_0)(z_\alpha / \omega)^2 = 1,420$$

based on power analysis and precision analysis, respectively. Obviously, a study of such size would need to recruit from very large areas over long periods, especially for the low prevalence, and is difficult to organize and likely to fail.

Instead, we may consider the proposed probability monitoring procedures to implement the inference based on a limited and affordable sample to achieve some statistical assurance. Suppose the maximum affordable sample size is $n = 600$. We may consider a monitoring procedure as each 100 more samples are accumulated, thus consisting of a sequence of $Q = 6$ accumulated sub-samples $\{s_1, s_2, \ldots, s_Q\}$ with sizes

$$n_1, n_2, \ldots, n_Q,$$

where

$$s_1 \subset s_2 \subset \cdots \subset s_Q$$

and

$$n_1 = 100, \quad n_2 = 200, \ldots, \quad n_Q = 600.$$

For the non-adaptive monitoring procedure, specify the boundary as $P_b = 0.01$ or 0.005, and the maximum of the possible cumulative probabilities that are less the boundary $P_b$ and the corresponding incident number $P_q^m$ and $r_q^m$ at each stage $q$ are listed. For the adaptive monitoring procedure, we consider the situation of early stopping due to futility/safety with same boundary $P_f = 0.05$ or 0.10 for each stage, and assume the observed number of incidents $r_q^a$ for each sub-sample $s_q$ is 0.8 of the maximum value that allows the study to continue, denote the corresponding cumulative probability as $P_q^a$, and determine the estimated incidence rate as

$$p_q = \min\left(p_0, \hat{p}_{q-1} + z_\varphi \sqrt{\frac{\hat{p}_{q-1}(1-\hat{p}_{q-1})}{n_{q-1}}}\right), \hat{p}_{q-1} = \frac{r_{q-1}^m/2}{n_{q-1}}, \varphi = 0.025$$

and

$$p_1 = p_0.$$

See Table 7.6 for more details.

Table 7.6 illustrates the non-adaptive and adaptive probability monitoring procedures for the real data. For non-adaptive monitoring, $P_q^*$ is the maximum of the possible cumulative probabilities that haven't crossed the boundary $P_b$ for sub-sample $s_q$, and $r_q^*$ is the corresponding number of incidents. For adaptive monitoring, we assume the observed number of incidents $r_q^a$ is 0.8 of the maximum of possible numbers of incidents whose cumulative probabilities are less than the futility boundary $P_f$ at each stage $q$, denote the corresponding cumulative probability as $P_q^a$, and determine the estimated incidence rate as

$$p_q = \min\left(p_0, \hat{p}_{q-1} + z_\varphi \sqrt{\frac{\hat{p}_{q-1}(1-\hat{p}_{q-1})}{n_{q-1}}}\right) \text{ where } \hat{p}_{q-1} = \frac{r_q^a}{n_{q-1}}, \varphi = 0.025$$

and

$$p_1 = p_0.$$

**TABLE 7.6**

Comparison of Non-Adaptive and Adaptive Probability Monitoring Procedures

| | | Non-Adaptive | | | | Adaptive | | | | | |
|---|---|---|---|---|---|---|---|---|---|---|---|
| | | $p_0 = 0.3$ | | $p_0 = 0.3$ | | $p_0 = 0.3$ | | | $p_0 = 0.3$ | | |
| | | $P_b = 0.01$ | | $P_b = 0.005$ | | $P_f = 0.05$ | | | $P_f = 0.10$ | | |
| $q$ | $n_q$ | $P_q^*$ | $r_q^*$ | $P_q^*$ | $r_q^*$ | $P_q^d$ | $r_q^d$ | $p_q$ | $P_q^a$ | $r_q^a$ | $p_q$ |
| 1 | 100 | 0.009 | 19 | 0.004 | 18 | 0.005 | 18 | 0.3 | 0.008 | 19 | 0.3 |
| 2 | 200 | 0.007 | 44 | 0.004 | 43 | 0.001 | 32 | 0.255 | 0.003 | 36 | 0.267 |
| 3 | 300 | 0.009 | 71 | 0.004 | 69 | 0.001 | 41 | 0.211 | 0.001 | 48 | 0.233 |
| 4 | 400 | 0.009 | 99 | 0.005 | 96 | 0.001 | 46 | 0.176 | 0.001 | 56 | 0.201 |
| 5 | 500 | 0.008 | 125 | 0.004 | 123 | 0.001 | 48 | 0.146 | 0.001 | 60 | 0.174 |
| 6 | 600 | 0.008 | 153 | 0.004 | 150 | 0.001 | 48 | 0.122 | 0.001 | 62 | 0.148 |

## 7.5 Concluding Remarks

Power analysis is frequently used to determine the sample size required in clinical study. As indicated in Chow et al. (2017), it is not the only way for sample size calculation in clinical research. As demonstrated in this chapter, there are other procedures for sample size calculation in clinical practice. These procedures include precision analysis, reproducibility analysis, and probability statement. The basic idea of power analysis is to choose a sample size which controls the type II error rate to achieve a desired power for detecting a clinically meaningful effect at a pre-specified level of significance, while precision analysis is based on the idea of controlling type I error rate to determine the sample size. Sample size calculation based on reproducibility analysis is to control margin and variability at the same time.

Although the above three traditional methods are commonly used for sample size calculation in clinical research, they may not be feasible in clinical research, especially for clinical trials with extremely low incidence rates and rare disease clinical trials with extremely low prevalence. In this chapter, we introduced the probability monitoring procedure proposed by Huang and Chow (2019) for the sample size determination for rare disease clinical development. Actually, Huang and Chow (2019) proposed two versions of probability monitoring procedures: One is non-adaptive version and the other is adaptive version. The non-adaptive version of probability monitoring procedure is simple yet inflexible, since it cannot stop the trial early, which may cause the waste of time and resources. On the other hand, the adaptive version is flexible but a little complicated, since it updates the expected clinical effect at each stage of the monitoring and allows stop the study early based on safety/futility and/or efficacy, which saves time and resources.

It should be noted that the proposed non-adaptive or adaptive version of probability monitoring procedure for sample size determination can be applied to complex innovative designs (CID) such as the $n$-of-1 trial design (Chow, 2019), adaptive trial design, and master protocol (platform trial design), which will be further discussed in Chapters 10 and 11, respectively.

# 8

## Real-World Data and Real-World Evidence

### 8.1 Introduction

Real-world data (RWD) refers to data relating to patient health status and/ or the delivery of health care routinely collected from a variety of sources. RWD sources include, but are not limited to, electronic health record (EHR), administrative claims and enrollment, personal digital health applications, public health databases, and emerging sources. Table 8.1 lists some examples of common RWD sources.

Among these data sources, EHR data source is probably the most important data source for RWD. EHR data source often includes data obtained from randomized and nonrandomized clinical trials (structured or unstructured) and patient experience data. The emerging sources include data sources that thematically capture data and information about patient physiology, biology, health, behavior, or their environment that have not been substantially assessed or validated. Thus, emerging data sources may include genomics, metabolomics, and proteomics. As indicated in the 21st Century Cure Act, real-world evidence (RWE) typically means data regarding the usage, or the potential benefit or risks of a drug derived from sources other than randomized clinical trials. However, RWD may contain data obtained from randomized or nonrandomized clinical studies. Thus, RWE is evidence derived from RWD through the application of research methods. For regulatory applications, RWE can further be defined as clinical evidence (e.g., substantial evidence (SE)) regarding the use and potential benefits or risks of a medical product derived from the analysis of RWD. This evidence can be used to study health status and measure a treatment's effectiveness, safety, and clinical benefit/risk in the real-world settings. In addition, this evidence provides valuable insights that traditional randomized controlled trials (RCTs) may not capture in natural settings or within relevant populations.

As the increasing availability of RWD/RWE, a commonly asked question is how RWD/RWE can be better integrated into drug development and regulatory review, especially for rare diseases drug development. As indicated in the 21st Century Cures Act passed by the United States Congress in December 2016 and subsequent the Prescription Drug User Fee Act (PDUFA VI), FDA should

**TABLE 8.1**

Examples of RWD Sources

| Data Source | Examples |
| --- | --- |
| Electronic health records (EHRs) | • Diagnoses, symptoms, and treatments<br>• Diagnostic test results (imaging, genetic tests, medical device data, etc.)<br>• Ordered/written prescriptions<br>• Demographics<br>• Patient characteristics<br>• Clinical data<br>• Patient experience data |
| Administrative enrollment and claims | • Diagnosis and procedure codes<br>• Outpatient pharmacy dispensing data<br>• Provider and facility information<br>• Plan/benefit information<br>• Dates of service<br>• Inpatient's lengths of stay |
| Personal digital health applications | • Patient experience data<br>• Data from personal medical device<br>• Sensor data from consumer device<br>• Socioeconomic and demographic data<br>• Over-the-counter medicine lists |
| Public health databases | • National death index<br>• EPA Air Quality Systems (AQS) data<br>• HRSA (Health Resources and Services Administration) Area Health Resource File (AHRF)<br>• AHRQ (Agency for Healthcare Research and Quality) Healthcare Cost and Utilization Project |
| Emerging sources | • Genomics<br>• Metabolomics<br>• Proteomics |

establish a program to evaluate the potential use of RWE (i) to support approval of new indications for a drug approved under section 505 (c) and (ii) to satisfy post-approval study requirements (21CCA, 2016; PDUFA VI, 2018). The program should be based on a framework as described in a recent FDA guidance on RWD/RWE (FDA, 2019). Along this line, it is of particular interest to the sponsors regarding how to incorporate RWE into evidence generation (i.e., substantial evidence for regulatory review and approval) in drug research and development. For this purpose, RWE will need to map to substantial evidence (SE), which is the current regulatory standard for the approval of drug products.

The remaining of this chapter is organized as follows. Section 8.2 compares the fundamental difference between RWE and SE (regulatory standard). Section 8.3 discusses how to incorporate RWE into SE in regulatory review and approval of drug products, and also includes some statistical considerations of integrity and validity of RWD in regulatory process. Section 8.4

provides statistical methods for the assessment of RWE in support of review and approval of regulatory submissions of drug products. Section 8.5 gives details about a simulation study. The last section of this chapter provides some concluding remarks.

## 8.2 Map of Real-World Evidence to Regulatory Standard

### 8.2.1 Substantial Evidence Versus Real-World Evidence

**Substantial Evidence** – For evaluation and marketing approval of a pharmaceutical entity, the sponsors are required to submit SE of effectiveness and safety accumulated adequate and well-controlled clinical trials to the FDA. SE can only be obtained through the conduct of randomized adequate and well-controlled clinical trials (21 CFR 314). As described in 21 CFR 314.126, adequate and well-controlled studies have the following characteristics: (i) a protocol and results report containing a clear objective statement and summary of proposed methods and analysis; (ii) use of a valid comparison with a control (e.g., placebo, dose, active, historical); (iii) a method of selecting patient that adequately assures they have the disease; (iv) a treatment assignment method that minimizes bias and ensures comparability between arms, ordinarily randomization; (v) measures to minimize subject, observer, and analysis bias, like blinding; (vi) well-defined and reliable methods for assessing patient response; and (vii) adequate analytical plan for assessing the effects of the drug. To provide a better understanding, the characteristics of an adequate and well-controlled clinical study are summarized in Table 8.2.

**Real-World Evidence** – As indicated earlier, RWE contains information on health care that is derived from multiple sources outside typical

**TABLE 8.2**

Characteristics of Adequate and Well-Controlled Study

| Characteristics | Requirement |
| --- | --- |
| Objectives | Clear |
| Methods of analysis | Appropriate statistical methods |
| Design | Valid for addressing scientific questions |
| Selection of subjects | Assurance of the disease under study |
| Assignment of subjects | Minimize bias |
| Participants of studies | Minimize bias |
| Assessment of responses | Well defined and reliable |
| Assessment of the effect | Accurate and reliable |

clinical research setting, including EHRs, claims and billing data, product and disease registries, and data gathered through personal devices and health application. Since RWD contains both randomized studies and nonrandomized trials (e.g., observational studies), RWE may be derived from RWD obtained from randomized studies and/or RWD obtained from observational studies.

In practice, it is recognized that RWE offers the opportunities as compared to RCTs to develop evidence that (i) includes broader populations/uses more typical of routine practice and (ii) includes effort on longer-term endpoints and endpoints more relevant to patients, providers, and payers. However, traditional RCTs will continue to be the gold standard for drug development. While concerns around validity and reliability can and will exist, RWE developed from observational studies can (i) provide an opportunity to develop robust evidence using high-quality data and sophisticated methods for producing causal-effect estimates when randomization is infeasible, (ii) enable longer follow-up to better understanding long-term outcomes, and (iii) be conducted in more cost-effective and efficient ways for certain types of clinical questions.

**A Comparison** – PDUFA VI requires FDA to enhance the use of RWE in regulatory decision-making, while the regulatory standard is still based on SE. Table 8.3 provides a comparison between RWE and SE in terms of (i) legal basis, (ii) bias and variability, (iii) sources where evidence is obtained from, (vi) clinical practice, (v) methods for assessment, and (vi) validity and integrity.

As can be seen from Table 8.3, we need to have good understanding of RWD before we fill the gaps between RWE and SE for potential use in regulatory review and approval of new indications of drug approved under section 505 (c). Corrigan-Curay (2018) suggested asking the following questions to

**TABLE 8.3**

A Comparison Between Substantial Evidence and Real-World Evidence

| Characteristic | Substantial Evidence | Real-World Evidence |
|---|---|---|
| Legal basis | CFR | 21st Century Cure Act |
| Bias | Bias is minimized | Selection bias |
| Variability | Expected and controllable | Expected, but not controllable |
| Evidence obtained from | Randomized clinical trials | Real-world data |
| Clinical practice | Reflect controlled clinical practice | Reflect real clinical practice |
| Methods for assessment | Statistical methods are well established | Statistical methods are not fully established |
| Validity and integrity | Accurate and reliable | Questionable |

CFR = Codes of Federal Regulations.

have a better understanding of the RWD, which will be used to derive RWE, for assessing the possible gaps between RWE and SE.

    i. Are there *consistent* measurements across systems/providers?

    ii. Is the frequency of assessment sufficient for *evidence* generation?

    iii. Are the data collected from a unique subset of patients, or its *representative*?

    iv. What is the *quality* of the data?

    v. Is it possible to capture in multiple databases, e.g., claims/EHRs for cross-*verification*?

    vi. How much of the data is *missing* and is it random?

The above list of questions cover representativeness, consistency, quality, verification, and integrity of RWD for the generation of RWE, which have an impact on the validity and reliability of RWE derived from RWD.

## 8.2.2 Map of RWE to Substantial Evidence

In order to map RWE to SE (current regulatory standard), we need to have good understanding of the RWD in terms of data relevancy/quality and its relationship with SE so that a fit-for-regulatory purpose RWE can be derived to map to the regulatory standard, which will be described in the subsequent sub-sections.

**Gap Analysis** – To map RWE to SE, it is suggested that a gap analysis be performed to determine the difference in evidence provided by RWD and data collected from randomized clinical trials (RCTs), which constitutes SE. Let $\mu_{SE}$ and $\sigma_{SE}$ be the expected mean response and the corresponding standard deviation of the response from RCT. Similarly, let $\mu_{RWD}$ and $\sigma_{RWD}$ be the expected mean response and the corresponding standard deviation of the response from RWD. Since the patient population of the RCT and the patient population of RWD are similar but different, it is reasonable to assume that $\mu_{RWD} = \mu_{SE} + \varepsilon$ and $\sigma_{RWD} = C\sigma_{SE}(C > 0)$, where $\varepsilon$ represents the shift in population mean and $C$ is the inflation factor of the population standard deviation. Thus, the (treatment) effect size of RWD adjusted for standard deviation can be expressed as follows:

$$E_{RWD} = \left| \frac{\mu_{RWD}}{\sigma_{RWD}} \right| = \left| \frac{\mu_{SE} + \varepsilon}{C\sigma_{SE}} \right| = |\Delta| \left| \frac{\mu_{SE}}{\sigma_{SE}} \right| = |\Delta| E_{SE}, \qquad (8.1)$$

where $\Delta = (1 + \varepsilon/\mu_{SE})/C$ and $E_{SE}$ and $E_{RWD}$ are the effect size of clinically meaningful importance observed from RCT and RWD (i.e., SE and RWE), respectively. $\Delta$ is referred to as a sensitivity index measuring the change in effect size between the patient population of the RCT and the patient population of RWD.

As it can be seen from (8.1), if $\varepsilon = 0$ and $C = 1$, $E_{SE} = E_{RWD}$. That is, the effect sizes of the two populations are identical. In this case, we claim that the results observed from the patient population of RWD can be generalized to the target patient population of the RCT. In other words, RWE generated from RWD can be used to support regulatory approval of the treatment under investigation. In other words, shift in population mean could be offset by the inflation/deflation of variability. As a result, the sensitivity index may remain unchanged while the target patient population has been shifted. Table 8.4 provides a summary of the impacts of various scenarios of population shift (i.e., change in $\varepsilon$) and change in inflation/deflation of population standard deviation (i.e., change in C, either inflation or deflation of variability).

As indicated by Chow and Shao (2005), in many clinical trials, the effect sizes of the two populations could be linked by baseline demographics or patient characteristics if there is a relationship between the effect sizes and the baseline demographics and/or patient characteristics (e.g., a covariate vector). In practice, however, such covariates may not exist or exist but not observable. In this case, the sensitivity index may be assessed by simply replacing $\varepsilon$ and C with their corresponding estimates (Chow and Shao, 2005). Intuitively, $\varepsilon$ and C can be estimated by

$$\hat{\varepsilon} = \hat{\mu}_{RWD} - \hat{\mu}_{SE} \text{ and } \hat{C} = \hat{\sigma}_{RWD}/\hat{\sigma}_{SE},$$

respectively. Thus, the sensitivity index can be estimated by

$$\hat{\Delta} = \frac{1 + \hat{\varepsilon}/\hat{\mu}_{SE}}{\hat{C}}.$$

In practice, the shift in population mean ($\varepsilon$) and/or the change in inflation/deflation of population standard deviation (C) could be random. If both $\varepsilon$ and C are fixed, the sensitivity index can be assessed based on the sample

**TABLE 8.4**

Changes in Sensitivity Index

| $\varepsilon/\mu$ (%) | Inflation of Variability | | Deflation of Variability | |
|---|---|---|---|---|
| | C (%) | Δ | C (%) | Δ |
| −20 | 120 | 0.667 | 80 | 1.000 |
| −10 | 120 | 0.750 | 80 | 1.125 |
| −5 | 120 | 0.792 | 80 | 1.188 |
| 0 | 120 | 0.833 | 80 | 1.250 |
| 5 | 120 | 0.875 | 80 | 1.313 |
| 10 | 120 | 0.917 | 80 | 1.375 |
| 20 | 120 | 1.000 | 80 | 1.500 |

means and sample variances obtained from the two populations. In the real-world problems, however, $\varepsilon$ and $C$ could be either fixed or random variables. In other words, there are three possible scenarios: (i) the case where $\varepsilon$ is random and $C$ is fixed, (ii) the case where $\varepsilon$ is fixed and $C$ is random, and (iii) the case where both $\varepsilon$ and $C$ are random. These possible scenarios are discussed in Lu et al. (2017).

Let $(\Delta_L, \Delta_U)$ be the range which is accepted by the regulatory agency. If $\hat{\Delta}$ falls within the regulatory acceptable range, we may claim RWE generated by RWD is equivalent to SE and conclude that RWE can be used in support of regulatory submission of the test treatment under investigation. Note that $(\Delta_L, \Delta_U)$ should be selected based on regulatory consideration with medical justification. As an example, if applying the concept of bioequivalence assessment, we can claim that the effect sizes of the two patient populations are equivalent if the confidence interval of $\Delta$ is within (80%, 120%) of $E_{SE}$.

**Data Relevancy** – A real-world data set is relevant if it is robust and representative of the population under study. Data relevancy and quality determine whether a real-world data set is able to address the regulatory question/concern in a clinical context of interest. In practice, data relevancy, which often focuses on selection bias, is usually assessed by asking the following questions:

i. Are the data in the data set representative of the population under study?
ii. Do critical data include exposures, covariates, and clinical outcomes?
iii. If more than one data source is used, are there permitted accurate links at the patient level?
iv. Are there sufficient patients and follow-up time in the data source to demonstrate the expected treatment effect, including adequate capture of potential safety events?

Thus, data relevancy usually refers to relevant target patient population (e.g., patient diagnosis), patient's adequate drug exposure, and the availability of outcomes.

**Data Quality or Data Reliability** – On the other hand, data quality, which often focuses on information bias, can be assessed by the following questions:

i. Are the data accurate?
ii. How complete are the data?
iii. Are the data transformation and data transfer transparent?
iv. If more than one data source is used, are measurements consistent across different data sources?

Thus, data quality or data reliability usually relates to data reliability which depends upon data integrity (e.g., whether there are missing data), accuracy of the data (e.g., whether there are selection or information biases of the data), and transparency in data transfer when there are multiple data sources.

**Fit-for-Regulatory Purpose Data** – The process of producing a fit-for-purpose real-world data set begins with the selection of one or more data sources. A final characterization of fit-for-purpose RWD should include quantitative summaries of data relevancy and data quality. Each fit-for-purpose RWD-supported submission should have sufficient documentations to transparently characterize the relevancy and quality of the fit-for-purpose RWD to the specific regulatory requirements/decisions. Such fit-for-purpose RWD-specific documentations usually consist of the following:

  i. Historical use and prior data management documentation of RWD sources;

 ii. Assessment of selection bias from data sources;

iii. Assessment of information bias from data sources;

 iv. Impacts of assumptions and procedures from data cleaning, transformation, de-identification, and linkage;

  v. Assessment of changes in key data element capture and coding over time;

 vi. Measurements of accuracy for critical data fields, such as consistency with source, sensitivity, and specificity of calculation and/or abstraction;

vii. Historical or verified validity measures of critical data fields;

viii. Assessments of data completeness by field and over time.

**Bayesian Approach for Data Borrowing** – Let $D_0$ and $D$ be the historical data available from similar previous studies and data from the current randomized clinical trial, respectively. Also, let $\pi(\theta)$ be the prior distribution of $\theta$, where $\theta$ is the parameter of interest. In this case, we may update the prior $\pi(\theta)$ by incorporating the information obtained from the historical data or relevant RWD as follows:

$$\pi(\theta|D, D_0) \propto L(\theta|D)\pi(\theta|D_0), \tag{8.2}$$

where $L(\theta|D)$ is the likelihood function given $D$. Under (8.2), Ibrahim and Chen (2000) proposed the concept of power prior (PP) distribution to raising the likelihood function of the historical data to a desired power under four commonly used classes of regression models. These regression models include (i) generalized linear model, (ii) generalized linear mixed model, (iii) semi-parametric proportional hazards model, and (iv) cure rate model for survival data.

Let $\pi_0(\theta)$ be the prior distribution of $\theta$ based on $D_0$. The idea is to update the prior based on $D_0$. Then, use the posterior as the new prior for obtaining statistical inference for $\theta$, which incorporates information $D_0$ into $D$. In other words, we have

$$\pi(\theta \,|\, D_0, \,\delta) \propto L(\theta \,|\, D_0)^\delta \,\pi_0(\theta),$$

where $L(\theta \,|\, D_0)$ is the likelihood function of $\theta$ given $D_0$ and $\delta \in [0,1]$ is a parameter that determines the amount of information $D_0$ that will be incorporated into $D$. Note that when $\delta = 1$, $D_0$ is fully utilized. When $\delta = 0$, no information from $D_0$ was borrowed. Thus, we have

$$\pi(\theta \,|\, D, \,D_0, \,\delta) \propto L(\theta \,|\, D)\pi(\theta \,|\, D_0, \,\delta) \propto L(\theta \,|\, D)L(\theta \,|\, D_0)^\delta \,\pi_0(\theta). \quad (8.3)$$

One of the major disadvantages of (8.3) is that $\delta$ is unknown in practice. To overcome the problem, alternatively, Duan et al. (2006) suggested the use of a modified power prior (MPP) by treating $\delta$ as a random variable following a distribution $\pi(\delta)$. This gives

$$\pi(\theta, \,\delta \,|\, D_0) = \frac{L(\theta \,|\, D_0)^\delta \,\pi_0(\theta)\pi(\delta)}{C(\delta)},$$

where $C(\delta) = \int_\Theta L(\theta \,|\, D_0)^\delta \,\pi_0(\theta)d\theta$, given $D_0$. Thus, the posterior distribution of $\theta$ and $\delta$ is given by

$$\pi(\theta, \,\delta \,|\, D, \,D_0) \propto L(\theta \,|\, D)\pi(\theta, \,\delta \,|\, D_0). \quad (8.4)$$

Since there exists no data to support the method of modified prior, Pan et al. (2017) proposed a method using Kolmogorov–Smirnov (KS) statistic to measure and calibrate the difference between current data and historical data. This method is referred to as the method of calibrated power prior (CPP). Let $D = (y_1, \,\ldots, \,y_n)$ be the historical data, $D_0 = (x_1, \,\ldots, \,x_m)$ be the current data, and $Z_{(1)} \leq \ldots Z_{(N)}$ be the ordered combined sample of $D_0$ and $D$ with a sample size of $N = n + m$. Thus, KS can be obtained as follows:

$$S_{KS} = \max_{i=1,\ldots,\,N} \left\{ \left| F\left(Z_{(i)}\right) - G\left(Z_{(i)}\right) \right| \right\},$$

where

$$F_m(t) = \Sigma I\left(x_j \leq t\right)/m \text{ and } G(t) = \Sigma I\left(y_i \leq t\right)/n$$

are the distribution functions of $D_0$ and $D$, respectively.

Since KS measures the difference between distributions of $D_0$ and $D$, a larger value of $S_{KS}$ is an indication that the distribution of $D_0$ is inconsistent with that of $D$. If we define

$$S = \max(m, n)^{1/4} S_{KS},$$

the relationship between $\delta$ and $S$ can then be described as follows:

$$\delta = g(S;\phi) = \frac{1}{1+\exp\{a+b\log(S)\}}, \tag{8.5}$$

where $\phi = (a, b)$ is a parameter that controls $\delta$ and $S$ correlation, $b > 0$. As it can be seen, a relatively smaller value of $\delta$ implies that the distribution of $D_0$ is significantly inconsistent with that of $D$.

Under the assumptions that $D_0$ and $D$ are normally distributed, i.e., $x_i \sim N(\mu_0, \sigma_0^2)$ and $y_j \sim N(\mu_0 + \gamma, \sigma_0^2)$, $i = 1,\ldots, m$ and $j = 1,\ldots, n$, parameters $a$ and $b$ in (3) can be determined by the following steps:

Step 1. Obtain estimates of population mean and population variance of $D_0$. That is, $\hat{\mu}_0 = \bar{x}$, where $\bar{x} = \sum_{i=1}^{m} x_i / m$ and $\hat{\sigma}_0^2 = \sum_{i=1}^{m} (x_i - \bar{x})/(m-1)$.

Step 2. Let $\gamma$ be the difference in mean between $D_0$ and $D$. Denote by $\gamma_c$ the mean difference which is negligible (in other words, $D_0$ and $D$ are consistent). Then, obtain the minimum mean difference, denoted by $\gamma_{\bar{c}}$, that will lead to the conclusion of inconsistences between the distributions of $D_0$ and $D$.

Step 3. Simulate M samples $(y_1, \ldots, y_n)$ from $N(\hat{\mu}_0 + \gamma, \hat{\sigma}_0^2)$. Then, for each sample, calculate KS between $D_0$ and $D$. Let $S^*(\gamma_c)$ be the median of the M KS statistics. Replace $\gamma_{\bar{c}}$ with $\gamma_c$, and repeat step 3.

Step 4. Let $S^*(\gamma_c)$ be the median of the M KS statistics. Parameters can be solved from the following equations:

$$\delta_c = g\{S^*(\gamma_c);\phi\}$$

$$\delta_{\bar{c}} = g\{S^*(\gamma_{\bar{c}});\phi\}.$$

If we choose $\delta_c$ close to 1 (say 0.98), $\delta_{\bar{c}}$ is constant close to 0 (say 0.01), and $a$ and $b$ can be obtained as follows:

$$a = \log\left(\frac{1-\delta_c}{\delta_c}\right) - \frac{\log\left\{\frac{(1-\delta_c)\delta_{\bar{c}}}{(1-\delta_{\bar{c}})\delta_c}\right\}\log\{S^*(r_c)\}}{\log\left\{\frac{S^*(r_c)}{S^*(r_{\bar{c}})}\right\}}$$

$$b = \frac{\log\left\{\dfrac{(1-\delta_c)\delta_{\bar{c}}}{(1-\delta_{\bar{c}})\delta_c}\right\}}{\log\left\{\dfrac{S^*(r_c)}{S^*(r_{\bar{c}})}\right\}}.$$

Once $a$ and $b$ have been determined, $\theta$'s CPP is given by

$$\pi(\theta \mid D_0, a, b) = L(\theta \mid D_0)^{\left[1+\exp\{a+b\log(S)\}\right]^{-1}} \pi_0(\theta). \tag{8.6}$$

Viele et al. (2014) proposed a method which is referred to as a test-then-pool approach to examine the difference between the current data (e.g., data collected from RCT) and historical data (e.g., RWD) by testing the following hypotheses:

$$H_0: p_h = p_c \text{ versus } H_a: p_h \neq p_c, \tag{8.7}$$

where $p_h$ and $p_c$ are the response rates of the historical data and current data, respectively. If we fail to reject the null hypothesis, historical data and current data can be combined for final analysis. It should be noted that the above method can only determine whether the historical data can be combined for a final analysis. This method is not useful to determine how much information can be borrowed for further analysis. To overcome the problem, Liu (2017) proposed the method of $p$-value-based power prior (PVPP) by testing the following hypotheses:

$$H_0: |\theta_h - \theta| > \eta \text{ versus } H_1: |\theta_h - \theta| < \eta. \tag{8.8}$$

Alternatively, Gravestock et al. (2017) proposed adaptive power priors with empirical Bayes (EBPP) by considering the estimate of $\delta$

$$\hat{\delta}(D_0, D) = \arg\max_{\delta\in[0,1]} L(\delta; D_0, D),$$

where

$$L(\delta; D_0, D) = \int L(\delta; D)\pi(\theta \mid \delta, D_0)d\theta = \frac{\int L(\delta; D)L(\theta; D_0)^{\delta}\pi(\theta)d\theta}{\int L(\theta; D_0)^{\delta}\pi(\theta)d\theta},$$

in which $\eta > 0$ is the pre-specified constant. Let p be the maximum of the two one-sided test p-values. Since under the test-then-pool approach, we can only take 0 or 1 for $\delta$, Gravestock et al. (2017) suggested considering the following continuous function of $p$:

$$\delta = \exp\left[\frac{k}{1-p}\ln(1-p)\right],$$

where $k$ is a pre-specified constant. A small $p$-value suggests that more information from the historical data or relevant RWD can be borrowed.

### 8.2.3 Potential Use of RWD/RWE in Clinical Studies

There is a wide spectrum of potential use of RWD/RWE in clinical studies, including (i) interventional randomized clinical trials, (ii) interventional nonrandomized trials, and (iii) non-interventional nonrandomized studies.

For traditional randomized trials, RWD may be used to assess enrollment criteria, trial feasibility, and site selection. In addition, EHR and claim data can be used to identify outcomes, while mobile technology can be sued to capture supportive endpoints (e.g., to assess ambulation). For trials in clinical practice settings, pragmatic RCTs may use eCRF and claims with/without EHR. If it is a single-arm study, we may use external control. For observational studies, we may use existing databases for case–control and/or retrospective cohort studies. It, however, should be noted that there are different challenges and opportunities for each approach.

**An Example for Extension of Indication** – A typical example is a recent approval of the extended indication of Ibrance (palbociclib, PAL) by the FDA in April 2019. The original indication of Ibrance is for treating HR (+) and HER2 (–) advanced or metastatic breast cancer in women. The sponsor submitted RWE derived from RWD of IQVIA (formerly Quintiles and IMS Health, Inc.) insurance database, Flatiron Health Breast Cancer database, and sponsor's own global database for Ibrance used in male breast cancer. RWE was derived from fit-for-regulatory purpose RWD, which consists of two cohorts of male metastatic breast cancer: the cohort of PAL plus endocrine therapy (ET) and the cohort of ET alone. The results of RWE indicated that median duration of treatment (first line) for PAL+ET is 8.5 months ($n = 37$) and ET alone is 4.3 months ($n = 214$). With all lines, the maximum response rate is 33.3% for PAL+ET ($n = 12$) and 12.5% for ET alone ($n = 8$).

The map of the RWE to SE is considered acceptable based on the rationale that (i) rarity of male breast cancer (about less than 1% of all breast cancers) and (ii) similarity of breast cancer biology between female and male.

**An Example for Post-Marketing Label Change** – Alteplase for acute ischemic stroke (AIS) was approved by the FDA at the dose of 0.9 mg/kg based on randomized trial, an National Institute of Neurological Disorders and Stroke (NINDS) trial. The drug was subsequently approved by Taiwan Food and Drug Administration (TFDA) in 2002. However, it is recognized that the relationship between the dose of recombinant tissue-type plasminogen activator (rt-PA) and its safety/efficacy for ischemic stroke has not been well evaluated in East Asian population.

Thus, Taiwan Thrombolytic Therapy for Acute Ischemic Stroke (TTT-AIS) study group conducted a prospective multicenter observational study based on fit-for-regulatory purpose RWD database from an Internet-based, academic, interactive thrombolysis therapy registration in Taiwan (Chao et al., 2014). RWE regarding different doses of rt-PA for acute stroke in Chinese patients in Taiwan is then derived. A total of 1,004 eligible patients were classified according to the dose of rt-PA received for managing acute ischemic: 0.9 mg/kg ($n = 422$), 0.8 mg/kg ($9n = 202$), 0.7 mg/kg ($9n = 199$), and 0.6 mg/kg ($9n = 181$). The safety outcome was symptomatic intracerebral hemorrhage and death within 3 months. The efficacy outcome was good functional outcome (modified Rankin's scale ≤1) at 3 months.

As indicated in Chao et al. (2014), there was a significant trend in symptomatic intracerebral hemorrhage with age ($p$-value = 0.002). A multivariate logistic regression analysis indicated that a dose of 0.9 mg/kg was a predictor of symptomatic intracerebral hemorrhage ($p$-value = 0.0109) and a dose ≤ 0.65 mg/kg was a predictor of good functional outcome ($p$-value = 0.0369). In patients aged from 71 to 80 years, there was a significant trend of increasing symptomatic intracerebral hemorrhage ($p$-value = 0.0130) and less good functional outcome ($p$-value = 0.0179) with increasing doses of rt-PA. There was also a trend of increasing mortality ($p$-value = 0.0971) at three months in these patients.

TFDA feels that this RWE does not support the dose of 0.9 mg/kg of rt-PA, which is optimal for all patients in East Asian population. In elderly patients (71–80 years), a lower dose of 0.6 mg/kg is associated with a better outcome. Thus, TFDA accepts post-marketing label change based on RWD/RWE.

## 8.3 Validity of Real-World Data

As indicated, RWD usually consists of data sets from randomized and/or nonrandomized (published or unpublished) studies. Thus, imbalances between the treatment group and the control group are likely to occur. In addition, studies with positive results are most likely to be published and accepted to the big data. In this case, even that the use of propensity score matching can help to reduce the bias due to some selection bias, the bias due to the fact that majority of data sets accepted to the big data are most likely those studies with positive results could be substantial and hence cannot be ignored. As a result, there is a bias of big data analytics regardless of whether the big data analytics are case–control studies, meta-analyses, or data mining in genomics studies. In this section, we attempt to assess the selection bias of accepting positive data sets into the RWD.

### 8.3.1 Bias of Real-World Data

Let $\mu$ and $\mu_B$ be the true mean of the target patient population with the disease under study and the true mean of the pooled historical data, respectively. Let $\varepsilon = \mu_B - \mu$, which depends upon the unknown percentage of data sets with positive results in the historical data. Now, let $\mu_P$ and $\mu_N$ be the true means of data sets of positive studies and nonpositive studies conducted in the target patient population, respectively. Also, let $r$ be the proportion of positive studies conducted on the target patient population, which is usually unknown. For illustration and simplicity purposes, we assume that there is no treatment-by-center interaction for those multicenter studies and there is no treatment-by-study interaction. In this case, we have

$$\mu = r\mu_P + (1-r)\mu_N, \tag{8.9}$$

where $\mu_P > \delta > \mu_N$, in which $\delta$ is the effect of clinical importance. In other words, a study with an estimated effect greater than $\delta$ is considered a *positive* study. As it can be seen from (8.9), if the historical data only contain data sets from positive studies, i.e., $r = 1$, (8.9) reduces to

$$\mu = \mu_P.$$

In other words, in this extreme case, the historical data does not contain any studies with nonpositive results. In practice, we would expect $\frac{1}{2} < r \leq 1$.

For a given historical data, $r$ can be estimated based on the number of positive studies in the historical data (i.e., $\hat{r}$). In practice, $\hat{r}$ usually overestimates the true $r$ because the published or positive studies are more likely to be included in historical data. Thus, we have

$$E(\hat{r}) - \Delta = ra.$$

Now, for simplicity, assume that all studies included in the historical data are of parallel design and all positive studies and nonpositive studies are of the same size $n_P$ and $n_N$, respectively. Let $x_{ij}$ be the response of the $i$th subject in the $j$th positive study, $i = 1,\ldots, n_P$ and $j = 1,\ldots, rn$, where $n$ is the total number of studies in the historical data. Also, let $y_{ij}$ be the response of the $i$th subject in the $j$th nonpositive study, $i = 1,\ldots, n_N$ and $j = 1,\ldots, (1-r)n$. Thus, the bias of $\hat{\mu}$ is given by

$$\text{Bias}\left(\hat{\mu}\right) = E\left(\hat{\mu}\right) - \mu = E\left[\hat{r}\hat{\mu}_P + (1-\hat{r})\hat{\mu}_N\right] - \mu$$

$$\approx (r+\Delta)\mu_P + (1-r-\Delta)\mu_N - \mu \quad , \tag{8.10}$$

$$= \Delta(\mu_P - \mu_N)$$

where

$$\hat{\mu}_P = \bar{x} = \frac{1}{rnn_P} \sum_{j=1}^{rn} \sum_{i=1}^{n_P} x_{ij}$$

and

$$\hat{\mu}_N = \bar{y} = \frac{1}{(1-r)nn_N} \sum_{j=1}^{(1-r)n} \sum_{i=1}^{n_N} y_{ij}.$$

Thus, we have

$$Bias\left(\hat{\mu}\right) = \Delta\left(\mu_P - \mu_N\right), \tag{8.11}$$

where $\mu_P > \delta > \mu_N$, in which $\delta$ is a pre-specified constant that indicates the effect of clinical importance. In practice, $\delta$ could be used to control the number of positive studies in extreme case; i.e., $r$ is 100%.

Regarding the power, by (8.9), the variance of $\hat{\mu}$ is given by

$$Var\left(\hat{\mu}\right) = \frac{1}{n}\hat{\sigma}^2 = r^2\left(\frac{\sigma_P^2}{rnn_P}\right) + (1-r)^2\left(\frac{\sigma_N^2}{(1-r)nn_N}\right)$$

$$= \frac{1}{n}\left[r\sigma_P^2/n_P + (1-r)\sigma_N^2/n_N\right] \geq 0,$$

where

$$\sigma_P^2 = \frac{1}{rnn_P} \sum_{j=1}^{rn} \sum_{i=1}^{n_P} \left(x_{ij} - \bar{x}\right)$$

and

$$\sigma_N^2 = \frac{1}{(1-r)nn_P} \sum_{j=1}^{(1-r)n} \sum_{i=1}^{n_P} \left(y_{ij} - \bar{y}\right),$$

$n_P$ and $n_N$ are the size of positive and nonpositive studies, respectively, and $n$ is the total number of studies accepted into the historical data. In addition, if we take derivative of the above, it leads to

$$\frac{\partial}{\partial r}\left[Var\left(\hat{\mu}\right)\right] = \frac{1}{n}\left(\sigma_P^2/n_P - \sigma_N^2/n_N\right).$$

As the between-group difference is larger in positive studies than that in nonpositive studies, it would be a reasonable assumption that $\sigma_P^2$ is larger than $\sigma_N^2$. Thus, if the sizes of positive and nonpositive are close, we have $\sigma_P^2/n_P > \sigma_N^2/n_N$, and $\hat{\sigma}^2$ is an increasing function of $r$. In this case, it is expected that the power of historical data analytics may decrease as $r$ increases if $n$ is fixed. The above discussion suggests that the power of the historical data analytics can be studied through the evaluation of the following probability:

$$P\left\{\hat{\sigma}_P^2/n_P > \hat{\sigma}_N^2/n_N \mid \mu_P, \mu_N, \sigma_P^2, \sigma_N^2, \text{ and } r\right\}$$

based on data available in the RWD.

### 8.3.2 Reproducibility Probability

The reproducibility probability is defined as an estimated power (EP) of the future trial using the data from the previous trials. Theoretically, different trials are independent; the probability of achieving a statistically significant result from the new trial would be identical with previous studies if these trials apply the same study design and hypothesis, regardless of the outcome of previous trials (Goodman, 1992). The probability of observing a significant clinical result when $H_1$ is indeed true is referred to as the power of the test procedure.

Suppose that the null hypothesis $H_0$ is rejected if and only if $|T| > c$, where $c$ is a positive known constant and $T$ is a test statistic. Considering a two-sided alternative hypothesis, the power would be

$$P\left(|T| > c \mid H_1\right) = P\left(|T| > c \mid \theta\right), \tag{8.12}$$

where $T$ is a test statistic, $H_a$ is the alternative hypothesis, and $\theta$ is an unknown parameter. Although trials are independent, the outcome of the historical trial can still be a reasonable reference for the later trial. According to Chow's study, reproducibility probability could be estimated through (i) the EP approach, (ii) the confidence bound (CB) approach, and (iii) the Bayesian approach (Shao and Chow, 2002). The CB approach is most conservative, followed by the Bayesian approach and then the EP approach. In this study, we use the EP approach and the CB approach to estimate reproducibility probability.

The EP approach is applied as follows. Consider a two-group parallel design with unequal variances. Let $x_{ij}$ be the $j$th subject in the $i$th group ($i = 1, 2$) and independently distributed as $N\left(\mu_i, \sigma_i^2\right)$, in which $\sigma_1^2 \neq \sigma_2^2$. Thus, when $n_1$ and $n_2$ are large, the statistics $T$ in (8.12) is given by

$$T = \frac{\bar{x}_1 - \bar{x}_2}{\sqrt{s_1^2/n_1 + s_2^2/n_2}}. \tag{8.13}$$

T approximately follows the normal distribution $N(\theta, 1)$ with

$$\theta = \frac{\mu_1 - \mu_2}{\sqrt{\sigma_1^2/n_1 + \sigma_2^2/n_2}}. \tag{8.14}$$

Thus, the reproducibility probability through the EP approach is obtained by replacing $\theta$ in (8.12) by the estimated statistic $T$

$$\hat{P} = \Phi\left[T(x) - z_{0.975}\right] + \Phi\left[-T(x) - z_{0.975}\right]. \tag{8.15}$$

The EP approach would provide an optimistic result that the adjustment of bias might be inadequate when applied in our proposed approach. The CB approach would provide a more conservative estimate of reproducibility probability.

Consider the previous case of two-group parallel design with unequal variances $\sigma_1^2$ and $\sigma_2^2$. When $n_1$ and $n_2$ are large, the statistics $T$ estimated by (8.13) approximately follow the distribution $N(\theta, 1)$, in which noncentral parameter $\theta$ is given by (8.14). Thus, the reproducibility probability through the CB approach is given by

$$\hat{P}_- = \Phi\left[\left|T(x)\right| - 2z_{0.975}\right]. \tag{8.16}$$

## 8.4 Statistical Methods for Assessment of Real-World Evidence

As indicated in the previous section, the validity of RWD could be assessed by $\Delta(\mu_P - \mu_N)$. We propose a reproducibility probability-based approach to estimate $\Delta(\mu_P - \mu_N)$ using RWD. In this approach, parameters $\Delta$ and $(\mu_P - \mu_N)$ are estimated separately.

### 8.4.1 Estimation of $\Delta$

As mentioned earlier, the proportion of positive trials in historical data, $\hat{r}$, is always overestimated. In other words, $\hat{r}$ is larger than the real proportion $r$. Following the concept of empirical power or reproducibility probability, we can estimate $r$ by the reproducibility probability of observing the positive result of a future study given that the observed mean response and the corresponding sample variance are as follows:

$$p = P\left\{\text{future study is positive} \mid \mu \equiv \hat{\mu}_B \text{ and } \sigma \equiv \hat{\sigma}_B\right\}.$$

The above expression can be interpreted as given the observed mean response $(\hat{\mu}_B)$ and the corresponding sample standard deviation $(\hat{\sigma}_B)$; we expect to see $p \times 100$ studies with positive results if we shall conduct the clinical trial under similar experimental conditions 100 times. Thus, intuitively, $p$ is a reasonable estimate for the unknown $r$. With historical data, we have

$$\hat{\mu}_B = \hat{\mu}_1 - \hat{\mu}_0 \text{ and } \hat{\sigma}_B = \sqrt{\frac{n_1 \hat{\sigma}_1^2 + n_2 \hat{\sigma}_0^2}{n_1 + n_0}}, \qquad (8.17)$$

where $\hat{\mu}_1$ and $\hat{\mu}_0$ are the pooled means of treatment and placebo group in historical data, $n_1$ and $n_0$ are the sizes of treatment and placebo group of pooled historical data, and $\hat{\sigma}_1^2$ and $\hat{\sigma}_0^2$ are the estimated variances of the treatment group and placebo group, respectively. Then, the statistic $T$ in (8.15) and (8.16) is given by

$$T = \sqrt{\frac{n_{m1} n_{m0}}{n_{m1} + n_{m0}}} \frac{\hat{\mu}_B}{\hat{\sigma}_B}, \qquad (8.18)$$

where $n_{m1}$ and $n_{m0}$ are the median sample sizes of treatment and placebo group in studies that are included in historical data. Thus, we have

$$\Delta = \hat{r} - \hat{p}. \qquad (8.19)$$

## 8.4.2 Estimation of $\mu_P - \mu_N$

Let $(L_P, U_P)$ and $(L_N, U_N)$ be the $(1-\alpha) \times 100\%$ confidence interval for $\mu_P$ and $\mu_N$, respectively. Under normality assumption and the assumption that $\sigma_P = \sigma_N$, we have

$$(L_P, U_P) = \hat{\mu}_P \pm z_{1-\alpha/2} \frac{\hat{\sigma}_P}{\sqrt{rnn_P}} \text{ and } (L_N, U_N) = \hat{\mu}_N \pm z_{1-\alpha/2} \frac{\hat{\sigma}_N}{\sqrt{(1-r)nn_N}}. \qquad (8.20)$$

When the selection bias indeed exists, it is reasonable to assume that positive studies and nonpositive studies are different. In other words, the mean effect of positive studies might be statistically different from that of nonpositive studies. In this case, we assume that $\mu_p > \delta > \mu_N$, and the confidence intervals of $\mu_P$ and $\mu_N$, i.e., $(L_P, U_P)$ and $(L_N, U_N)$, would not overlap each other. At extreme cases, $U_N$ is close to $L_P$. Thus, we have

$$\hat{\mu}_P - z_{1-\frac{\alpha}{2}} \frac{\hat{\sigma}_P}{\sqrt{n_P}} \approx \hat{\mu}_N + z_{1-\frac{\alpha}{2}} \frac{\hat{\sigma}_N}{\sqrt{n_N}}.$$

This leads to

$$\hat{\mu}_P - \hat{\mu}_N \approx z_{1-\frac{\alpha}{2}} \frac{\hat{\sigma}_P}{\sqrt{n_P}} + z_{1-\frac{\alpha}{2}} \frac{\hat{\sigma}_N}{\sqrt{n_N}} = z_{1-\frac{\alpha}{2}} \left( \frac{\hat{\sigma}_P}{\sqrt{n_P}} + \frac{\hat{\sigma}_N}{\sqrt{n_N}} \right).$$

By representing Distance$(\mu_P, \mu_N)$ as the estimation of $\mu_P - \mu_N$, we have

$$\text{Distance}(\mu_P, \mu_N) = z_{1-\frac{\alpha}{2}}\left(\frac{\hat{\sigma}_P}{\sqrt{n_P}} + \frac{\hat{\sigma}_N}{\sqrt{n_N}}\right). \tag{8.21}$$

### 8.4.3 Remarks

In some extreme cases, there are only data from positive studies, which are achievable. The study with the least effect size would be used to estimate $L_N$ and $U_N$.

These two parameters, $\Delta$ and $(\mu_P - \mu_N)$, are corresponding to two assumptions: (i) The positive study is more likely to be published, and (ii) the positive study and negative study have different distributions, which are reasonable considerations for selection bias in historical trial data. Based on the assumptions of the proposed approach, we suggest a two-step procedure to determine whether it is proper to apply the approach.

*Step 1.* Calculate the proportion of positive studies in historical data, $\hat{r}$, through (8.8) and compare it with the designed power of each study included in the historical data set. If the proportion of positive studies is larger than power of most studies (mostly larger than all the studies in practice), then conduct step 2.

*Step 2.* Calculate the mean difference of positive studies and nonpositive studies, $\hat{\mu}_P - \hat{\mu}_N$, and compare it with the theoretical distance given by (8.14). The adjustment could be made when

$$\hat{\mu}_P - \hat{\mu}_N < z_{1-\frac{\alpha}{2}}\left(\frac{\hat{\sigma}_P}{\sqrt{n_P}} + \frac{\hat{\sigma}_N}{\sqrt{n_N}}\right).$$

Through this two-step procedure, the estimated bias would be able to reduce bias that overestimates $M_1$. The EP approach is considered as a conservative adjustment, which is reasonable when the bias is not extreme. When $\hat{r}$ is extremely larger than most power of studies included, the CB approach would be more suitable.

## 8.5 Simulation Study

In this section, we designed simulation experiments to examine the performance of the proposed bias correction procedure by comparing the corrected treatment effect with the true effect size of the target patient population. The following procedure is repeated 1,000 times. For each combination of parameters in these scenarios, 100 repetitions are performed. The simulation study and data analysis are conducted by SAS 9.4.

### 8.5.1 Parameter Specifications

We assume that each trial contains two groups (treatment and control) of patients, and the responses of two groups follow the normal distribution $N(2,5)$ and $N(0,5)$ separately. Thus, 100 subjects per group would have a power of 80% to detect a clinically meaningful difference of 2. Sample sizes of 49, 63, 79, 100, and 133 per group are simulated to achieve 50%, 60%, 70%, 80%, and 90% power. Positive studies and nonpositive studies are categorized through a $t$-test procedure.

To assess the performance of the proposed bias correction procedure, bias is defined as the relative difference between the estimated effect size $\hat{\mu}_B$ from pooled data and true effect size $\mu$, which would be 2 in this experiment, as follows: Bias $= (\hat{\mu}_B - \mu)/\mu$. Adjusted bias is defined as $(\hat{\mu}_B - \mu - \varepsilon)/\mu$, in which $\varepsilon$ is the estimated bias correction factor by the proposed method.

To assess the performance of the proposed approach, we consider the following scenarios:

Scenario 1: Performance of the proposed bias adjustment approach using the EP or the CB approach.

Scenario 2: Performance of the proposed bias adjustment approach when negative studies are absent.

Scenario 3: Performance of the proposed bias adjustment approach for a small number of historical studies.

In scenario 1, the goal of the simulation study is to assess the performance of the proposed bias adjustment approach and compare the performance of the EP approach or the CB approach. Table 8.5 and Figure 8.1 summarize the amount of bias and adjustment. As expected, the selection bias increases with the observed proportion of positive study in historical data. The EP approach can reduce the bias slightly when $r$ is larger than the designed power by 15%. When the proportion $r$ is close to the designed power ($\pm 10\%$), the EP approach is not suitable as it may increase the bias. The CB approach yields the similar performance. The amount of adjustment made by the CB approach is larger than that of the EP approach. When the proportion of positive study, $r$, is nearly 15% larger than the designed power, the adjusted bias is shown to be very close to 0. Both the proposed approach can achieve a conservative estimate of $M_1$ by adjusting the bias toward the right direction, and the adjustment of the EP approach is shown to be more moderate than that of the CB approach.

Table 8.6 presents the performances of the EP approach and the CB approach when historical data contains only positive studies. In this situation, with a fixed $\hat{r} = 100\%$, bias becomes smaller when designed power increases. By considering the study with the least effect size as the negative study, the EP approach showed a slight reduction in bias. Compared with Table 8.1, the performance of the EP approach shows no difference, and bias is also slightly

**TABLE 8.5**

Performance of Proposed Bias Adjustment Approach at Different Power

| Power | R | Bias | Adjusted Bias | |
|---|---|---|---|---|
| | | | EP Approach | CB Approach |
| 0.5080 | 0.70 | 0.2597 | 0.2571 | 0.0473 |
| | 0.75 | 0.3340 | 0.3304 | 0.1012 |
| | 0.80 | 0.3945 | 0.3872 | 0.1388 |
| | 0.85 | 0.4617 | 0.4481 | 0.1739 |
| | 0.90 | 0.5082 | 0.4812 | 0.1664 |
| | 0.95 | 0.5888 | 0.5449 | 0.1483 |
| 0.6122 | 0.70 | 0.1324 | 0.1355 | −0.0529 |
| | 0.75 | 0.1904 | 0.1910 | −0.0109 |
| | 0.80 | 0.2573 | 0.2560 | 0.0359 |
| | 0.85 | 0.3128 | 0.3053 | 0.0639 |
| | 0.90 | 0.3726 | 0.3565 | 0.0803 |
| | 0.95 | 0.4256 | 0.3918 | 0.0464 |
| 0.7102 | 0.70 | 0.0386 | 0.0477 | −0.1230 |
| | 0.75 | 0.0934 | 0.1002 | −0.0825 |
| | 0.80 | 0.1543 | 0.1589 | −0.0383 |
| | 0.85 | 0.2111 | 0.2129 | −0.0013 |
| | 0.90 | 0.2625 | 0.2553 | 0.0139 |
| | 0.95 | 0.2975 | 0.2725 | −0.0349 |
| 0.8074 | 0.70 | −0.0351 | −0.0190 | −0.1740 |
| | 0.75 | 0.0149 | 0.0290 | −0.1343 |
| | 0.80 | 0.0640 | 0.0758 | −0.0978 |
| | 0.85 | 0.1156 | 0.1229 | −0.0651 |
| | 0.90 | 0.1510 | 0.1498 | −0.0620 |
| | 0.95 | 0.2112 | 0.1995 | −0.0620 |
| 0.9035 | 0.70 | −0.1024 | −0.0770 | −0.2128 |
| | 0.75 | −0.0680 | −0.0465 | −0.1881 |
| | 0.80 | −0.0256 | −0.0076 | −0.1542 |
| | 0.85 | 0.0129 | 0.0261 | −0.1314 |
| | 0.90 | 0.0715 | 0.0806 | −0.0896 |
| | 0.95 | 0.1121 | 0.1111 | −0.0974 |

EP: Estimated power.
CB: Confidence bound.

reduced from 1.6% to 6.4%: The reduction is increased with the increase in the amount of bias. The CB approach also yields the similar outcomes, as shown in Table 8.1. When the designed power is around 80%, the adjusted bias is distributed around 0. Among all tested designed powers, the simulation shows that both approaches are still reliable when the historical data is constituted by only positive studies.

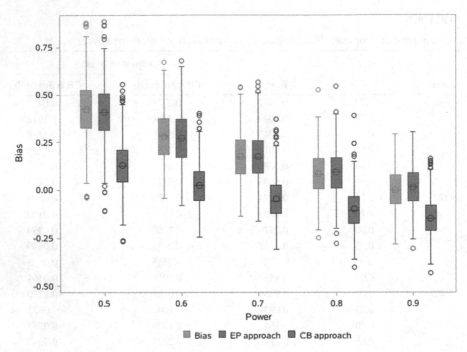

**FIGURE 8.1**
Performance of the proposed bias adjustment approach at different power.

**TABLE 8.6**

Performance of the Proposed Bias Adjustment Approach When Negative Studies are Absent

| Power | Bias | Adjusted Bias | |
| --- | --- | --- | --- |
| | | EP Approach | CB Approach |
| 0.5080 | 0.6384 | 0.5747 | 0.1704 |
| 0.6122 | 0.5047 | 0.4597 | 0.1184 |
| 0.7102 | 0.3602 | 0.3228 | 0.0242 |
| 0.8074 | 0.2751 | 0.2519 | 0.0064 |
| 0.9035 | 0.1444 | 0.1283 | −0.0702 |

Table 8.7 shows the result when the historical data is constituted by a small number of positive studies, which is a usual condition in practice. In this scenario, we assume that historical data sets are constituted by 2–10 positive studies. The designed power of each study is set to be 80%. Both approaches are shown to be capable of bias reduction. However, as shown in Figure 8.2, the bias of historical data that is constituted by a small number of studies is highly uncertain; in other words, small historical data is more likely to produce extreme outcomes. In this case, the CB approach may increase the

**TABLE 8.7**

Performance of Proposed Bias Adjustment Approach for Different Number of Studies

| Number of Studies | Bias | Adjusted Bias | |
| --- | --- | --- | --- |
| | | EP Approach | CB Approach |
| 2 | 0.2660 | 0.1988 | –0.1577 |
| 3 | 0.2751 | 0.2310 | –0.0957 |
| 4 | 0.2663 | 0.2255 | –0.0850 |
| 5 | 0.2500 | 0.2109 | –0.0898 |
| 6 | 0.2636 | 0.2290 | –0.0553 |
| 7 | 0.2625 | 0.2297 | –0.0503 |
| 8 | 0.2283 | 0.1911 | –0.0942 |
| 9 | 0.2544 | 0.2228 | –0.0529 |
| 10 | 0.2483 | 0.2158 | –0.0594 |

variability of bias. With the number of studies included in the historical data accumulated, the EP approach shows better performances. This is because the bias of historical data becomes larger when more positive studies are included.

In summary, both the EP approach and the CB approach are capable of bias reduction when the bias indeed exists. The EP approach is more conservative,

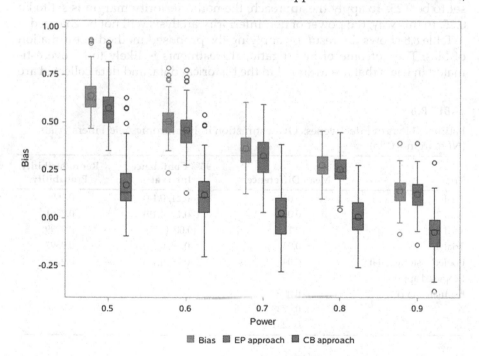

**FIGURE 8.2**
Performance of the proposed bias adjustment approach when negative studies are absent.

which is appropriate when the bias is not extreme. The CB approach would significantly reduce the bias when the EP of each study that is included in the historical data is significantly lower than the proportion of positive studies observed. Either approach is not suitable when the number of studies is too small or the situation that the proportion of positive studies is close to the designed power.

### 8.5.2 An Example

We illustrate an application of the proposed approach to adjust the non-inferiority margin using a historical data of hyaluronic acid gel clinical trials. Hyaluronic acid fillers are a kind of medical device for the correction of soft tissue defects and have been used for aesthetic purposes since the 1990s. The historical data we used is constituted by four non-inferiority trials of follow-on hyaluronic acid fillers with a similar study design. All these studies were approved by independent ethics committees in each study site. Restylane is used as an active control in all these trials, and the treatment group is different in each trial. Hyaluronic acid fillers are injected into the deep dermis to correct moderate to deep nasolabial folds (NLFs). The wrinkle severity rating scale (WSRS) is used to evaluate the severity of NLFs.

To demonstrate the method, we consider the mean change in WSRS from baseline as the primary effectiveness parameter. Non-inferiority margin is set to be 0.25. To apply the approach, the non-inferiority margin is set to be 0.25. In this way, the power of non-inferiority analysis will not be changed.

Table 8.8 shows the result by applying the proposed method of estimation of bias. The outcome of investigational treatments is likely to be overestimated in trials that are included in the historical data, and data collected are

### TABLE 8.8

Estimated Bias of Effectiveness Overestimation in Hyaluronic Acid Fillers Trials (*NI* margin = 0.25)

| Trial | Mean Difference | 95% Confidence Interval | Reproducibility Probability |
|---|---|---|---|
| Trial 1 | −0.03 | (−0.21, 0.14) | 0.6699 |
| Trial 2 | 0.00 | (−0.19, 0.19) | 0.7561 |
| Trial 3 | 0.14 | (0.00, 0.28) | 0.9999 |
| Trial 4 | 0.03 | (−0.13, 0.19) | 0.9397 |
| Pooled historical data | 0.04 | (−0.05, 0.12) | 0.9262 |
| Proposed approach | | | |
| Estimated bias | 0.0091 | | |
| Δ | 0.0738 | | |
| $\mu_P - \mu_N$ | 0.1231 | | |

*Notes:*
Mean difference = Investigational treatment − Active control.
Reproducibility probability is calculated through the EP approach.

all positive studies. In this case, the bias should be considered. The reproducibility probability of pooled data is found to be 92.62%. Thus, the estimated bias is 0.0091. It should be noticed that in this analysis, the effectiveness of active control is not available. In this case, the estimated bias can't be used to adjust the non-inferiority margin. However, the estimated bias could be applied in sensitivity analysis.

Note that this work example is used to illustrate the proposed statistical methodology, and the studies about non-inferiority margin in hyaluronic acid fillers should not be selected merely according to the result of this example. More considerations in the study design of hyaluronic acid fillers trial could refer to the study of Rzany et al. (2017).

### 8.5.3 Discussion

In this section, we discussed a reproducibility probability-based approach to estimate the selection bias of $M_1$ using historical trials whose IPD is available. In this approach, we defined a formula expression of selection bias in historical trial data, which is constituted by the overestimated fraction of proportion of positive studies, $\Delta$, and theoretical distance between positive study and negative study, $(\mu_P - \mu_N)$. We use two methods, EP approach and CB approach, to estimate the reproducibility probability. Simulation study shows that the proposed approach in general yields satisfactory performances. The reproducibility probability-based bias adjustment approach could provide a conservative estimate of effect size for the specification of non-inferiority margin. Also, this method could be applied in sensitivity analysis.

In the simulation study, we noticed that the number of studies included in historical data is an essential requirement of the proposed approach. When the proportion of positive studies is larger than the designed power of most studies that are included in the historical data, selection bias is considered. To quantify the bias, the number of included studies should be enough to get a stable estimate of $\hat{r}$. Otherwise, without a sufficient number of studies, the estimated $\hat{r}$ would be easily affected by the outcome of a single study. For example, suppose a set of historical data is constituted by four positive studies, and $\hat{r}$ would be 100%. When a non-positive study is added, $\hat{r}$ would be 80%. In this case, the estimate of bias may not be reliable. We also notice that when the proportion of positive studies, $\hat{r}$, is lower than or close to the designed power of most studies in historical data, the proposed approach tends to enlarge the bias. This is because the bias is not obvious in this situation that the proposed approach would make an overadjustment. Another reason that may lead to the failure of the proposed approach is no difference between positive studies and nonpositive studies, which is corresponding to the second assumption of the proposed approach as aforementioned. In this case, formula (8.13) would produce an overestimated $(\mu_P - \mu_N)$, which may raise unexpected bias.

## 8.6 Concluding Remarks

Although RWE offers the opportunities to develop robust evidence using high-quality data and sophisticated methods for producing causal-effect estimates, randomization is feasible. In this chapter, we have demonstrated that the assessment of treatment effect (RWE) based on RWD could be biased due to potential selection and information biases of RWD. Although fit-for-purpose RWE may meet regulatory standards under certain assumptions, it is not the same as SE (current regulatory standard). In practice, it is then suggested that when there are gaps between fit-for-purpose RWE and SE, we should make efforts to fill the gaps for an accurate and reliable assessment of the treatment effect.

As indicated by Corrigan-Curay (2018), there is a value of using RWE to support regulatory decisions in drug review and approval process. However, while incorporating RWE into evidence generation, many factors must be considered at the same time before we can map RWE to SE (current regulatory standard) for regulatory review and approval. These factors include, but are not limited to, (i) efficacy or safety; (ii) relationship to available evidence; (iii) clinical context, e.g., rare, severe, life-threatening, or unmet medical need; and (iv) natural of endpoint/concerns about bias. In addition, leveraging RWE to support new indications and label revisions can help accelerate high-quality RWE earlier in the product lifecycle, providing more relevant evidence to support higher-quality and higher-value care for patients. Incorporating RWE into product labeling can lead to better-informed patient and provider decisions with more relevant information. For this purpose, it is suggested characterizing RWD quality and relevancy for regulatory purposes. Ultimate regulatory acceptability, however, will depend upon how robust these studies can be. That is, how well they minimize the potential for bias and confounding.

# 9

## Innovative Approach for Rare Diseases Drug Development

### 9.1 Introduction

As indicated in Chapter 1, rare disease is defined as a disorder or condition that affects less than 200,000 persons in the United States. Most rare diseases are genetically related and thus are present throughout the person's entire life, even if symptoms do not immediately appear (FDA, 2015b, 2019a). To encourage the development of rare disease drug products, FDA provides several incentives to expedite development programs for rare diseases. These expedited programs include (i) fast-track designation, (ii) breakthrough therapy designation, (iii) priority review designation, and (iv) accelerated approval for regulatory review and approval of rare disease drug products. FDA, however, emphasizes that the agency will not create a statutory standard for the approval of orphan drugs that is different from the standard for the approval of drugs in common conditions (FDA, 2015b, 2015c, 2019a). For the approval of drug products, however, FDA requires that substantial evidence regarding effectiveness and safety of the drug products be provided. Substantial evidence can only be obtained through the conduct of adequate and well-controlled investigations (21 CFR 314.126(a)). For rare diseases drug development, it is then of particular interest to the investigator regarding how to conduct clinical trials with a limited number of subjects available for obtaining substantial evidence in support of regulatory review and approval of the drug product under investigation.

For the approval of drug products, a traditional approach is to conduct a randomized clinical trial (RCT) by testing a null hypothesis of no treatment effect (i.e., ineffectiveness) against an alternative hypothesis that there is a treatment effect (i.e., effectiveness). Effectiveness is concluded if we reject the null hypothesis at a pre-specified level of significance (say 5%). To ensure there is high probability of correctly detecting a clinically meaningful difference (or treatment effect), a pre-study power calculation (i.e., power analysis for sample size calculation) is often made to ensure that there are sufficient number of subjects for achieving a desired power for correctly detecting a clinically meaningful treatment effect if such a treatment effect truly exists. In rare disease clinical trials, however, power calculation may not be feasible because there are only a limited number of subjects available for the rare diseases under study. In this case, innovative thinking and approach for rare diseases drug development are needed for providing substantial evidence meeting the same standard as drugs in common conditions with certain statistical assurance.

In what follows, some out-of-the-box innovative thinking designs for rare disease drug (clinical) development are described. These out-of-the-box innovative thinking designs include, but are not limited to, (i) probability monitoring procedure for sample size calculation/justification (Huang and Chow, 2019); (ii) demonstrating not-ineffectiveness rather than demonstrating effectiveness based on a limited number of subjects available of the small patient population of the rare disease under study (Chow and Huang, 2019a); (iii) utilizing (borrowing) real-world data (RWD), which generates real-world evidence (RWE) in support of regulatory approval of the rare disease drug development; and (iv) the use of complex innovative designs such as *n*-of-1 trial design, adaptive trial design, master protocols, and Bayesian approach.

In the next section, some basic considerations in rare disease clinical trials are outlined. An innovative concept for demonstrating not-ineffectiveness rather than demonstrating effectiveness with a limited number of subjects available for a rare disease under study is introduced in Section 9.3. In Section 9.4, an innovative approach utilizing a two-stage adaptive seamless trial design in conjunction with the concept of RWD/RWE is proposed for regulatory consideration. Also included in this section are the probability monitoring procedure for sample size calculation and an example concerning rare disease development. Some concluding remarks are given in the last section of this chapter.

## 9.2 Basic Statistical Considerations

For the approval of drug products, FDA requires that substantial evidence regarding the effectiveness and safety of the drug products under development be provided. This, however, is a major challenge for rare disease drug product development due to the fact that there are only a limited number of subjects available. Thus, some basic statistical principles are necessarily considered, which are described below.

**Historical Data and Ethical Consideration** – Due to the small sample available in rare diseases drug development, FDA encourages the sponsors to evaluate the existing natural history knowledge in rare diseases drug development programs (FDA, 2015b, 2019a). Natural history studies help in defining the disease population, selecting clinical endpoints, and developing new or optimized biomarkers in early rare disease drug development. Most importantly, natural history studies provide an external control group for clinical trials. In practice, natural history studies could be conducted either prospectively or retrospectively based on electronic health records (EHRs) such as existing medical records and/or patient charts (FDA, 2015b, 2019a).

Regarding ethical considerations, as indicated in Chapter 2, an institutional review board (IRB) ensures that ethical requirements are fulfilled, including biases are eliminated, ethical issues between investigators and participants are balanced, and the research does not exploit individuals or groups. In practice, critical and emerging ethical issues are challenges to rare disease clinical research. These challenges include relatively few participants, the need for multi-site studies, innovative designs, and the need to protect privacy in a contained research environment.

**Small Sample Requirement** – In rare diseases clinical trials, power analysis for sample size calculation is not feasible, especially in the situations where there are (i) clinical trials with extremely low incidence rates and (ii) clinical trials with rare disease drug development. This is because (i) it may require a large sample size for detecting a relatively small difference and (ii) eligible patients may not be available for a small target patient population. In this case, an innovative method based on a probability monitoring procedure as described in Chapter 7 is useful. The concept of a probability monitoring procedure (either adaptive version or non-adaptive version) is to select an appropriate sample size for controlling the probability of crossing safety and/or efficacy boundaries. For rare disease clinical development, an adaptive probability monitoring procedure is recommended if a multiple-stage adaptive trial design is used.

**Endpoint Selection** – In rare diseases clinical trials, the selection of appropriate study endpoints is critical for an accurate and reliable evaluation of safety and effectiveness of a test treatment under investigation. Sample size calculation depends upon not only (i) data types (e.g., continuous, binary response, or time-to-event data) of the primary endpoint, but also (ii) absolute change, relative change, and/or responder defined based on either absolute difference or relative difference. In practice, it is usually not clear which study endpoint can best inform the disease status and measure the treatment effect. Different study endpoints such as clinical endpoint, surrogate endpoint, or biomarker may not translate one another although they may be highly correlated with one another. Thus, it is of interest to develop an innovative endpoint, namely, therapeutic index based on a utility function to combine and utilize information collected from all study endpoints. More details regarding the development of therapeutic index have been previously discussed in Chapter 4.

**Innovative Trial Design** – As indicated in Chapter 2, small patient population is a challenge to rare disease clinical trials. Thus, there is a need for innovative trial designs in order to obtain substantial evidence with a small number of subjects available for achieving the same standard for regulatory approval. These innovative trial designs include, but are not limited to, the $n$-of-1 trial design, an adaptive trial design, master protocols, and a Bayesian design. The use of innovative trial design is not only to fix the dilemma of samples available (e.g., $n$-of-1 trial design and master protocol) but also to allow flexibility to effectively, accurately, and reliably assess the treatment effect and increase the probability of success of the intended clinical trials (e.g., adaptive trial design in conjunction with Bayesian approach for borrowing RWD). More details regarding these innovative trial designs for potential use in rare diseases clinical development are discussed in Chapters 10–12.

**Bayesian Approach for Data Borrowing** – To meet the statutory requirements of the 21st Century Cures Act, the FDA has provided draft guidance outlining its policies for using RWE to support regulatory decision regarding safety and/or efficiency. The sponsors and applicants are encouraged to use RWD in order to generate RWE as part of a regulatory submission in a simple, uniform format.

According to the FDA's RWE program (FDA, 2018c), RWD are data relating to patient's health status and/or the delivery of health care routinely collected from a variety of sources, such as EHR, claims data, patient registries, and patient-generated data; however, RWE is the clinical evidence about the usage and potential benefits or risks of a medical product derived from the analysis of RWD. RWD can be helpful to (i) generate hypotheses tests, (ii) identify biomarkers, (iii) assess trial feasibility, (iv) inform Bayesian prior, (v) identify prognostic indicators or patient characteristic for enrichment or stratification, and (vi) assemble cohorts in rare diseases drug development.

The FDA has used RWE to monitor and evaluate post-market drug safety (e.g., through the Sentinel system). However, the use of RWE to support treatment effectiveness is generally limited to oncology or rare disease single-arm trials. Challenges exist, such as (i) populations may not be comparable, (ii) diagnostic criteria or outcomes may not be equivalent, (iii) outcomes are assessed differently, and (iv) variability is found in the follow-up procedures.

Obviously, RWD and RWE are attractive in rare disease drug developments. Bayesian approach could be useful for data borrowing in rare disease drug development. For example, (i) an informative prior could be constructed based on RWD that are well evaluated and verified, and (ii) a power prior may be constructed based on the likelihood of historical RWD and an initial prior. However, due to the above-mentioned challenges and the difficulty in evaluating and verifying the comparison of the borrowed data and study data, Bayesian data borrowing approach is a popular area with issues and deserves further research.

## 9.3 Innovative Thinking for Rare Disease Drug Development

To overcome the problem of small patient population for rare diseases, some out-of-the-box innovative thinking regarding design (sample size requirement and complex innovative trial design) and analysis (e.g., demonstration of effectiveness and the utilization of RWD/RWE) are necessarily considered.

**Probability Monitoring Procedure for Sample Size** – For rare disease clinical development, it is recognized that a pre-study power analysis for sample size calculation is not feasible due to the fact that there are only a limited number of subjects available for the intended trial, especially when the anticipated treatment effect is relatively small and/or the variability is relatively large. In this case, alternative methods such as precision analysis (or confidence interval approach), reproducibility analysis, and probability monitoring approach may be considered for providing substantial evidence with certain statistical assurance (Chow, Shao et al., 2017). It, however, should be noted that the resultant sample sizes from these different analyses could be very different with different levels of statistical assurance achieved. Thus, for rare disease clinical trials, it is suggested that an appropriate sample size should be selected for achieving certain statistical assurance under a valid trial design. To overcome the problem, Huang and Chow (2019) proposed a probability monitoring procedure for sample size calculation/justification, which can substantially reduce the required sample size for achieving certain statistical assurance.

As an example, an appropriate sample size may be selected based on a probability monitoring approach such that the probability of crossing safety boundary is controlled at a pre-specified level of significance. Suppose an investigator plans to monitor the safety of a rare disease clinical trial sequentially at several times, $t_i$, $i = 1, \ldots, K$. Let $n_i$ and $P_i$ be the sample size and the probability of observing an event at time $t_i$. Thus, an appropriate sample size can be selected such that the following probability of crossing safety stopping boundary is less than a pre-specified level of significance:

$$p_k = P\{\text{across safety stopping boundry} \mid n_k, P_k\} < \alpha, \, k = 1, \ldots, K. \quad (9.1)$$

Note that the concept of the probability monitoring approach should not be mixed up the concepts with those based on power analysis, precision analysis, and reproducibility analysis. Statistical methods for data analysis should reflect the desired statistical assurance under the trial design (Table 9.1).

**TABLE 9.1**

Stopping Boundaries for Two-Stage Efficacy Designs

| One-sided $\alpha$ | | 0.005 | 0.010 | 0.015 | 0.020 | 0.025 | 0.030 |
|---|---|---|---|---|---|---|---|
| 0.025 | $\alpha_1$ | 0.2050 | 0.1832 | 0.1564 | 0.1200 | 0.0250 | – |
| 0.05 | $\alpha_2$ | 0.3050 | 0.2928 | 0.2796 | 0.2649 | 0.2486 | 0.2300 |

*Source:* Chang (2007).

**Demonstrating Not-Ineffectiveness Versus Demonstrating Effectiveness** – For the approval of a new drug product, the sponsor is required to provide substantial evidence regarding safety and efficacy of the drug product under investigation. In practice, a typical approach is to conduct adequate and well-controlled clinical studies, and test the following point hypotheses:

$$H_0: \textit{ineffectiveness versus } H_a: \textit{effectiveness.} \tag{9.2}$$

The rejection of the null hypothesis of *ineffectiveness* is in favor of the alternative hypothesis of effectiveness. Most researchers interpret that the rejection of the null hypothesis is the demonstration of the alternative hypothesis of effectiveness. It, however, should be noted that "in favor of effectiveness" does not imply "the demonstration of effectiveness." In practice, hypotheses (9.2) should be

$$H_0: \textit{ineffectiveness versus } H_a: \textit{not-ineffectiveness.} \tag{9.3}$$

In other words, the rejection of $H_0$ would lead to the conclusion of "*not $H_0$,*" which is $H_a$, i.e., "not-ineffectiveness," as given in (9.3). As can be seen from $H_a$ in (9.2) and (9.3), the concept of *effectiveness* (9.2) and the concept of *not-ineffectiveness* (3) are not the same (Chow and Huang, 2019a). Not-ineffectiveness does not imply effectiveness in general. Thus, the traditional approach for clinical evaluation of the drug product under investigation can only demonstrate "*not-ineffectiveness,*" but not "*effectiveness.*" The relationship between demonstrating "effectiveness" (9.2) and demonstrating "not-ineffectiveness" (9.3) is illustrated in Figure 9.1. As it can be seen from Figure 9.1, "not-ineffectiveness" consists of two parts, namely, the portion of "inconclusiveness" and the portion of "effectiveness."

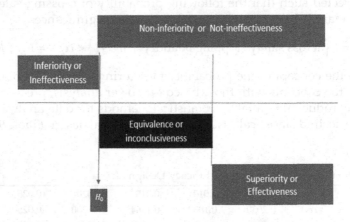

**FIGURE 9.1**
Demonstrating *"effectiveness"* or *"not-ineffectiveness."*

For a placebo-controlled clinical trial comparing a test treatment ($T$) and a placebo control ($P$), let $\theta = \mu_T - \mu_R$ be the treatment effect of the test treatment as compared to the placebo, where $\mu_T$ and $\mu_R$ are the mean responses of the test treatment and the placebo, respectively. For a given sample, e.g., test results from a previous or pilot study, let $(\theta_L, \theta_U)$ be a $(1-\alpha) \times 100\%$ confidence interval of $\theta$. In this case, hypotheses (9.2) become

$$H_0: \theta \le \theta_L \text{ versus } H_a: \theta > \theta_U, \tag{9.4}$$

while hypotheses (9.3) are given by

$$H_0: \theta \le \theta_L \text{ versus } H_a: \theta > \theta_L. \tag{9.5}$$

Hypotheses (9.4) are similar to the hypotheses set in Simon's two-stage optimal design for cancer research. At the first stage, Simon suggested testing whether the response rate has exceeded a pre-specified undesirable response rate. If yes, then proceed to test whether the response rate has achieved a pre-specified desirable response rate. Note that Simon's hypotheses testing actually is an interval hypotheses testing problem. On the other hand, hypotheses (9.5) is a typical one-sided test for non-inferiority of the test treatment as compared to the placebo. Thus, the rejection of inferiority leads to the conclusion of non-inferiority, which consists of equivalence (the area of inconclusiveness, i.e., $\theta_L < \theta < \theta_U$) and superiority (i.e., effectiveness). For a given sample size, the traditional approach for the clinical evaluation of the drug product under investigation can only demonstrate that the drug product is *not ineffective* when the null hypothesis is rejected. For demonstrating that the drug product is truly effective, we need to perform another test to rule out the possibility of inconclusiveness (i.e., to reduce the probability of inconclusiveness).

In practice, however, we typically test point hypotheses of equality at the $\alpha = 5\%$ level of significance. The rejection of the null hypothesis leads to the conclusion that there is a treatment effect. An adequate sample size is then selected to have a desired power (say 80%) to determine whether the observed treatment effect is clinically meaningful and hence claim that the effectiveness is demonstrated. For testing point hypotheses of no treatment effect, many researchers prefer testing the null hypothesis at the $\alpha = 1\%$ rather than at the $\alpha = 5\%$ level of significance in order to account for the possibility of inconclusiveness. In other words, if the observed $p$-value falls between 1% and 5%, we claim the test result is *inconclusive*. It should be noted that the concept of point hypotheses testing for no treatment effect is very different from interval hypotheses testing (9.2) and one-sided hypotheses testing for non-inferiority (9.3). In practice, however, point hypotheses testing, interval hypotheses testing, and one-sided hypotheses for testing non-inferiority have been mixed up and used in pharmaceutical research and development.

As a result, the rejection of the null hypothesis of ineffectiveness cannot directly imply that the drug product is effective unless the probability of inconclusiveness, denoted by $p_{IC}$, is negligible, i.e.,

$$p_{IC} = P\{\text{inconclusiveness}\} < \varepsilon, \tag{9.6}$$

where $\varepsilon$ is a pre-specific number, which is agreed upon between clinician and regulatory reviewer (Chow and Huang, 2019; Chow, 2020).

Note that the concept of demonstrating "not-ineffectiveness" is similar to that of establishing non-inferiority. One can test for superiority (i.e., effectiveness) once the non-inferiority has been established without paying any statistical penalties.

**The Use of RWD/RWE** – The 21st Century Cures Act passed by the United States Congress in December 2016 requires that the FDA shall establish a program to evaluate the potential use of RWE, which is derived from RWD to (i) support the approval of new indication for a drug approved under section 505 (c) and (ii) satisfy post-approval study requirements. RWD refers to data relating to patient health status and/or the delivery of health care routinely collected from a variety of sources. RWD sources include, but are not limited to, EHR, administrative claims and enrollment, personal digital health applications, public health databases, and emerging sources. Although RWE offers the opportunities to develop robust evidence using high-quality data and sophisticated methods for producing causal-effect estimates, randomization is feasible. In this chapter, we have demonstrated that the assessment of treatment effect (RWE) based on RWD could be biased due to potential selection and information biases of RWD. Although fit-for-purpose RWE may meet regulatory standards under certain assumptions, it is not the same as substantial evidence (current regulatory standard). In practice, it is then suggested that when there are gaps between fit-for-purpose RWE and substantial evidence, we should make efforts to fill the gaps for an accurate and reliable assessment of the treatment effect.

In order to map RWE to substantial evidence (current regulatory standard), we need to have good understanding of the RWD in terms of data relevancy/quality and its relationship with substantial evidence so that a fit-for-regulatory purpose RWE can be derived to map to regulatory standard.

As indicated by Corrigan-Curay (2018), there is a value of using RWE to support regulatory decisions in drug review and approval process. However, while incorporating RWE into evidence generation, many factors must be considered at the same time before we can map RWE to substantial evidence (current regulatory standard) for regulatory review and approval. These factors include, but are not limited to, (i) efficacy or safety; (ii) relationship to available evidence; (iii) clinical context, e.g., rare, severe, life-threatening, or unmet medical need; and (iv) nature of endpoint/concerns about bias. In addition, leveraging RWE to support new indications and label revisions can help accelerate high-quality RWE earlier in the product lifecycle, providing more relevant evidence to support higher-quality and higher-value care for patients. Incorporating RWE into product labeling can lead to better-informed patient and provider decisions with more relevant information. For this purpose, it is suggested characterizing RWD quality and relevancy for regulatory purposes. Ultimate regulatory acceptability, however, will depend upon how robust these studies can be. That is, how well they minimize the potential for bias and confounding.

**Complex Innovative Design** – As indicated earlier, small patient population is a challenge to rare disease clinical trials. Thus, there is a need for innovative trial designs in order to obtain substantial evidence with a small number of subjects available for achieving the same standard for regulatory approval. In this sub-section, several innovative trial designs, including $n$-of-1 trial design, an adaptive trial design, master protocols, and a Bayesian design, are discussed.

One of the major dilemmas for rare diseases clinical trials is the unavailability of patients with the rare diseases under study. In addition, it is unethical to consider a placebo control in the intended clinical trial. Thus, it is suggested an $n$-of-1 crossover design be considered. An $n$-of-1 trial design is to apply $n$ treatments (including placebo) in an individual at different dosing periods with sufficient washout in between dosing periods. A complete $n$-of-1 trial design is a crossover design consisting of all possible combinations of treatment assignment at different dosing periods.

Another useful innovative trial design for rare disease clinical trials is an adaptive trial design. In its draft guidance on adaptive clinical trial design, FDA defines an adaptive design as a study that includes a *prospectively* planned opportunity for the modification of one or more specified aspects of the study design and hypotheses based on the analysis of (usually interim) data from subjects in the study (FDA, 2010, 2019c). The FDA guidance has been served as an official document describing the potential use of adaptive designs in clinical trials since it was published in 2019. It, however, should be noted that the FDA draft guidance on adaptive clinical trial design is currently being revised in order to reflect pharmaceutical practice and FDA's current thinking.

Woodcock and LaVange (2017) introduced the concept of master protocol for studying multiple therapies, multiple diseases, or both in order to answer more questions in a more efficient and timely fashion. Master protocols include the following types of trials: umbrella, basket, and platform. The type of umbrella trial is to study multiple targeted therapies in the context of a single disease, while the type of basket trial is to study a single therapy in the context of multiple diseases or disease subtypes. The platform is to study multiple targeted therapies in the context of a single disease in a perpetual manner, with therapies allowed to enter or leave the platform on the basis of decision algorithm. As indicated by Woodcock and LaVange (2017), if designed correctly, master protocols offer a number of benefits, including streamlined logistics, improved data quality, collection and sharing, as well as the potential to use innovative statistical approaches to study design and analysis. Master protocols may be a collection of sub-studies or a complex statistical design or platform for rapid learning and decision-making.

Under the assumption that historical data (e.g., previous studies or experience) are available, Bayesian methods for borrowing information from different data sources may be useful. These data sources could include, but are not limited to, natural history studies and expert's opinion regarding prior distribution about the relationship between endpoints and clinical outcomes. The impact of borrowing on results can be assessed through the conduct of sensitivity analysis. One of the key questions of particular interest to the investigator and regulatory reviewer is that how much to borrow in order to (i) achieve desired statistical assurance for substantial evidence, and (ii) maintain the quality, validity, and integrity of the study.

## 9.4 Innovative Approach for Rare Disease Drug Development

**Chow's Proposed Innovative Approach** – Combining the out-of-the-box innovative thinking designs regarding rare disease drug development described in the previous section, Chow and Chang (2019) and Chow (2020) proposed the following innovative approach utilizing a two-stage adaptive approach in conjunction with the use of RWD/RWE for rare diseases drug development. This innovative approach is briefly summarized below.

*Step 1*. Select a small sample size $n_1$ at stage 1 as deemed appropriate by the principal investigator (PI) based on both medical and nonmedical considerations. Note that $n_1$ may be selected based on the probability monitoring procedure.

*Step 2*. Test hypotheses (9.3) for not-ineffectiveness at the $\alpha_1$-level, a pre-specified level of significance. If fails to reject the null hypothesis of ineffectiveness, then stop the trial due to futility. Otherwise, proceed to the next stage.

Note that an appropriate value of $\alpha_1$ can be determined based on the evaluation of the trade-off with the selection of $\alpha_2$ for controlling the overall type I error rate at the significance level of $\alpha$. The goal of this step is to establish non-inferiority (i.e., not-ineffectiveness) of the test treatment with a limited number of subjects available at the $\alpha_1$-level of significance based on the concept of probability monitoring procedure for sample size justification and performing a non-inferiority (not-ineffectiveness) test with a significance level of $\alpha_1$.

*Step 3a*. Recruit additional $n_2$ subjects at stage 2. Note that $n_2$ may be selected based on the probability monitoring procedure. Once the non-inferiority (not-ineffectiveness) has been established at stage 1, sample size re-estimation may be performed for achieving the desirable statistical assurance (say 80% power) for the establishment of effectiveness of the test treatment under investigation at the second stage (say $N_2$, sample size required at stage 2).

***Step 3b.*** Obtain (borrow) $N_2 - n_2$ data from previous studies (RWD) if the sample size of $n_2$ subjects is not enough for achieving desirable statistical assurance (say 80% power) at stage 2. Note that data obtained from the $n_2$ subjects are from RCT, while data obtained from the other $N_2 - n_2$ are from RWD.

***Step 4.*** Combine data from both step 3a (data obtained from RCT) and step 3b (data obtained from RWD) at stage 2, and perform a statistical test to eliminate the probability of inconclusiveness. That is, perform a statistical test to determine whether the probability of inconclusiveness has become negligible at the $\alpha_2$-level of significance. For example, if the probability of inconclusiveness is less than a pre-specified value (say 5%), we can then conclude the test treatment is effective.

In summary, for review and approval of rare diseases drug products, Chow and Huang (2019a), Chow and Chang (2019) and Chow (2020) proposed first to demonstrate not-ineffectiveness with limited information (patients) available at a pre-specified level of significance of $\alpha_1$. Then, after the not-ineffectiveness of the test treatment has been established, collect additional information (RWD) to rule out the probability of *inconclusiveness* for the demonstration of effectiveness at a pre-specified level of significance of $\alpha_2$ under the two-stage adaptive seamless trial design.

**Statistical Properties of Chow's Proposal** – In this section, we explore more technical details about the statistical properties of the proposed innovative approach. Consider testing hypotheses $H_0 : \theta \leq \theta_L$ versus $H_a : \theta > \theta_U$ in (9.4). We divide this test into two stages. At stage 1, we test for *not-ineffectiveness*, i.e., $H_0 : \theta \leq \theta_L$ versus $H_a : \theta > \theta_L$ in (9.5). At stage 2, we test for *effectiveness* after the *not-ineffectiveness* has been established at stage 1, i.e.,

$$H_0 : \theta \leq \theta_U \text{ versus } H_a : \theta > \theta_U \text{ given that } \theta > \theta_L. \tag{9.7}$$

We then select significance levels of $\alpha_1$ and $\alpha_2$ at stages 1 and 2 according to what described in Section 9.3 for a well-controlled overall type I error rate at a pre-specified level $\alpha$. The hypothesis of not-ineffectiveness is then tested with a significance level of $\alpha_1$ at stage 1, and the hypothesis of effectiveness is tested with a significance level of $\alpha_2$ at stage 2 by ruling out the probability of inconclusiveness.

Without the loss of generality, assume the study is a single arm with a continuous endpoint. Assume the true value of $\theta$ is $\theta_T$ with $\theta_T > \theta_L$ and the population variance is $\sigma^2$. Suppose the targeted power is $1 - \beta$. Under normal assumption or with normal approximation, the required sample size based on a pre-study power analysis at stage 1 is given by

$$N_1 = \left( \frac{z_{a_1} + z_\beta}{\theta_T - \theta_L} \right)^2 \sigma^2. \tag{9.8}$$

Suppose in the rare disease clinical trial, it is not feasible to recruit $N_1$ patients for the study, and suppose that only $r_1$ proportion of the required sample size $N_1$ is available for the study; i.e., only $n_1 = r_1 N_1$ subjects are available for testing not-ineffectiveness at stage 1. In this situation, with a limited sample size of $n_1$, it may be difficult to achieve statistically significant conclusion of not-ineffectiveness even when the true $\theta$ is effectiveness using traditional hypothesis testing procedure. Instead, we may carry out the probability monitoring procedure proposed by Huang and Chow (2019) to justify the selected sample size of $n_1$ by achieving some statistical assurance of establishing not-ineffectiveness. Note that a total sample of $n_1$ subjects with observations $X_{n_1} = \{X_1, X_2, ..., X_{n_1}\}$ will be obtained at stage 1. To carry out the probability monitoring procedure, we may specify a series of time points and a probability boundary $P_b$, e.g., $P_b = 0.05$. At each time point where a cumulative sub-sample of $X_{n_1}$ is collected, a test statistic can be constructed and a probability of crossing efficacy boundary can be calculated and compared to $P_b$. The monitoring procedure continues until the boundary $P_b$ is crossed or there are no more samples to collect. We then conclude not-ineffectiveness and proceed to stage 2 if $P_b$ is never crossed; otherwise, the trial is stopped due to ineffectiveness. Note that in practice, the probability boundary $P_b$ is often selected based on clinical judgment in conjunction with statistical justification for desirable statistical assurance.

Now suppose that the not-ineffectiveness has been established at stage 1. At stage 2, we want to test the hypothesis of effectiveness. By power analysis, the required sample size for stage 2 can be obtained as

$$N_2 = \left( \frac{z_{\alpha_2} + z_\beta}{\theta_T - \theta_U} \right)^2 \sigma^2. \tag{9.9}$$

By comparing (9.9) with (9.8), it can be verified that $N_2$ can be larger than $N_1$ if the true value $\theta_T > \theta_U$. Similarly, assume only $r_2$ proportion of the required sample size $N_2$ is available; i.e., only $n_2 = r_2 N_2$ subjects are available for testing the effectiveness at stage 2, and the observations set is $X_{n_2} = \{X_{n_1+1}, X_{n_1+2}, ..., X_{n_1+n_2}\}$. To achieve the desirable statistical assurance for the rare disease clinical trial, one plausible way is to borrow data from previous studies or utilize RWD (obtained from adequate and well-controlled clinical studies). Assume that the RWD is from adequate and well-controlled clinical trials with the same population. We then borrow a real-world sample with size $K = N_2 - n_2$ using Bayesian approach and denote the data set as $Y_K = \{Y_1, Y_2, ..., Y_K\}$. Thus, we have a combined data set from the random study at stage 2 and previous studies $Z_{N_2} = \{X_{n_2}, Y_K\} = \{X_{n_1+1}, X_{n_1+2}, ..., X_{n_1+n_2}, Y_1, Y_2, ..., Y_K\}$ for data analysis at stage 2. Note that $Z_{N_2}$ is independent of $X_{n_1}$. Thus, the determination of the combination of $\alpha_1$ and $\alpha_2$ based on Section 9.3 can control the overall type I error at $\alpha$. Note that conditional on the establishment of not-ineffectiveness at stage 1, we can rule out the inconclusiveness to conclude the effectiveness. Note that inconclusiveness means that $\theta \in (\theta_L, \theta_U)$. Given the sample $Z_{N_2}$, we may construct a test statistic $T_2$ as the sample mean, derive its asymptotic or approximated distribution, and define the probability of inconclusiveness $P_I$ as the probability that $T_2 \in (\theta_L, \theta_U)$, i.e.,

$$P_I = \Pr\left( T_2 \in (\theta_L, \theta_U) \right). \tag{9.10}$$

Given a pre-specified threshold $\delta$, we may decide that the probability of inconclusiveness is negligible if $P_I < \delta$ and conclude the effectiveness; otherwise, the effectiveness is rejected. Note that in practice, the threshold $\delta$ is often selected based on clinical judgment in conjunction with statistical justification for desirable statistical assurance.

**An Example** – In this section, Chow proposed innovative two-stage approach in rare disease drug development through an example based on a real study of evaluation of treatments applicable to fairly advanced (say stage 3) primary biliary cirrhosis, which has very low prevalence (Lilford et al., 1995). Suppose that a new drug (perhaps a biological modifier) is proposed as an alternative to conventional ursodeoxycholic treatment for advanced primary biliary cirrhosis with very low prevalence. The five-year mortality rate among patients with this stage of primary ciliary cirrhosis is $p_L = 30\%$ with the current treatment. A single-arm trial is planned to investigate the effect of the drug due to the low prevalence, with the expectation of improvement in mortality $p$ from $p_L = 30\%$ to $p_U = 27\%$. Thus, we want to evaluate the hypotheses

$$H_0: p \geq p_L \text{ versus } H_a: p < p_U. \tag{9.11}$$

According to the two-stage approach, we test for *not-ineffectiveness* at stage 1, i.e.,

$$H_0: p \geq p_L \text{ versus } H_a: p < p_L; \tag{9.12}$$

and test for *effectiveness* after the *not-ineffectiveness* has been established at stage 2, i.e.,

$$H_0: p \geq p_U \text{ versus } H_a: p < p_U \text{ given that } p < p_L \tag{9.13}$$

by ruling out the probability of inconclusiveness, i.e.,

$$P_I = \Pr\left( T_2 \in (p_U, p_L) \right), \tag{9.14}$$

where $T_2$ is the estimated mortality (sample mean) at stage 2.

Consider the significance level of $\alpha = 5\%$, power of $1 - \beta = 90\%$, and select an appropriate combination of $\alpha_1$ and $\alpha_2$ (see, e.g., Chang, 2007). Assume that the true mortality based on the new treatment is $p_T = 24\%$. The sample sizes based on power analysis for stages 1 and 2 are

$$N_1 = p_L\left(1-p_L\right)\left(z_{a_1} + z_\beta\right)^2 \Big/ \left(p_T - p_L\right)^2$$

$$N_2 = p_U\left(1-p_U\right)\left(z_{a_2} + z_\beta\right)^2 \Big/ \left(p_T - p_U\right)^2. \tag{9.15}$$

We list in Table 9.2 several possible combinations of $(N_1, N_2)$ and the total sample size $N = N_1 + N_2$ according to the combinations of $(\alpha_1, \alpha_2)$.

Without the loss of generality, we select $\alpha_1 = 0.020$ and $\alpha_2 = 0.265$; thus, $N_1 = 649$, $N_2 = 799$, and the total required sample size is . In rare disease drug development, a study of such size would need to recruit from very large areas over long periods. Suppose only 20% of the required total sample size $N$ is available for the study, i.e., $n = 290$. We need to allocate the sample sizes $n_1$ and $n_2$ for stages 1 and 2, where $n_1 + n_2 = n$. There might be an optimal ratio $\rho = n_1/n_2$ to achieve the study goal. To simplify, it is reasonable to set the ratio as $\rho = N_1/N_2 = 0.812$; thus, $n_1 = \dfrac{\rho}{1+\rho}N = 130$ and $n_2 = \dfrac{1}{1+\rho}N = 160$. Note that $r_1 = n_1/N_1 = 0.2$ and $r_2 = n_2/N_2 = 0.2$ are equal.

For stage 1, given the limited sample size of $n_1 = 130$, it is difficult to achieve statistically significant conclusion of not-ineffectiveness even when the true $p_T$ is effectiveness using traditional hypothesis testing procedure. Instead, we carry out the non-adaptive probability monitoring procedure proposed by Huang and Chow (2019) based on the limited sample to achieve some statistical assurance of establishing not-ineffectiveness. Specifically, we consider a monitoring procedure as each 20 more samples (except the last accumulation, which is 10 more samples) are accumulated, thus consisting of a sequence of $Q=7$ accumulated sub-samples $\{s_{11}, s_{12},..., s_{1Q}\}$ with sizes $n_{11}, n_{12}, ..., n_{1Q}$, where $s_{11} \subset s_{12} \subset ... \subset s_{1Q}$ and $n_{11} = 20$, $n_{12} = 40$, $n_{13} = 60$, $n_{14} = 80$, $n_{15} = 100$, $n_{16} = 120$, $n_{1Q} = 130$. Denote the number of accumulated incidents and the cumulative probability as $r_q$ and $P_q, q = 1,..., Q$, where $P_q = B(r_q; n_q, p_L)$ and $B$ represent the cumulative binomial distribution. Without the loss of generality, suppose the probability boundary $P_b$ could be 0.01, 0.025, 0.05, or 0.1. Note that in practice, the probability boundary $P_b$ is often selected based on clinical judgment in conjunction with statistical justification for desirable statistical assurance. We figure out

**TABLE 9.2**

Illustration of Possible Sample Combination $(N_1, N_2)$

| $\alpha_1$ | 0.005 | 0.010 | 0.015 | 0.020 | 0.025 | 0.030 | 0.035 |
|---|---|---|---|---|---|---|---|
| $\alpha_2$ | 0.305 | 0.293 | 0.280 | 0.265 | 0.249 | 0.230 | 0.208 |
| $N_1$ | 868 | 760 | 695 | 649 | 613 | 584 | 559 |
| $N_2$ | 703 | 731 | 763 | 799 | 842 | 894 | 961 |
| $N$ | 1,571 | 1,491 | 1,458 | 1,448 | 1,455 | 1,478 | 1,520 |

*Note:* $N = N_1 + N_2$ according to the combination $(\alpha_1, \alpha_2)$ for the two-stage approach. Total one-sided type I error rate is $\alpha = 0.05$.

**TABLE 9.3**

Probability Monitoring Procedures

| | | $p_L = 0.3$ | | $p_L = 0.3$ | | $p_L = 0.3$ | | $p_L = 0.3$ | |
| | | $P_b = 0.01$ | | $P_b = 0.025$ | | $P_b = 0.05$ | | $P_b = 0.1$ | |
| $q$ | $n_q$ | $P_q^*$ | $r_q^*$ | $P_q^*$ | $r_q^*$ | $P_q^*$ | $r_q^*$ | $P_q^*$ | $r_q^*$ |
|---|---|---|---|---|---|---|---|---|---|
| 1 | 20 | 0.008 | 1 | 0.008 | 1 | 0.035 | 2 | 0.035 | 2 |
| 2 | 40 | 0.009 | 5 | 0.024 | 6 | 0.024 | 6 | 0.055 | 8 |
| 3 | 60 | 0.006 | 10 | 0.014 | 11 | 0.029 | 12 | 0.010 | 14 |
| 4 | 80 | 0.008 | 15 | 0.016 | 16 | 0.030 | 17 | 0.087 | 19 |
| 5 | 100 | 0.009 | 20 | 0.016 | 21 | 0.048 | 23 | 0.076 | 24 |
| 6 | 120 | 0.009 | 25 | 0.016 | 26 | 0.042 | 28 | 0.096 | 30 |
| 7 | 130 | 0.007 | 27 | 0.020 | 29 | 0.049 | 31 | 0.073 | 32 |

*Note:* $P_q^*$ is the maximum of the possible cumulative probabilities that haven't crossed the boundary $P_b$ for sub-sample $s_q$, and $r_q^*$ is the corresponding number of incidents.

maximum of the possible cumulative probabilities $P_q^m$ that are less the boundary $P_b$ and the corresponding incident number $r_q^m$ at each stage $q$, which are listed in Table 9.3

Given a real sample, the probability monitoring procedure continues as the observed $P_q$ never crosses the boundary $P_b$ (equivalently, $P_q \leq P_q^m$ and $r_q \leq r_q^m$ in Table 9.3), or there are no more samples to collect. We then conclude not-ineffectiveness and proceed to stage 2 if $P_b$ is never crossed; otherwise, the trial is stopped due to ineffectiveness.

Now suppose that the not-ineffectiveness has been established at stage 1 based on the probability monitoring procedure with a sample of size $n_1 = 130$. At stage 2, we want to test the hypothesis of effectiveness. Note that only $n_2 = 160$ subjects are available instead of the required size $N_2 = 799$. To achieve the desirable statistical assurance, we borrow a sample with size $K = N_2 - n_2 = 639$ using Bayesian approach from previous studies or utilize RWD, which is obtained from adequate and well-controlled clinical studies with the same population. Thus, we have a combined data set from the random study at stage 2 and previous studies for evaluating the hypothesis of effectiveness by ruling out the inconclusiveness.

Specifically, we construct a test statistic $T_2$ as the sample mean based on the combined data at stage 2. Under the normal approximation, $T_2$ follows the following normal distribution:

$$T_2 \sim \left( \hat{p}, \frac{\hat{p}(1-\hat{p})}{N_2} \right), \quad (9.16)$$

where $\hat{p}$ is the observed sample mean from the combined data at stage 2. Based on (9.10), the probability of inconclusiveness $P_I$ is

$$P_I = \Pr\left(T_2 \in \left(p_U, p_L\right)\right) = \Phi\left(\frac{p_L - \hat{p}}{\sqrt{\hat{p}(1-\hat{p})/N_2}}\right) - \Phi\left(\frac{p_U - \hat{p}}{\sqrt{\hat{p}(1-\hat{p})/N_2}}\right), \quad (9.17)$$

where $\Phi(\cdot)$ is the cumulative distribution function of the standard normal distribution. Given a pre-specified threshold $\delta$, we can then decide that the probability of inconclusiveness is negligible if $P_I < \delta$ and conclude the effectiveness; otherwise, the effectiveness is rejected. Note that in practice, the threshold $\delta$ is often selected based on clinical judgment in conjunction with statistical justification for desirable statistical assurance.

## 9.5 Concluding Remarks

As discussed, for rare disease drug development, power analysis for sample size calculation may not be feasible due to the fact that there is small patient population. FDA draft guidance emphasizes that the same standards for regulatory approval will be applied to rare diseases drug development despite the small patient population. Thus, often there is insufficient power for rare disease drug clinical investigation. In this case, alternatively, it is suggested that sample size calculation or justification should be performed based on precision analysis, reproducibility analysis, or probability monitoring approach for achieving certain statistical assurance.

In practice, it is a dilemma for having the same standard with less subjects in rare disease drug development. Thus, it is suggested that innovative design and statistical methods should be considered and implemented for obtaining substantial evidence regarding effectiveness and safety in support of regulatory approval of rare disease drug products. In this chapter, several innovative trial designs such as complete $n$-of-1 trial design, adaptive seamless trial design, trial design utilizing the concept of master protocols, and Bayesian trial design are introduced. The corresponding statistical methods and sample size requirement under respective study designs are derived. These study designs are useful in speeding up rare disease development process and identifying any signal, pattern or trend, and/or optimal clinical benefits of the rare disease drug products under investigation.

Due to the small patient population in rare disease clinical development, the concept of generalizability probability can be used to determine whether the clinical results can be generalized from the targeted patient population (e.g., adults) to a different but similar patient population (e.g., pediatrics or elderly) with the same rare disease. In practice, the generalizability probability can be evaluated through the assessment of sensitivity index between the targeted patient population and the different patient population (Lu et al., 2017). The degree of generalizability probability can then be used to judge whether the intended trial has provided substantial evidence regarding the effectiveness and safety for the different patient population (e.g., pediatrics or elderly).

In practice, although an innovative and yet complex trial design may be useful in rare disease drug development, it may introduce operational bias to the trial and consequently increase the probability of making errors. It is then suggested that quality, validity, and integrity of the intended trial utilizing an innovative trial design should be maintained.

# 10

# The n-of-1 Trial Design and Its Application

## 10.1 Introduction

Biological products as therapeutic agents are made from a variety of living cells or living organisms. The application of biologics ranges from treating disease and medical conditions to preventing and diagnosing diseases (Endrenyi et al., 2017). Under the Biologic Price, Competition, and Innovation (BPCI) Act passed by the United Congress in 2009, the development of biosimilar (test) products compared to an innovative biological (reference) product provides a more affordable alternative to general patient population. As indicated in the BPCI Act, a biosimilar product is defined as a biological product that is highly similar to the reference product notwithstanding minor differences in clinically inactive components, and there are no clinically meaningful differences in terms of safety, purity, and potency. If a proposed biosimilar product has been demonstrated to be highly similar to the reference product, the proposed biosimilar product will be approved by the regulatory authority. An approved biosimilar product can be used as a substitute for the reference product for treating patients with the diseases under study or with the diseases approved for the reference product.

In practice, as more and more biosimilar products become available in the marketplace, interesting questions have been raised. The questions are that whether the approved biosimilar products can be used interchangeable with the reference product, and whether the approved biosimilar products can be used interchangeable with other approved biosimilar products. To address this question, as indicated in the BPCI Act, a valid trial design should allow the evaluation of (i) the risk in terms of safety and diminished efficacy of alternating or switching between the use of the biological product and the use of the reference product, (ii) the risk of using the reference product without such alternation or switch, and (iii) the relative risk between switching/alternation and without switching/alternation. For this purpose, a crossover design is necessarily employed.

FDA (2017a) interprets the risk of switch and alternation as the result of single switch and multiple switches, respectively. Thus, in this chapter, we will focus on design and analysis of switching designs in biosimilar drug development. In its recent draft guidance, FDA recommends a $2x(m + 1)$

crossover design as the switching designs for the assessment of the risk between switching/alternation and without switching/alternation, where $m$ is the number of switches. Alternatively, Chow, Song et al. (2017) suggested a complete $n$-of-1 trial design be considered. As indicated by Chow, Song et al. (2017), FDA recommended $2x(m + 1)$ switching designs are special cases of complete $n$-of-1 trial design with $m$ switches.

In this chapter, relative merits and limitations of FDA-recommended $2 \times (m+1)$ crossover designs for biosimilar switching studies as compared to the complete $n$-of-1 trial design will be examined. Sample size requirement and statistical methods for data analysis under these switching designs will also be studied. In the next section, the concept of interchangeable biosimilar products is briefly outlined. In Section 3, the relative efficiency of the FDA-recommended $2 \times (m+1)$ crossover switching designs as compared to the complete $n$-of-1 trial design is studied. Section 10.4 discusses power calculation and statistical analysis under these switching designs. In the last section, brief concluding remarks are given.

## 10.2 Interchangeable Biosimilar Products

As described in Section 351(k)(4) of the Public Health Services (PHS) Act, the term "interchangeable" or "interchangeability" refers that "the biological product may be substituted for the reference product without the intervention of the health care provider who prescribed the reference product." A proposed biosimilar product is considered *interchangeable* provided that (i) if it is highly similar to the reference product and can be expected to produce the same clinical result as the reference product in any given patient, and (ii) if the biological product is administered more than once to an individual, the risk in terms of safety or diminished efficacy of alternating or switching between the use of the biological product and the use of the reference product is not greater than the risk of using the reference product without such alternation or switching.

**Interpretation of Interchangeability** – Regarding the second part of (i), in practice, it is often difficult (if not impossible) to demonstrate that a proposed interchangeable biosimilar product can produce the same clinical result as the reference product in *any* given patient. In other words, for *every* patient, we need to demonstrate that the proposed interchangeable biosimilar product will produce the same clinical result as the reference product before interchangeability between the interchangeable biosimilar product and the reference product can be claimed. However, statistically, it is possible to demonstrate that the proposed biosimilar product can produce the same clinical result as the reference product in any given patient *with certain assurance*. In this case, the concept of individual bioequivalence

(IBE) for addressing drug switchability proposed by the FDA in early 2000 may be useful (FDA, 2001, 2003). This is because the assessment of IBE will not only evaluate the difference in treatment effect within each individual, but also take into consideration the effect due to subject-by-treatment interaction, which is known to have an impact on drug interchangeability (Chow and Liu, 2008a).

**Selection of Study Population** – In the design of interchangeability study, it is ideal to select a sufficiently sensitive study population for safety. There might be a limitation to design a pivotal phase III biosimilarity study to show interchangeability at the same time since the necessary study population for the phase III pivotal study may not be a sufficiently sensitive patient population for the interchangeability study.

**Switching and Alternation** – Let T and R stand for a proposed biosimilar (test) product and an innovative biological (reference) product, respectively. Most researchers interpret switching as switch from (T to R), (R to T), (R to R) or (T to T) and alternation as switch from (R to T to R), (R to R to R), (T to R to T), or (T to T to T). In other words, switching is generally referred to as switch from one product (R or T) to another (R or T), where T could be a different interchangeable biosimilar product, which has been demonstrated to be highly similar to the same reference product. For the alternation, it may begin with one product (R or T) and after a few switches return to the same product, where T could be a different interchangeable biosimilar product, which has been demonstrated to be highly similar to the same reference product.

In its recent draft guidance on interchangeability, FDA distinguishes the concepts of switching and alternation for interchangeable biosimilar products (FDA, 2017a). Switching is generally referred to as a single switch from one product to another such as (R to T) and (R to R). On the other hand, alternation is referred as multiple switches, such as (R to T to R) and (R to R to R) for two switches and (R to T to R to T) and (R to R to R to R) for three switches.

**Criteria for Risk Evaluation of Switching and Alternation** – For the evaluation of the potential risk of switching and/or alternation based on the assessment of biosimilarity between products, several criteria for the assessment of bioequivalence/biosimilarity are useful. First, one may consider the well-established criteria for the assessment of bioequivalence (Chow and Liu, 2008a) and biosimilarity (Chow, 2013). Alternatively, the scaled average bioequivalence (SABE) criterion proposed by Haidar et al. (2008) for highly variable drug products may be considered as most biosimilar products are considered highly variable drug products. Chow et al. (2015) proposed a new scaled criterion for drug interchangeability (SCDI) based on the criterion for the assessment of IBE by taking the variability due to subject-by-drug interaction into consideration. The SCDI criterion is found to be superior to the classical bioequivalence criterion and SABE in many cases especially when there are notable large variabilities due to the reference product and/or the subject-by-drug interaction.

## 10.3 Switching Designs

As indicated earlier, an interchangeable biosimilar product is expected to produce the same clinical results as the reference product in any given patient with the disease under study. To determine whether the proposed interchangeable biosimilar product can produce the same clinical results in any given patient, a switching design with the nature of crossover within individual subjects is necessarily employed. In what follows, two useful switching designs are described: One is a two-sequence crossover design recommended by the FDA (2017), and the other one is a complete $n$-of-1 trial design proposed by Chow et al. (2017).

### 10.3.1 The $2 \times (m + 1)$ Crossover Design

In its recent guidance on interchangeability, the FDA recommends a $2 \times (m+1)$ crossover design, where $m$ is the number of switches. With single switch, i.e., $m = 1$, FDA recommends a crossover design consisting of two sequences of RT and RR, denoted by (RT, RR), be used. As it can be seen, (RT, RR) design allows the evaluation of the effect of the switch from R to T and the effect of the switch from R to R (i.e., no switch). In addition, the relative risk between the switch from R to T and no switch (i.e., the switch from R to R) can also be assessed. Note that the $2 \times 2$ crossover design with single switch, i.e., (RT, RR), is a partial design of the $4 \times 2$ Balaam design, i.e., (RR, TT, RT, TR).

When $m = 2$ (i.e., there are two switches), FDA suggests a $2 \times 3$ crossover design that consists of the two sequences of RTR and RRR, denoted by (RTR, RRR), be used. The $2 \times 3$ crossover design with two switches allows the evaluation of the effect of the switch from R to T, the effect of the switch from T to R, and the effect of the switch from R to R (i.e., no switch). In addition, the relative risk between the switch from R to T and no switch (i.e., the switch from R to R) and the relative risk between the switch from T to R and no switch can also be assessed.

When $m = 3$ (i.e., there are three switches), FDA suggests a $2 \times 4$ crossover design that consists of the two sequences of RTRT and RRRR, denoted by (RTRT, RRRR), be used. Similar to the $2 \times 3$ crossover design with two switches, the $2 \times 4$ crossover design with three switches allows the evaluation of the effect of the switch from R to T, the effect of the switch from T to R, and the effect of the switch from R to R (i.e., no switch). In addition, the relative risk between the switch from R to T and no switch (i.e., the switch from R to R) and the relative risk between the switch from T to R and no switch can also be assessed.

### 10.3.2 Complete $n$-of-1 Trial Design

In recent years, the $n$-of-1 trial design has become a very popular design for the evaluation of the difference in treatment effect within the same individual when $n$ treatments are administered at different dosing periods. Thus, $n$-of-1 trial design is in fact a crossover design. Following similar ideas of switching designs with single switch and/or multiple switches, Chow et al. (2017) proposed the use of so-called complete n-of-1 trial design for the assessment of relative risk between switching/alternation and without switching/alternation.

The construction of a complete n-of-1 trial design depends upon $m$, the number of switches. For example, if $m = 1$ (single switch), the complete n-of-1 trial design will consist of $m + 1 = 2$ periods. At each dosing period, there are two choices (i.e., either R or T). Thus, there are a total of $2^{m+1} = 2^2 = 4$ sequences (i.e., combinations of R and T). This results in a $4 \times 2$ Balaam design, i.e., (RR, TT, RT, TR). When (two switches), the complete n-of-1 trial design will consist of $m + 1 = 3$ periods. At each dosing period, there are two choices (i.e., either R or T). Thus, there are a total of $2^{m+1} = 2^3 = 8$ sequences. This results in an $8 \times 3$ crossover design. S$m = 2$imilarly, where there are three switches (i.e., $m = 3$), the complete n-of-1 trial design will consist of $m + 1 = 4$ periods. At each dosing period, there are two choices (i.e., either R or T). Thus, there are a total of $2^{m+1} = 2^4 = 16$ sequences (i.e., combinations of R and T). This results in a $16 \times 4$ crossover design. To provide a better understanding, Table 10.1 lists a complete n-of-1 trial design with $m = 1$ (single switch), $m = 2$ (two switches), and $m = 3$ (three switches) that may be useful for biosimilar switching studies.

As it can be seen from Table 10.1, the switching designs with single switch (i.e., (RT, RR)), two switches (i.e., (RTR, RRR)), and three switches

**TABLE 10.1**

Complete n-of-1 Trial Design with $m = 1$, 2, and 3

| Group | Period I | Period II | Period III | Period IV |
|-------|----------|-----------|------------|-----------|
| 1 | R | R | R | R |
| 2 | R | T | R | R |
| 3 | T | T | R | R |
| 4 | T | R | R | R |
| 5 | R | R | T | R |
| 6 | R | T | T | T |
| 7 | T | R | T | R |
| 8 | T | T | T | T |
| 9 | R | R | R | T |
| 10 | R | R | T | T |
| 11 | R | T | R | T |
| 12 | R | T | T | R |
| 13 | T | R | R | T |
| 14 | T | R | T | T |
| 15 | T | T | R | T |
| 16 | T | T | T | R |

*Note:* $m = 1$ (single switch with two periods); $m = 2$ (two switches with three periods); $m = 3$ (three switches with four periods).

(i.e., (RTRT, RRRR)) are partial designs of the $n$-of-1 trial designs with single switch (two periods), two switches (three periods), and three switches (four periods), respectively.

## 10.4 Statistical Model and Analysis

The switching designs discussed in the previous section can be generally described as a $K \times J$ crossover design. For example, for FDA-recommended switching designs with two switches, $K = 2$ and $J = 3$, while for the complete $n$-of-1 trial design with two switches, $K = 8$ and $J = 3$. Thus, the switching designs discussed in the previous section can be described in a statistical model under a general $K \times J$ ($K$-sequence and $J$-period) crossover design comparing two treatments (i.e., R and T).

### 10.4.1 Statistical Model

Let $Y_{ijk}$ be the response of the $i$th subject in the $k$th sequence at the $j$th period. Thus, $Y_{ijk}$ can be described in the following model:

$$Y_{ijk} = \mu + G_k + S_{ik} + P_j + D_{d(j,k)} + C_{d(j-1,k)} + e_{ijk} \tag{10.1}$$

$$i = 1, 2, \ldots, n_k; j = 1, 2, \ldots, J; k = 1, 2, \ldots, K; d = T \text{ or } R,$$

where $\mu$ is the overall mean; $G_k$ is the fixed $k$th sequence effect; $S_{ik}$ is the random effect for the $i$th subject within the $k$th sequence with mean 0 and variance $\sigma_S^2$; $P_j$ is the fixed effect for the $j$th period; $D_{d(j,k)}$ is the drug effect for the drug at the $k$th sequence in the $j$th period; $C_{d(j-1,k)}$ is the carry-over effect, where $C_{d(0,k)} = 0$; and $e_{ijk}$ is the random error with mean 0 and variance $\sigma_e^2$. Under the model, it is assumed that $S_{ik}$ and $e_{ijk}$ are mutually independent.

Under Model (1), denote $\beta$ as a parameter vector, $\left(\mu, G_1, G_2, \ldots, G_K, P_1, P_2, \ldots, P_J, D_T, D_R, C_T, C_R\right)'$, which contains all unknown parameters in the model. Let $X$ be the design matrix of the $K \times J$ crossover design. Thus, $\beta$ can be estimated by $\hat{\beta} = (X'X)^{-1} X' \bar{Y}$, where $\bar{Y}$ is the vector of observed cell means. Thus, statistical inference of the treatment effect after switch can then be assessed simply by the following steps:

Step 1: Set up the design matrix for the $K \times J$ crossover design.

Step 2: Find $(X'X)^{-1} X'$, where $X$ is the $K \times J$ crossover design matrix. We then obtain $\hat{D}_R$ and $\hat{D}_T$.

Step 3: The estimates of $\theta_{ij} = D_i - D_j$ can be obtained by the difference of the corresponding coefficients between $\hat{D}_i$ and $\hat{D}_j$.

Step 4: The estimates of carry-over effects $\lambda_{ij} = C_i - C_j$, $i \neq j$ can be similarly obtained.

## 10.4.2 Analysis of FDA-Recommended Switching Design with Three Switches

As an example, consider FDA-recommended switching design with three switches, i.e., (RTRT, RRRR), which is a $2 \times 4$ crossover design with $K = 2$ and $J = 4$. Under the design, $\beta$ is given by

$$\left(\mu, G_1, G_2, P_1, P_2, P_3, P_4, D_T, D_R, C_T, C_R\right)'$$

$$= \left(\beta_{11}, \beta_{21}, \beta_{31}, \beta_{41}, \beta_{12}, \beta_{22}, \beta_{32}, \beta_{42}\right)',$$

which contains all unknown parameters in the model. Under the model, the design matrix is given by

$$X = \begin{pmatrix} 1 & 1 & 1 & 1 & 1 & 1 & 1 & 1 \\ 1 & 1 & 1 & 1 & 0 & 0 & 0 & 0 \\ 0 & 0 & 0 & 0 & 1 & 1 & 1 & 1 \\ 1 & 0 & 0 & 0 & 1 & 0 & 0 & 0 \\ 0 & 1 & 0 & 0 & 0 & 1 & 0 & 0 \\ 0 & 0 & 1 & 0 & 0 & 0 & 1 & 0 \\ 0 & 0 & 0 & 1 & 0 & 0 & 0 & 1 \\ 0 & 0 & 0 & 0 & 0 & 1 & 0 & 1 \\ 1 & 1 & 1 & 1 & 1 & 0 & 1 & 0 \\ 0 & 0 & 0 & 0 & 0 & 0 & 1 & 0 \\ 0 & 1 & 1 & 1 & 1 & 1 & 0 & 1 \end{pmatrix}, \beta = \begin{pmatrix} \beta_{11} \\ \beta_{21} \\ \beta_{31} \\ \beta_{41} \\ \beta_{12} \\ \beta_{22} \\ \beta_{32} \\ \beta_{42} \end{pmatrix}.$$

The aim is to construct the method of moment estimator in linear form of the observed cell means $\tilde{Y} = \beta'\bar{Y}$, where $\bar{Y}$ is the vector of observed cell means, where

$$\bar{Y} = \left(\bar{Y}_{.11}, \bar{Y}_{.21}, \bar{Y}_{.31}, \bar{Y}_{.41}, \bar{Y}_{.12}, \bar{Y}_{.22}, \bar{Y}_{.32}, \bar{Y}_{.42}\right)', \quad \bar{Y}_{.jk} = \frac{\sum_{i=1}^{n_k} Y_{ijk}}{n_k}.$$

Based on $\beta$s for estimating $D_T - D_R$, we have the following unbiased estimator for drug effect:

$$\tilde{D} = \frac{1}{2}\left[\left(2\bar{Y}_{.11} - \bar{Y}_{.21} - \bar{Y}_{.41}\right) - \left(2\bar{Y}_{.12} - \bar{Y}_{.22} - \bar{Y}_{.42}\right)\right]$$

$$E\left(\tilde{D}\right) = D_T - D_R, \quad \text{Var}\left(\tilde{D}\right) = \frac{3}{2}\sigma_e^2\left(\frac{1}{n_1} + \frac{1}{n_2}\right).$$

When $n_1 = n_2 = n$, $\text{Var}\left(\tilde{D}\right) = \frac{3\sigma_e^2}{n}$. To test the bioequivalence based on the average bioequivalence method,

$$H_0: |D_T - D_R| > \theta \quad \text{versus} \quad H_a: |D_T - D_R| \leq \theta,$$

where $\theta$ is the pre-specified margin. The null hypothesis will be rejected and the bioequivalence will be demonstrated when the statistic

$$T_D = \frac{\tilde{D} - \theta}{\hat{\sigma}_e^2\sqrt{\frac{3}{n}}} > t\left[\frac{\alpha}{2}, n_1 + n_2 - 5\right].$$

The corresponding confidence interval at $\alpha$-significance level is

$$\tilde{D} \pm t\left[\frac{\alpha}{2}, n_1 + n_2 - 5\right]\hat{\sigma}_e^2\sqrt{\frac{3}{2}\left(\frac{1}{n_1} + \frac{1}{n_2}\right)}.$$

## 10.4.3 Analysis of Complete *n*-of-1 Design with Three Switches

Under the general model (10.1), following similar ideas, it can be verified that

$$E\left(\tilde{D}\right) = D_T - D_R, \quad \text{Var}\left(\tilde{D}\right) = \frac{\sigma_e^2}{11n},$$

where $n$ is the number of subjects enrolled in each sequence (assuming equally allocated).

The null hypothesis will be rejected and the bioequivalence will be demonstrated when the statistic

$$T_D = \frac{\tilde{D} - \theta}{\hat{\sigma}_e^2\sqrt{\frac{1}{11n}}} > t\left[\frac{\alpha}{2}, 16n - 5\right].$$

The corresponding confidence interval at $\alpha$-significance level is

$$\tilde{D} \pm t\left[\frac{\alpha}{2}, 16n - 5\right]\hat{\sigma}_e^2\sqrt{\frac{1}{11n}}.$$

Under the model without the first-order carry-over effect, we can derive a different unbiased estimator $\tilde{D}$ such that $E(\tilde{D}) = D_T - D_R$, $\mathrm{Var}(\tilde{D}) = \frac{\sigma_e^2}{12n}$.

### 10.4.4 A Comparison

For a fixed sample size, e.g., $n = 48$, the number of patients that can be randomly assigned to each sequence in RRRR/RTRT design is 24, whereas the number of patients that can be randomly assigned to each sequence in complete $n$-of-1 design with four dosing periods is 3. Correspondingly, the variances of drug effects for (RTRT, RRRR) design and the complete $n$-of-1 design when adjusting for the carry-over effect are $\frac{\sigma_e^2}{8}$ and $\frac{\sigma_e^2}{33}$, respectively. The relative efficiency between two study designs is 24.24%, indicating that the efficiency of (RTRT, RRRR) design is 24.24% of the complete $n$-of-1 trial design. When ignoring the carry-over effect, the variance of drug effects in (RTRT, RRRR) design is $\frac{\sigma_e^2}{12}$, and the variance of drug effects in the complete $n$-of-1 design is $\frac{\sigma_e^2}{36}$. The relative efficiency between two study designs increases to 33.33%.

Therefore, the partial design (RTRT, RRRR) is less efficient than the complete $n$-of-1 design. When the washout is sufficient, the relative efficiency of partial design increases, but is still less than the complete design.

## 10.5 Sample Size Requirement

The sample size determination under the fixed power and significance level is derived based on the following hypothesis testing:

$$H_0: |D_T - D_R| > \theta \quad \text{versus} \quad H_1: |D_T - D_R| \le \theta.$$

According to the ±20% rule, the bioequivalence is concluded if the average bioavailability of the test drug effect is within ±20% of that of the reference drug effect with a certain assurance. Therefore, $\theta$ is usually represented by $\nabla \mu_R$, where $\nabla = 20\%$, and the hypothesis testing can be rewritten as follows:

$$H_0: \ \mu_T - \mu_R \langle -\nabla \mu_R \text{ or } \mu_T - \mu_R \rangle \nabla \mu_R \text{ versus } H_a: \ -\nabla \mu_R \le \mu_T - \mu_R \le \nabla \mu_R.$$

The power function can be written as

$$P(R) = F_v \left( \left[ \frac{\nabla - R}{CV \sqrt{b/n}} \right] - t(\alpha, v) \right) - F_v \left( t(\alpha, v) - \left[ \frac{\nabla + R}{CV \sqrt{b/n}} \right] \right) \text{ (see details in }$$

Appendix),

where $R = \dfrac{\mu_T - \mu_R}{\mu_R}$ is the relative change; $CV = \dfrac{S}{\mu_R}$; $\mu_T$, $\mu_R$ are the average bioavailability of the test and reference formulations, respectively; $S$ is the squared root of the mean square error from the analysis of variance table for each crossover design; $[-\nabla \mu_R, \nabla \mu_R]$ is the bioequivalence limits; $t(\alpha, v)$ is the upper $\alpha$th quantile of a $t$-distribution with $v$ degrees of freedom; $F_v$ is the cumulative distribution function of the $t$-distribution; and $b$ is the constant value for the variance of drug effect.

Accordingly, the exact sample size formula when $R = 0$ is

$$n \ge b \left[ t(\alpha, v) + t \left( \frac{\beta}{2}, v \right) \right]^2 [CV / \nabla]^2;$$

the approximate sample size formula when $R > 0$ is

$$n \ge b \left[ t(\alpha, v) + t(\beta, v) \right]^2 [CV / (\nabla - R)]^2 \text{ (see details in Appendix)}.$$

Setting $\nabla = 0.2$, to achieve the 80% or 90% power under 5% significance level, the required sample sizes for (RTRT, RRRR) design and complete $n$-of-1 design are shown in Tables 10.2 and 10.3.

Another way to determine the sample size is based on testing the ratio of drug effect between the biosimilar product and the reference product. Considering $\delta \in (0.8, 1.25)$ is the bioequivalence range of $\mu_T / \mu_R$, the hypothesis changes to

$$H_0: \frac{\mu_T}{\mu_R} \Big\langle 0.8 \text{ or } \frac{\mu_T}{\mu_R} \Big\rangle 1.25 \text{ versus } H_a: 0.8 \le \frac{\mu_T}{\mu_R} \le 1.25.$$

In case of the skewed distribution, the hypotheses are transformed to the logarithmic scale,

$$H_0: \log \mu_T - \log \mu_R \langle \log(0.8) \text{ or } \log \mu_T - \log \mu_R \rangle \log(1.25) \text{ versus}$$

$$H_a: \log(0.8) \le \log \mu_T - \log \mu_R \le \log(1.25)$$

**TABLE 10.2**

Number of Subjects Required for (RTRT, RRRR) Design ($\nabla$ = 0.2, $\alpha$ = 5%)

| Power (%) | CV (%) | 0% | 5% | R 10% | 15% |
|---|---|---|---|---|---|
| 80 | 10 | 32 | 40 | 80 | 300 |
| | 12 | 44 | 52 | 112 | 432 |
| | 14 | 56 | 68 | 152 | 588 |
| | 16 | 72 | 88 | 196 | 764 |
| | 18 | 88 | 112 | 244 | 968 |
| | 20 | 108 | 136 | 300 | 1,192 |
| | 22 | 128 | 164 | 364 | 1,440 |
| | 24 | 152 | 196 | 432 | 1,716 |
| | 26 | 180 | 228 | 508 | 2,012 |
| | 28 | 208 | 264 | 588 | 2,332 |
| | 30 | 236 | 300 | 672 | 2,676 |
| | 32 | 268 | 344 | 764 | 3,044 |
| | 34 | 304 | 388 | 864 | 3,436 |
| | 36 | 340 | 432 | 968 | 3,852 |
| | 38 | 376 | 480 | 1,076 | 4,292 |
| | 40 | 416 | 532 | 1,192 | 4,752 |
| 90 | 10 | 40 | 52 | 108 | 416 |
| | 12 | 52 | 72 | 152 | 596 |
| | 14 | 68 | 96 | 208 | 812 |
| | 16 | 88 | 124 | 268 | 1,056 |
| | 18 | 112 | 152 | 340 | 1,336 |
| | 20 | 136 | 188 | 416 | 1,648 |
| | 22 | 164 | 228 | 504 | 1,996 |
| | 24 | 192 | 268 | 596 | 2,372 |
| | 26 | 224 | 316 | 700 | 2,784 |
| | 28 | 260 | 364 | 812 | 3,228 |
| | 30 | 300 | 416 | 932 | 3,704 |
| | 32 | 340 | 472 | 1,056 | 4,216 |
| | 34 | 380 | 532 | 1,192 | 4,756 |
| | 36 | 428 | 596 | 1,336 | 5,332 |
| | 38 | 476 | 664 | 1,488 | 5,940 |
| | 40 | 524 | 736 | 1,648 | 6,584 |

**TABLE 10.3**

Number of Subjects Required for the Complete $n$-of-1 Design ($\nabla = 0.2$, $\alpha = 5\%$)

| Power (%) | CV (%) | R | | | |
|---|---|---|---|---|---|
| | | 0% | 5% | 10% | 15% |
| 80 | 10 | 16 | 16 | 16 | 48 |
| | 12 | 16 | 16 | 16 | 64 |
| | 14 | 16 | 16 | 32 | 80 |
| | 16 | 16 | 16 | 32 | 96 |
| | 18 | 16 | 16 | 32 | 128 |
| | 20 | 16 | 32 | 48 | 160 |
| | 22 | 32 | 32 | 48 | 176 |
| | 24 | 32 | 32 | 64 | 224 |
| | 26 | 32 | 32 | 64 | 256 |
| | 28 | 32 | 48 | 80 | 288 |
| | 30 | 32 | 48 | 96 | 336 |
| | 32 | 48 | 48 | 96 | 384 |
| | 34 | 48 | 48 | 112 | 432 |
| | 36 | 48 | 64 | 128 | 480 |
| | 38 | 48 | 64 | 144 | 528 |
| | 40 | 64 | 80 | 160 | 592 |
| 90 | 10 | 16 | 16 | 16 | 64 |
| | 12 | 16 | 16 | 32 | 80 |
| | 14 | 16 | 16 | 32 | 112 |
| | 16 | 16 | 16 | 48 | 144 |
| | 18 | 16 | 32 | 48 | 176 |
| | 20 | 32 | 32 | 64 | 208 |
| | 22 | 32 | 32 | 64 | 256 |
| | 24 | 32 | 48 | 80 | 304 |
| | 26 | 32 | 48 | 96 | 352 |
| | 28 | 48 | 48 | 112 | 400 |
| | 30 | 48 | 64 | 128 | 464 |
| | 32 | 48 | 64 | 144 | 512 |
| | 34 | 48 | 80 | 160 | 592 |
| | 36 | 64 | 80 | 176 | 656 |
| | 38 | 64 | 96 | 192 | 736 |
| | 40 | 80 | 96 | 208 | 800 |

Then, the sample size formulas for different $\delta$ are given below (see details in Appendix),

$$n \geq b\left[t(\alpha, v) + t\left(\frac{\beta}{2}, v\right)\right]^2 [CV / \ln 1.25]^2 \text{ if } \delta = 1$$

$$n \geq b\left[t(\alpha, v) + t(\beta, v)\right]^2 \left[CV / (\ln 1.25 - \ln \delta)\right]^2 \text{ if } 1 < \delta < 1.25$$

$$n \geq b\left[t(\alpha, v) + t(\beta, v)\right]^2 \left[CV / (\ln 0.8 - \ln \delta)\right]^2 \text{ if } 0.8 < \delta < 1.$$

To achieve 80% or 90% power under 5% significance level, the required sample sizes for (RTRT, RRRR) and complete $n$-of-1 designs are calculated in Tables 10.4 and 10.5, respectively.

**TABLE 10.4**

Number of Subjects Required for (RTRT, RRRR) Design – Multiplicative Model with $\alpha = 5\%$ and Similarity Margin of (0.8, 1.25)

| Power (%) | CV (%) | $\delta$ | | | | | | | |
|---|---|---|---|---|---|---|---|---|---|
| | | 0.85 | 0.90 | 0.95 | 1.00 | 1.05 | 1.1 | 1.15 | 1.2 |
| 80 | 10 | 104 | 30 | 16 | 14 | 16 | 26 | 56 | 226 |
| | 12 | 148 | 42 | 22 | 18 | 20 | 36 | 80 | 324 |
| | 14 | 200 | 56 | 28 | 24 | 26 | 48 | 108 | 438 |
| | 16 | 260 | 72 | 34 | 30 | 34 | 60 | 140 | 572 |
| | 18 | 330 | 90 | 44 | 36 | 42 | 76 | 176 | 724 |
| | 20 | 406 | 110 | 52 | 44 | 52 | 94 | 216 | 892 |
| | 22 | 492 | 132 | 64 | 52 | 62 | 112 | 260 | 1,080 |
| | 24 | 584 | 156 | 74 | 62 | 72 | 134 | 310 | 1,284 |
| | 26 | 684 | 184 | 88 | 72 | 86 | 156 | 364 | 1,508 |
| | 28 | 794 | 212 | 102 | 84 | 98 | 180 | 420 | 1,748 |
| | 30 | 910 | 244 | 116 | 96 | 112 | 206 | 482 | 2,006 |
| | 32 | 1,036 | 276 | 132 | 108 | 128 | 234 | 548 | 2,282 |
| | 34 | 1,170 | 312 | 148 | 122 | 144 | 264 | 620 | 2,576 |
| | 36 | 1,310 | 350 | 166 | 136 | 160 | 296 | 694 | 2,888 |
| | 38 | 1,460 | 388 | 184 | 152 | 178 | 330 | 772 | 3,216 |
| | 40 | 1,618 | 430 | 204 | 168 | 198 | 366 | 856 | 3,564 |
| 90 | 10 | 142 | 40 | 20 | 16 | 20 | 34 | 76 | 310 |
| | 12 | 204 | 56 | 28 | 22 | 28 | 48 | 110 | 446 |
| | 14 | 276 | 76 | 36 | 28 | 36 | 64 | 148 | 606 |

*(Continued)*

**TABLE 10.4 (*Continued*)**

Number of Subjects Required for (RTRT, RRRR) Design – Multiplicative Model with $\alpha = 5\%$ and Similarity Margin of (0.8, 1.25)

| Power (%) | CV (%) | 0.85 | 0.90 | 0.95 | 1.00 | 1.05 | 1.1 | 1.15 | 1.2 |
|---|---|---|---|---|---|---|---|---|---|
| | 16 | 360 | 98 | 48 | 36 | 46 | 84 | 192 | 792 |
| | 18 | 456 | 122 | 60 | 46 | 58 | 104 | 242 | 1,002 |
| | 20 | 562 | 150 | 72 | 56 | 70 | 128 | 298 | 1,236 |
| | 22 | 680 | 182 | 86 | 66 | 84 | 154 | 360 | 1,494 |
| | 24 | 808 | 216 | 102 | 78 | 100 | 184 | 428 | 1,778 |
| | 26 | 948 | 254 | 120 | 92 | 116 | 216 | 502 | 2,088 |
| | 28 | 1,098 | 294 | 140 | 106 | 136 | 250 | 582 | 2,420 |
| | 30 | 1,260 | 336 | 160 | 120 | 154 | 286 | 668 | 2,778 |
| | 32 | 1,434 | 382 | 180 | 136 | 176 | 324 | 760 | 3,160 |
| | 34 | 1,618 | 430 | 204 | 154 | 198 | 366 | 856 | 3,568 |
| | 36 | 1,814 | 482 | 228 | 172 | 222 | 410 | 960 | 3,998 |
| | 38 | 2,022 | 538 | 254 | 192 | 246 | 456 | 1,070 | 4,456 |
| | 40 | 2,240 | 596 | 282 | 212 | 274 | 506 | 1,186 | 4,936 |

**TABLE 10.5**

Number of Subjects Required for the Complete *n*-of-1 – Multiplicative Model with $\alpha = 5\%$ and Similarity Margin of (0.8, 1.25)

| Power (%) | CV (%) | 0.85 | 0.90 | 0.95 | 1.00 | 1.05 | 1.1 | 1.15 | 1.2 |
|---|---|---|---|---|---|---|---|---|---|
| 80 | 10 | 32 | 16 | 16 | 16 | 16 | 16 | 16 | 64 |
| | 12 | 48 | 16 | 16 | 16 | 16 | 16 | 32 | 80 |
| | 14 | 64 | 16 | 16 | 16 | 16 | 16 | 32 | 112 |
| | 16 | 80 | 32 | 16 | 16 | 16 | 16 | 48 | 144 |
| | 18 | 96 | 32 | 16 | 16 | 16 | 32 | 48 | 192 |
| | 20 | 112 | 32 | 16 | 16 | 16 | 32 | 64 | 224 |
| | 22 | 128 | 48 | 32 | 16 | 16 | 32 | 64 | 272 |
| | 24 | 144 | 48 | 32 | 32 | 32 | 48 | 80 | 320 |
| | 26 | 176 | 48 | 32 | 32 | 32 | 48 | 96 | 368 |
| | 28 | 208 | 64 | 32 | 32 | 32 | 48 | 112 | 432 |
| | 30 | 224 | 64 | 32 | 32 | 32 | 64 | 128 | 496 |
| | 32 | 256 | 80 | 48 | 32 | 32 | 64 | 144 | 560 |
| | 34 | 288 | 80 | 48 | 32 | 48 | 80 | 160 | 640 |

*(Continued)*

**TABLE 10.5 (*Continued*)**

Number of Subjects Required for the Complete *n*-of-1 – Multiplicative Model with $\alpha = 5\%$ and Similarity Margin of (0.8, 1.25)

| Power (%) | CV (%) | 0.85 | 0.90 | 0.95 | 1.00 | 1.05 | 1.1 | 1.15 | 1.2 |
|---|---|---|---|---|---|---|---|---|---|
| | | | | | $\delta$ | | | | |
| | 36 | 320 | 96 | 48 | 48 | 48 | 80 | 176 | 704 |
| | 38 | 368 | 96 | 48 | 48 | 48 | 96 | 192 | 784 |
| | 40 | 400 | 112 | 64 | 48 | 64 | 96 | 224 | 880 |
| 90 | 10 | 48 | 16 | 16 | 16 | 16 | 16 | 32 | 80 |
| | 12 | 64 | 16 | 16 | 16 | 16 | 16 | 32 | 112 |
| | 14 | 80 | 32 | 16 | 16 | 16 | 16 | 48 | 160 |
| | 16 | 96 | 32 | 16 | 16 | 16 | 32 | 48 | 208 |
| | 18 | 112 | 32 | 16 | 16 | 16 | 32 | 64 | 256 |
| | 20 | 144 | 48 | 32 | 16 | 32 | 32 | 80 | 304 |
| | 22 | 176 | 48 | 32 | 32 | 32 | 48 | 96 | 368 |
| | 24 | 208 | 64 | 32 | 32 | 32 | 48 | 112 | 448 |
| | 26 | 240 | 64 | 32 | 32 | 32 | 64 | 128 | 512 |
| | 28 | 272 | 80 | 48 | 32 | 48 | 64 | 144 | 592 |
| | 30 | 320 | 96 | 48 | 32 | 48 | 80 | 176 | 688 |
| | 32 | 352 | 96 | 48 | 48 | 48 | 80 | 192 | 768 |
| | 34 | 400 | 112 | 64 | 48 | 64 | 96 | 224 | 880 |
| | 36 | 448 | 128 | 64 | 48 | 64 | 112 | 240 | 976 |
| | 38 | 496 | 144 | 64 | 48 | 64 | 112 | 272 | 1,088 |
| | 40 | 544 | 160 | 80 | 64 | 80 | 128 | 304 | 1,200 |

To compare the sample sizes between two designs, for instance, we have a summary table (see Table 10.6) of the number of subjects for the additive models between (RTRT, RRRR) design and complete *n*-of-1 trial design with 80% power, CV = 20%, and $\theta = 5\%$ and 10%.

Table 10.7 summarizes the number of subjects for the multiplicative models between (RTRT, RRRR) design and complete *n*-of-1 trial design with 80% power, CV = 20%, and $\delta = 0.90$ and 1.00.

**TABLE 10.6**

Summary for Additive Model

| | R | |
|---|---|---|
| | 5% | 10% |
| (RTRT, RRRR) | 136 | 300 |
| Complete *n*-of-1 design | 32 | 48 |

**TABLE 10.7**

Summary for Multiplicative Model

|  | $\delta$ | |
| --- | --- | --- |
|  | 0.90 | 1.00 |
| (RTRT, RRRR) | 110 | 44 |
| Complete $n$-of-1 design | 32 | 16 |

## 10.6 Concluding Remarks

The drug interchangeability under switching and alternation can be assessed in both 2 × 4 parallel crossover design (RTRT, RRRR) and the complete $n$-of-1 design with three switches (four dosing periods). (RTRT, RRRR) design, as a partial design of the complete $n$-of-1 design, is suitable for evaluating the switch from R to R, switch from R to T, and switch from T to T. The analysis for assessing alternations of R to R to R, R to T to R, and T to R to T can also be conducted under this design. However, different from the partial design, the complete $n$-of-1 design brings a broader framework for switching and alternation. With 16 sequences, outcomes under all possible switches and alternations can be utilized to demonstrate the drug interchangeability. Compared to (RTRT, RRRR) design, the complete $n$-of-1 design contains more information for testing and reference drug, which provides the opportunity to comprehensively assess the drug biosimilarity and interchangeability.

The (RTRT, RRRR) design is less efficient than the complete $n$-of-1 design, in terms of the variance of drug effect and relative efficiency. Adjusting for the first-order carry-over effect that may exist during the clinical trial, the variance of the difference between the testing drug effect and reference drug effect in (RTRT, RRRR) design is greater than that in the complete $n$-of-1 design. The relative efficiency in the partial design is 24.24% of that in the complete design (for a fixed sample size $n = 48$). When the washout period is long enough so that the carry-over effect can be ignored, the complete $n$-of-1 design yields a smaller variance of drug effect and higher efficiency in the assessment of drug interchangeability. To achieve the expected power for evaluating the biosimilarity/interchangeability under two common hypotheses, the partial design requires a larger sample size than the complete design. Because of the smaller sample size required in the complete $n$-of-1 design, conducting the drug trial under the complete design may save money and time, and be more efficient than the partial design. However, too many sequences in the complete design will cause too few subjects to be allocated to each sequence. In case of the withdrawal of subjects, the conduction of complete design may face missing data issue, which will affect the evaluations more than that in the partial design.

In this chapter, only one testing drug and one reference drug are involved in the evaluation for both designs. In practice, however, the testing drug and the reference drug may be different. In the situation where more than one testing drug and reference drug is included, the above analysis can be generalized to evaluate the drug interchangeability but is more complicated. The complete *n*-of-1 design is expected to perform better in terms of the relative efficiency and required less sample size than the partial design since in the situation of more than two drugs.

## Appendix

### Hypothesis Based on Additive Effect

$$H_0: \mu_T - \mu_R \langle -\nabla \mu_R \text{ or } \mu_T - \mu_R \rangle \nabla \mu_R \text{ versus } H_a: -\nabla \mu_R \le \mu_T - \mu_R \le \nabla \mu_R$$

Since Power $= P(\text{Reject } H_0 | H_a \text{ is true})$, the power function can be derived as

$$P\left[ \frac{-(\mu_T - \mu_R) - \nabla \mu_R}{\text{sd}(\bar{Y}_T - \bar{Y}_R)} + t(\alpha, v) \le \frac{\bar{Y}_T - \bar{Y}_R - (\mu_T - \mu_R)}{\text{sd}(\bar{Y}_T - \bar{Y}_R)} \le \frac{-(\mu_T - \mu_R) + \nabla \mu_R}{\text{sd}(\bar{Y}_T - \bar{Y}_R)} \right.$$

$$\left. - t(\alpha, v) \mid -\nabla \mu_R \le \mu_T - \mu_R \le \nabla \mu_R \right]$$

$$= P\left[ \frac{\bar{Y}_T - \bar{Y}_R - (\mu_T - \mu_R)}{\text{sd}(\bar{Y}_T - \bar{Y}_R)} \le \frac{-(\mu_T - \mu_R) + \nabla \mu_R}{\text{sd}(\bar{Y}_T - \bar{Y}_R)} - t(\alpha, v) \mid -\nabla \mu_R \le \mu_T - \mu_R \le \nabla \mu_R \right]$$

$$- P\left[ \frac{\bar{Y}_T - \bar{Y}_R - (\mu_T - \mu_R)}{\text{sd}(\bar{Y}_T - \bar{Y}_R)} \le \frac{-(\mu_T - \mu_R) + \nabla \mu_R}{\text{sd}(\bar{Y}_T - \bar{Y}_R)} - t(\alpha, v) \mid -\nabla \mu_R \le \mu_T - \mu_R \le \nabla \mu_R \right]$$

$$= F_v\left[ \left( \nabla - \frac{\mu_T - \mu_R}{\mu_R} \right) \middle/ \frac{S}{\mu_R} \sqrt{\frac{b}{n}} - t(\alpha, v) \right] - F_v\left[ \left( -\nabla - \frac{\mu_T - \mu_R}{\mu_R} \right) \middle/ \frac{S}{\mu_R} \sqrt{\frac{b}{n}} + t(\alpha, v) \right]$$

$$= F_v\left[ \left[ \frac{\nabla - R}{\text{CV} \sqrt{\dfrac{b}{n}}} \right] - t(\alpha, v) \right] - F_v\left[ t(\alpha, v) - \left[ \frac{\nabla + R}{\text{CV} \sqrt{\dfrac{b}{n}}} \right] \right].$$

Thus, for sample size formula, we have
When $R = 0$,

$$P\left[\frac{-\nabla}{CV\sqrt{\frac{b}{n}}} + t(\alpha, v) < \frac{(\bar{Y}_T - \bar{Y}_R)/\mu_R}{CV\sqrt{\frac{b}{n}}} < \frac{\nabla}{CV\sqrt{\frac{b}{n}}} - t(\alpha, v)\right] \geq 1 - \beta.$$

Since a central $t$-distribution is symmetric about 0, the lower and upper endpoints are also symmetric about 0; thus,

$$\frac{\nabla}{CV\sqrt{\frac{b}{n}}} - t(\alpha, v) = -\left[\frac{-\nabla}{CV\sqrt{\frac{b}{n}}} + t(\alpha, v)\right].$$

Let $t(\alpha, v)$ be the upper $\alpha$-quantile of the t-distribution and $t\left(\frac{\beta}{2}, v\right)$ is the upper $\frac{\beta}{2}$ quantile of the $t$-distribution, then

$$\left|\frac{\nabla}{CV\sqrt{\frac{b}{n}}} - t(\alpha, v)\right| \geq t\left(\frac{\beta}{2}, v\right)$$

$$\Rightarrow n \geq b\left[t(\alpha, v) + t\left(\frac{\beta}{2}, v\right)\right]^2 [CV/\nabla]^2.$$

When $R > 0$,

$$P\left[\frac{-\nabla - R}{CV\sqrt{\frac{b}{n}}} + t(\alpha, v) < \frac{(\bar{Y}_T - \bar{Y}_R)/\mu_R - R}{CV\sqrt{\frac{b}{n}}} < \frac{\nabla - R}{CV\sqrt{\frac{b}{n}}} - t(\alpha, v)\right] \geq 1 - \beta.$$

In this case, the lower and upper endpoints are not symmetric about 0 because

$$-\left[\frac{\nabla - R}{CV\sqrt{\frac{b}{n}}} - t(\alpha, v)\right] = \frac{-\nabla + R}{CV\sqrt{\frac{b}{n}}} + t(\alpha, v) > \frac{-\nabla - R}{CV\sqrt{\frac{b}{n}}} + t(\alpha, v).$$

Thus, to derive the sample size in a less conservative way, the power function can be written as

$$P\left[\frac{(\bar{Y}_T - \bar{Y}_R)/\mu_R - R}{CV\sqrt{\dfrac{b}{n}}} < \frac{\nabla - R}{CV\sqrt{\dfrac{b}{n}}} - t(\alpha, \upsilon)\right] \geq 1 - \beta$$

$$\frac{\nabla - R}{CV\sqrt{\dfrac{b}{n}}} - t(\alpha, \upsilon) \geq t(\beta, \upsilon)$$

$$\Rightarrow n \geq b\left[t(\alpha, \upsilon) + t(\beta, \upsilon)\right]^2 \left[CV/(\nabla - R)\right]^2.$$

## Hypothesis Based on Multiplicative Effect

$$H_0\colon \log\mu_T - \log\mu_R \langle \log(0.8) \text{ or } \log\mu_T - \log\mu_R \rangle \log(1.25) \text{ versus}$$

$$H_a\colon \log(0.8) \leq \log\mu_T - \log\mu_R \leq \log(1.25).$$

Similarly, the power function according to the hypothesis testing can be written as

$$F_\upsilon\left(\left[\frac{\log(1.25) - \log(\mu_T/\mu_R)}{CV\sqrt{\dfrac{b}{n}}}\right] - t(\alpha, \upsilon)\right) - F_\upsilon\left(t(\alpha, \upsilon) - \left[\frac{\log(0.8) - \log(\mu_T/\mu_R)}{CV\sqrt{\dfrac{b}{n}}}\right]\right).$$

Denote $\delta = \dfrac{\mu_T}{\mu_R}$, when $\delta = 1$,

$$P\left[\frac{\log(0.8)}{CV\sqrt{\dfrac{b}{n}}} + t(\alpha, \upsilon) < \frac{\log\bar{Y}_T - \log\bar{Y}_R}{CV\sqrt{\dfrac{b}{n}}} < \frac{\log(1.25)}{CV\sqrt{\dfrac{b}{n}}} - t(\alpha, \upsilon)\right] \geq 1 - \beta.$$

In this situation, the lower and upper endpoints are not symmetric about 0, so applying a less conservative method, the sample size can be derived based on the following procedure:

$$\frac{\log(1.25)}{CV\sqrt{\dfrac{b}{n}}} - t(\alpha, \upsilon) \approx -\left[\frac{\log(0.8)}{CV\sqrt{\dfrac{b}{n}}} + t(\alpha, \upsilon)\right].$$

Then,

$$\left| \frac{\log(1.25)}{\text{CV}\sqrt{\dfrac{b}{n}}} - t(\alpha, v) \right| \geq t\left( \frac{\beta}{2}, v \right)$$

$$\Rightarrow n \geq b\left[ t(\alpha, v) + t\left( \frac{\beta}{2}, v \right) \right]^2 \left[ \text{CV} / \ln 1.25 \right]^2.$$

When $1 < \delta < 1.25$,

$$P\left[ \frac{(\log \bar{Y}_T - \log \bar{Y}_R) - (\log \delta)}{\text{CV}\sqrt{\dfrac{b}{n}}} < \frac{\log(1.25) - \log \delta}{\text{CV}\sqrt{\dfrac{b}{n}}} - t(\alpha, v) \right] \geq 1 - \beta$$

$$\frac{\log(1.25) - \log \delta}{\text{CV}\sqrt{\dfrac{b}{n}}} - t(\alpha, v) \geq t(\beta, v)$$

$$\Rightarrow n \geq b\left[ t(\alpha, v) + t(\beta, v) \right]^2 \left[ \text{CV} / (\ln 1.25 - \ln \delta) \right]^2.$$

When $0.8 < \delta < 1$,

$$P\left[ \frac{(\log \bar{Y}_T - \log \bar{Y}_R) - \log \delta}{\text{CV}\sqrt{\dfrac{b}{n}}} > \frac{\log(0.8) - \log \delta}{\text{CV}\sqrt{\dfrac{b}{n}}} + t(\alpha, v) \right] \geq 1 - \beta$$

$$\Rightarrow n \geq b\left[ t(\alpha, v) + t(\beta, v) \right]^2 \left[ \text{CV} / (\ln 0.8 - \ln \delta) \right]^2.$$

# 11

## Two-Stage Adaptive Seamless Trial Design

### 11.1 Introduction

In the past decade, adaptive design methods in clinical research have attracted much attention because they offer the principal investigators not only the flexibility for identifying clinical benefit of a test treatment under investigation, but also efficiency for speeding up the development process. The FDA adaptive design draft guidance defines an adaptive design as a clinical study that includes a prospectively planned opportunity for the modification of one or more specified aspects of the study design and hypotheses based on the analysis of data (usually interim data) from subjects in the study (FDA, 2010, 2019c). As it is recognized by many investigators/researchers, the use of adaptive design methods in clinical trials may allow the researchers to correct assumptions used at the planning stage and select the most promising option early. In addition, adaptive designs make the use of cumulative information of the ongoing trial, which provide the investigator an opportunity to react earlier to surprises regardless of positive or negative results. Thus, the adaptive design approaches may increase the probability of success of the drug development.

Despite the possible benefits for having a second chance to modify the trial at interim when utilizing an adaptive design, it can be more problematic operationally due to bias that may have introduced to the conduct of the trial. As indicated by the FDA draft guidance, operational biases may occur when adaptations in trial and/or statistical procedures are applied after the review of interim (unblinded) data. As a result, it is a concern whether scientific integrity and validity of trial are warranted. Chow and Chang (2011) indicated that trial procedures include, but are not limited to, inclusion/exclusion criteria, dose/dose regimen and treatment duration, endpoint selection and assessment, and/or laboratory testing procedures employed. On the other hand, statistical procedures are referred to as study design, statistical hypotheses (which can reflect study objectives), endpoint selection, power analysis for sample size calculation, sample size re-estimation, and/or sample size adjustment, randomization schedules, and statistical analysis plan (SAP). With respect to these trials and statistical procedures, commonly employed adaptations at interim include, but are not limited to, sample size re-estimation at interim analysis;

adaptive randomization with unequal treatment allocation (e.g., change from 1:1 ratio to 2:1 ratio); deleting, adding, or modifying treatment arms after the review of interim data; shifting in patient population due to protocol amendment, different statistical methods; changing study endpoints (e.g., change response rate and/or survival to time-to-disease progression in cancer trials); and changing hypotheses/objectives (e.g., switch a superiority hypothesis to a non-inferiority hypothesis). Therefore, the use of the adaptive design methods in clinical trials seems promising because of their potential flexibility for identifying any possible clinical benefit, signal, and/or trend regarding efficacy and safety of the test treatment under investigation. However, major adaptations may have an impact on the integrity and validity of the clinical trials, which may raise some critical concerns to the accurate and reliable evaluation of the test treatment under investigation. These concerns include (i) the control of the overall type I error rate at a pre-specified level of significance, (ii) the correctness of the obtained $p$-values, and (iii) the reliability of the obtained confidence interval. Most importantly, major (significant) adaptations may have resulted in a totally different trial that is unable to address the scientific/medical questions the original study intended to answer.

As indicated by Chow (2011), a seamless trial design is defined as a trial design that combines two independent trials into a single study that can addresses study objectives from individual studies. An adaptive seamless design is referred to as a seamless trial design that would use data collected before and after the adaptation in the final analysis. In practice, a two-stage seamless adaptive design typically consists of two stages (phases): a learning (or exploratory) phase (stage 1) and a confirmatory phase (stage 2). The objective of the learning phase is not only to obtain information regarding the uncertainty of the test treatment under investigation but also to provide the investigator the opportunity to stop the trial early due to safety and/or futility/efficacy based on accrued data or to apply some adaptations like adaptive randomization at the end of stage 1. The objective of the second stage is to confirm the findings observed from the first stage. A two-stage seamless adaptive trial design has the following advantages: (i) It may reduce lead time between studies (the traditional approach) and (ii) it provides the investigator the second chance to redesign the trial after the review of accumulated date at the end of stage 1. Most importantly, data collected from both stages are combined for a final analysis in order to fully utilize all data collected from the trial for a more accurate and reliable assessment of the test treatment under investigation.

As indicated in Chow and Tu (2008) and Chow (2011), in practice, two-stage seamless adaptive trial designs can be classified into the following four categories depending upon study objectives and study endpoints at different stages.

Table 11.1 indicates that there are four different types of two-stage seamless adaptive designs depending upon whether study objectives and/or study endpoints at different stages are the same. For example, Category I designs (i.e., SS designs) include those designs with the same study objectives and

**TABLE 11.1**

Types of Two-Stage Seamless Adaptive Designs

| Study Objectives | Study<br>Same (S) | Endpoint<br>Different (D) |
|---|---|---|
| Same (S) | I = SS | II = SD |
| Different (D) | III = DS | IV = DD |

*Source:* Chow (2011).

same study endpoints, while Category II and Category III designs (i.e., SD and DS designs) are referred to those designs with the same study objectives but different study endpoints and different study objectives but the same study endpoints, respectively. Category IV designs (i.e., DD designs) are the study designs with different study objectives and different study endpoints. In practice, different study objectives could be treatment selection for stage 1 and efficacy confirmation for stage 2. On the other hand, different study endpoints could be biomarker, surrogate endpoints, or a clinical endpoint with a shorter duration at the first stage versus clinical endpoint at the second stage. Note that a group sequential design with one planned interim analysis is often considered an SS design.

In practice, typical examples for a two-stage adaptive seamless design include a two-stage adaptive seamless phase I/II design and a two-stage adaptive seamless phase II/III design. For the two-stage adaptive seamless phase I/II design, the objective at the first stage may be for biomarker development and the study objective for the second stage is usually to establish early efficacy. For a two-stage adaptive seamless phase II/III design, the study objective is often for treatment selection (or dose finding), while the study objective at the second stage is for efficacy confirmation. In this chapter, our focus will be placed on Category II designs. The results can be similarly applied to Category III and Category IV designs.

It should be noted that the terms "seamless" and "phase II/III" were not used in the FDA draft guidance as they have sometimes been adopted to describe various design features (FDA, 2010, 2019c). In this chapter, a two-stage adaptive seamless phase II/III design only refers to a study containing stage 1 (an exploratory phase for phase II trial) and stage 2 (a confirmatory phase for phase III study), while data collected at both stages (phases) will be used for the final analysis.

One of the questions that are commonly asked when applying a two-stage adaptive seamless design in clinical trials is sample size calculation/allocation. For the first kind (i.e. Category I, SS) of two-stage seamless designs, the methods based on individual *p*-values as described in Chow and Chang (2011) can be applied. However, for other kinds (i.e., Category II to Category IV) of two-stage seamless trial designs, standard statistical methods for group sequential design are not appropriate and hence should not be applied directly.

For Category II–IV trial designs, power analysis and/or statistical methods for data analysis are challenging to the biostatistician. For example, a commonly asked question is "How do we control the overall type I error rate at a pre-specified level of significance?" In the interest of stopping trial early, "How to determine stopping boundaries?" is a challenge to the investigator and the biostatistician. In practice, it is often of interest to determine whether the typical O'Brien–Fleming type of boundaries is feasible. Another challenge is "How to perform a valid analysis that combines data collected from different stages?" To address these questions, Chow (2011) discussed the concept of a multiple-stage transitional seamless adaptive design, which takes into consideration of different study objectives and study endpoints.

## 11.2 Properties of Two-Stage Adaptive Design

As compared to the traditional approach (i.e., having two separate studies), a two-stage seamless adaptive design is preferred in terms of controlling type I error rate and power. For comparison of controlling the overall type I error rate, consider a two-stage adaptive trial design that combines a phase II trial and a phase III study. Let $\alpha_{II}$ and $\alpha_{III}$ be the corresponding type I error rate for the phase II trial and the phase III study, respectively. Thus, for the traditional approach, the overall type I error rate is given by $\alpha = \alpha_{II}\alpha_{III}$. In the two-stage adaptive seamless phase II/III design, on the other hand, the actual desired alpha is given by $\alpha = \alpha_{III}$. Thus, as compared to the traditional approach, the $\alpha$ for a two-stage adaptive phase II/III design is actually $1 / \alpha_{II}$ times larger. Similarly, let $Power_{II}$ and $Power_{III}$ be the power for the phase II trial and the phase III study, respectively. Then, the power for the traditional approach is $Power = Power_{II} * Power_{III}$. In the two-stage phase II/III adaptive design, the power is given by $Power = Power_{III}$. Thus, as compared to the traditional approach, the power for a two-stage phase II/III adaptive design is $1 / Power_{II}$ times larger.

A two-stage seamless adaptive trial design has the following advantages: First, it may help in reducing lead time between studies for the traditional approach. In practice, the lead time between the end of the phase II trial and kick-off the phase III study is estimated about 6–12 months. This is because that usually the phase III study will not be initiated until the final clinical report of the phase II trial is completed. After the completion of a clinical study, it will usually take about 4–6 months to clean and lock the database, programming and data analysis, and final report. Besides, before we kick off the phase III trial, protocol development, site selection/initiation, and IRB review/approval will also take some time. Thus, the use of a two-stage phase II/III adaptive trial design will definitely reduce the lead time between studies. In addition, the nature of adaptive trial design will also allow the investigator to make a go/no-go decision early (i.e., at the end of the first stage). In terms of the sample size required, a two-stage phase II/III adaptive design may require a smaller sample size as compared to the traditional approach. Most importantly, a two-stage phase II/III adaptive trial design allows us to fully utilize data collected from both stages for a combined analysis, which will provide a more accurate and reliable assessment of the test treatment under investigation.

In what follows, an overview of statistical methods for analysis of different types (i.e. Category I–IV) of two-stage designs is provided (see also Chow and Lin, 2015). In addition, a case study concerning the evaluation of a test treatment for treating patients with hepatitis C infection of a clinical study utilizing a Category IV adaptive design is presented.

## 11.3 Analysis for Category I Adaptive Designs

Category I design with the same study objectives and the same study endpoints at different stages is considered similar to a typical group sequential design with one planned interim analysis. Thus, standard statistical methods for group sequential design are often employed. It, however, should be noted that with various adaptations that are applied, these standard statistical methods may not be appropriate. In practice, many interesting methods for Category I designs are available in the literature. These methods include (i) Fisher's criterion for combining independent $p$-values (Bauer and Kohne, 1994; Bauer and Rohmel, 1995; Posch and Bauer, 2000), (ii) weighted test statistics (Cui et al., 1999), (iii) the conditional error function approach (Liu and Chi, 2001; Proschan and Hunsberger, 1995), and (iv) conditional power approaches (Li et al., 2005).

Among these methods, Fisher's method for combining $p$-values provides great flexibility in selecting statistical tests for individual hypotheses based on sub-samples. Fisher's method, however, lacks flexibility in the choice of boundaries (Muller and Schafer 2001). For Category I adaptive designs, many related issues have been studied. For example, Rosenberger and Lachin (2003) explored the potential use of response-adaptive randomization. Chow et al. (2005) examined the impact of population shift due to protocol amendments. Li et al. (2005) studied a two-stage adaptive design with a survival endpoint, while Hommel et al. (2005) studied a two-stage adaptive design with correlated data. An adaptive design with a bivariate endpoint was studied by Todd (2003). Tsiatis and Mehta (2003) showed that there exists a more powerful group sequential design for any adaptive design with the sample size adjustment.

For illustration purpose, in what follows, we will introduce the method based on the sum of $p$-values (MSP) by Chang (2007) and Chow and Chang (2011). The MSP follows the idea of considering a linear combination of the $p$-values from different stages.

### 11.3.1 Theoretical Framework

Similar to statistical methods described in Section 2.4 of Chapter 2, consider a clinical trial with $K$-stage design. The objective of the trial can be formulated as the following intersection of the individual hypothesis tests from the interim analyses:

$$H_0: H_{01} \cap \ldots \cap H_{0K}, \tag{11.1}$$

where $H_{0i}, i = 1, \ldots, K$ is the null hypothesis to be tested at the $i$th interim analysis. Note that there are some restrictions on $H_{0i}$; i.e., rejection of any $H_{0i}, i = 1, \ldots, K$ will lead to the same clinical implication (e.g., drug is efficacious); hence, all $H_{0i}, i = 1, \ldots, K$ are constructed for testing the same endpoint within a trial. Otherwise, the global hypothesis cannot be interpreted.

In practice, the hypotheses of interest to be tested can be written as

$$H_{0i}: \eta_{i1} \geq \eta_{i2} \text{ versus } H_{ai}: \eta_{i1} < \eta_{i2}, \tag{11.2}$$

where $\eta_{i1}$ and $\eta_{i2}$ are the responses of the two treatment groups at the $i$th stage. Under $H_0$, when $\eta_{i1} = \eta_{i2}$, the $p$-value $p_i$ for the sub-sample at the $i$th stage is assumed uniformly distributed on [0, 1]. Under the null hypothesis, Chang (2007) proposed the following linear combination of the $p$-values:

$$T_k = \sum_{i=1}^{K} w_{ki} p_i, i = 1, \ldots, K, \tag{11.3}$$

where $w_{ki} > 0$ and $K$ is the number of interim analyses planned. When $w_{ki} = 1$, (11.1) becomes

$$T_k = \sum_{i=1}^{K} p_i, i = 1, \ldots, K. \tag{11.4}$$

It can be verified that the error rate at the $k$th stage is given by

$$\pi_k = \psi_k(\alpha_k). \tag{11.5}$$

Consequently, the experiment-wise type I error rate is the sum of $\pi_k, k = 1, \ldots, K$, can be obtained as

$$\alpha = \sum_{k=1}^{K} \pi_k. \tag{11.6}$$

Note that stopping boundaries can be determined with appropriate choices of $\alpha_k$ (see also Section 2.4). The adjusted $p$-value calculation is the same as the one in a classic group sequential design (Jennison and Turnbull, 2006). The key idea is that when the test statistic at the $k$th stage $T_k = t = \alpha_k$ (i.e., just on the efficacy stopping boundary), the $p$-value is equal to alpha spent $\sum_{i=1}^{k} \pi_i$. This is true regardless of which error spending function is used and consistent with the $p$-value definition of the traditional design. As indicated in Chang (2007), the adjusted $p$-value corresponding to an observed test statistic $T_k = t$ at the $k$th stage can be defined as

$$p(t;k) = \sum_{i=1}^{k-1} \pi_i + \psi_k(t), \quad k = 1,..,K. \tag{11.7}$$

Note that $p_i$ is the stage-wise (unadjusted) $p$-value from a sub-sample at the $i$th stage, while $p(t;k)$ are the adjusted $p$-values calculated from the test statistic, which are based on the cumulative sample up to the $k$th stage where the trial stops, equations (11.6) and (11.7) are valid regardless of how $p_i$ are calculated.

### 11.3.2 Two-Stage Design

In this section, for simplicity, we will consider the MSP and apply the general framework to the two-stage designs as outlined in Chang (2007) and Chow and Chang (2011), which are suitable for the following adaptive designs that allow (1) early efficacy stopping, (2) early stopping for both efficacy and futility, and (3) early futility stopping. These adaptive designs are briefly described below.

**Early Efficacy Stopping** – For simplicity, consider $K = 2$ (i.e., a two-stage design), which allows for early efficacy stopping (i.e., $\beta_1 = 1$). By (11.5), the type I error rates to spend at stages 1 and 2 are given by

$$\pi_1 = \psi_1(\alpha_1) = \int_0^{\alpha_1} dt_1 = \alpha_1, \tag{11.8}$$

and

$$\pi_2 = \psi_2(\alpha_2) = \int_{\alpha_1}^{\alpha_2} \int_{t}^{\alpha_1} dt_2 \, dt_1 = \frac{1}{2}(\alpha_2 - \alpha_1)^2, \tag{11.9}$$

respectively. Using equations (11.8) and (11.9), (11.6) becomes

$$\alpha = \alpha_1 + \frac{1}{2}(\alpha_2 - \alpha_1)^2. \tag{11.10}$$

Solving for $\alpha_2$, we obtain

$$\alpha_2 = \sqrt{2(\alpha - \alpha_1)} + \alpha_1. \tag{11.11}$$

where $\alpha_1$ is the stopping probability (error spent) at the first stage under the null hypothesis condition and $\alpha - \alpha_1$ is the error spent at the second stage. As a result, if the test statistic $t_1 = p_1 > \alpha_2$, it is certain that $t_2 = p_1 + p_2 > \alpha_2$. Therefore, the trial should stop when $p_1 > \alpha_2$ for futility.

Based on the relationship among $\alpha_1$, $\alpha_2$, and $\alpha$ as given in (11.10), various stopping boundaries can be considered with appropriate choices of $\alpha_1$, $\alpha_2$, and $\alpha$. For illustration purpose, Table 11.2 provides some examples of the stopping boundaries from equations (11.10) and (11.11).

By combining (11.7)–(11.11), the adjusted $p$-value is given by

$$p(t;k) = \begin{cases} t & \text{if } k = 1 \\ \alpha_1 + \dfrac{1}{2}(t - \alpha_1)^2 & \text{if } k = 2' \end{cases} \tag{11.12}$$

where $t = p_1$ if the trial stops at stage 1 and $t = p_1 + p_2$ if the trial stops at stage 2. Table 11.3 provides some examples of the stopping boundaries from equations (11.12).

**Early Efficacy or Futility Stopping** – For this case, it is obvious that if $\beta_1 \geq \alpha_2$, the stopping boundary is the same as it is for the design with early

**TABLE 11.2**

Stopping Boundaries for Two-Stage Efficacy Designs

| One-sided $\alpha$ | | | | | | | |
|---|---|---|---|---|---|---|---|
| | $\alpha_1$ | 0.005 | 0.010 | 0.015 | 0.020 | 0.025 | 0.030 |
| 0.025 | $\alpha_2$ | 0.2050 | 0.1832 | 0.1564 | 0.1200 | 0.0250 | – |
| 0.05 | $\alpha_2$ | 0.3050 | 0.2928 | 0.2796 | 0.2649 | 0.2486 | 0.2300 |

*Source:* Chang (2007).

**TABLE 11.3**

Stopping Boundaries for Two-Stage Efficacy and Futility Designs

| One-Sided $\alpha$ | | $\beta_1 = 0.15$ | | | | |
|---|---|---|---|---|---|---|
| 0.025 | $\alpha_1$ | 0.005 | 0.010 | 0.015 | 0.020 | 0.025 |
| | $\alpha_2$ | 0.2154 | 0.1871 | 0.1566 | 0.1200 | 0.0250 |
| | | $\beta_1 = 0.2$ | | | | |
| 0.05 | $\alpha_1$ | 0.005 | 0.010 | 0.015 | 0.020 | 0.025 |
| | $\alpha_2$ | 0.3333 | 0.3155 | 0.2967 | 0.2767 | 0.2554 |

*Source:* Chang (2007).

efficacy stopping. However, futility boundary $\beta_1$ when $\beta_1 \geq \alpha_2$ is expected to affect the power of the hypothesis testing. Therefore,

$$\pi_1 = \int_0^{\alpha_1} dt_1 = \alpha_1, \tag{11.13}$$

and

$$\pi_2 = \begin{cases} \displaystyle\int_{\alpha_1}^{\beta_1} \int_{t_1}^{\alpha_2} dt_2\, dt_1 & \text{for } \beta_1 \leq \alpha_2 \\[3ex] \displaystyle\int_{\alpha_1}^{\alpha_2} \int_{t_1}^{\alpha_2} dt_2\, dt_1 & \text{for } \beta_1 > \alpha_2 \end{cases}. \tag{11.14}$$

Thus, it can be verified that

$$\alpha = \begin{cases} \alpha_1 + \alpha_2(\beta_1 - \alpha_1) - \dfrac{1}{2}(\beta_1^2 - \alpha_1^2) & \text{for } \beta_1 < \alpha_2 \\[3ex] \alpha_1 + \dfrac{1}{2}(\alpha_2 - \alpha_1)^2 & \text{for } \beta_1 \geq \alpha_2 \end{cases}. \tag{11.15}$$

Similarly, under (11.15), various boundaries can be obtained with appropriate choices of $\alpha_1$, $\alpha_2$, $\beta_1$, and $\alpha$ (Table 11.4). The adjusted $p$-value is given by

$$p(t;k) = \begin{cases} t & \text{if } k = 1 \\[2ex] \alpha_1 + t(\beta_1 - \alpha_1) - \dfrac{1}{2}(\beta_1^2 - \alpha_1^2) & \text{if } k = 2 \text{ and } \beta_1 < \alpha_2, \\[3ex] \alpha_1 + \dfrac{1}{2}(t - \alpha_1)^2 & \text{if } k = 2 \ \beta_1 \geq \alpha_2 \end{cases} \tag{11.16}$$

where $t = p_1$ if the trial stops at stage 1 and $t = p_1 + p_2$ if the trial stops at stage 2.

**TABLE 11.4**

Stopping Boundaries for Two-Stage Futility Design

| One-Sided $\alpha$ | $\beta_1$ | 0.1 | 0.2 | 0.3 | $\geq 0.4$ |
|---|---|---|---|---|---|
| 0.025 | $\alpha_2$ | 0.3000 | 0.2250 | 0.2236 | 0.2236 |
| 0.05 | $\alpha_2$ | 0.5500 | 0.3500 | 0.3167 | 0.3162 |

*Source:* Chang (2007).

**Early Futility Stopping** – A trial featuring early futility stopping is a special case of the previous design, where $\alpha_1 = 0$ in equation (11.15). Hence, we have

$$\alpha = \begin{cases} \alpha_2\beta_1 - \dfrac{1}{2}\beta_1^2 & \text{for } \beta_1 < \alpha_2 \\[2mm] \dfrac{1}{2}\alpha_2^2 & \text{for } \beta_1 \geq \alpha_2 \end{cases}. \tag{11.17}$$

Solving for $\alpha_2$, it can be obtained that

$$\alpha_2 = \begin{cases} \dfrac{\alpha}{\beta_1} + \dfrac{1}{2}\beta_1 & \text{for } \beta_1 < \sqrt{2\alpha} \\[2mm] \sqrt{2\alpha} & \text{for } \beta_1 \geq \alpha_2 \end{cases}. \tag{11.18}$$

Examples of the stopping boundaries generated using equation (11.18) are presented in Table 11.4. The adjusted $p$-value can be obtained from equation (11.16), where $\alpha_1 = 0$, i.e.,

$$p(t;k) = \begin{cases} t & \text{if } k = 1 \\[2mm] \alpha_1 + t\beta_1 - \dfrac{1}{2}\beta_1^2 & \text{if } k = 2 \text{ and } \beta_1 < \alpha_2 \\[2mm] \alpha_1 + \dfrac{1}{2}t^2 & \text{if } k = 2 \; \beta_1 \geq \alpha_2 \end{cases}. \tag{11.19}$$

### 11.3.3 Conditional Power

Conditional power with or without clinical trial simulation is often considered for sample size re-estimation in adaptive trial designs. As discussed earlier, since the stopping boundaries for the most existing methods are based on either $z$-scale or $p$-value, to link a $z$-scale and a $p$-value, we will consider $p_k = 1 - \Phi(z_k)$ or inversely, $z_k = \Phi^{-1}(1 - p_k)$, where $z_k$ and $p_k$ are the normal $z$-score and the $p$-value from the sub-sample at the $k$th stage, respectively. It should be noted that $z_2$ has asymptotically normal distribution with $N(\delta / se(\hat{\delta}_2), 1)$ under the alternative hypothesis, where $\hat{\delta}_2$ is the estimation of treatment difference in the second stage and

$$se(\hat{\delta}_2) = \sqrt{2\hat{\sigma}^2 / n_2} \approx \sqrt{2\sigma^2 / n_2}.$$

The conditional power can be evaluated under the alternative hypothesis when rejecting the null hypothesis $H_0$. That is,

$$z_2 \geq B(\alpha_2, p_1). \tag{11.20}$$

Thus, the conditional probability given the first-stage naïve $p$-value, $p_1$ at the second stage is given by

$$P_C(p_1,\delta) = 1 - \Phi\left( B(\alpha_2,p_1) - \frac{\delta}{\sigma}\sqrt{\frac{n_2}{2}} \right), \alpha_1 < p_1 \le \beta_1. \qquad (11.21)$$

As an example, for the method based on the product of stage-wise $p$-values (MPP), the rejection criterion for the second stage is

$$p_1 p_2 \le \alpha_2, \text{ i.e., } z_2 \ge \Phi^{-1}(1-\alpha_2/p_1).$$

Therefore,

$$B(\alpha_2,p_1) = \Phi^{-1}(1-\alpha_2/p_1).$$

Similarly, for MSP, the rejection criterion for the second stage is

$$p_1 + p_2 \le \alpha_2, \text{ i.e., } z_2 = B(\alpha_2,p_1) = \Phi^{-1}(1-\max(0,\alpha_2-p_1)).$$

On the other hand, for the inverse normal method (21), the rejection criterion for the second stage is

$$w_1 z_1 + w_2 z_2 \ge \Phi^{-1}(1-\alpha_2),$$

i.e.,

$$z_2 \ge (\Phi^{-1}(1-\alpha_2) - w_1\Phi^{-1}(1-p_1))/w_2,$$

where $w_1$ and $w_2$ are the prefixed weights satisfying the condition of $w_1^2 + w_2^2 = 1$. Note that the group sequential design and CHW (Cui-Hung-Wang) method (Cui et al., 1999) are special cases of the inverse normal method. Since the inverse normal method requires two additional parameters ($w_1$ and $w_2$), for simplicity, we will only compare the conditional powers of MPP and MSP. For a valid comparison, the same $\alpha_1$ is used for both methods. As it can be seen from equation (Lehmacher and Wassermer, 1999), the comparison of the conditional power is equivalent to the comparison of function $B(\alpha_2,p_1)$. Equating the two $B(\alpha_2,p_1)$, we have

$$\frac{\hat{\alpha}_2}{p_1} = \tilde{\alpha}_2 - p_1, \qquad (11.22)$$

where $\hat{\alpha}_2$ and $\tilde{\alpha}_2$ are the final rejection boundaries for MPP and MSP, respectively. Solving (11.22) for $p_1$, we obtain the critical point for $p_1$.

$$\eta = \frac{\tilde{\alpha}_2 \mp \sqrt{\tilde{\alpha}_2^2 - 4\tilde{\alpha}_2}}{2}. \tag{11.23}$$

Equation (11.23) indicates that when $p_1 < \eta_1$ or $p_2 > \eta_2$, MPP has a higher conditional power than MSP. When $\eta_1 < p_1 < \eta_2$, MSP has a higher conditional power than MPP. As an example, for one-sided test at $\alpha = 0.025$, if we choose $\alpha_1 = 0.01$ and $\beta_1 = 0.3$, then $\hat{\alpha}_2 = 0.0044$, and $\tilde{\alpha}_2 = 0.2236$, which result in $\eta_1 = 0.0218$ and $\eta_2 = 0.2018$ by equation (11.23).

Note that the unconditional power $P_w$ is nothing but the expectation of conditional power, i.e.,

$$P_w = E_\delta[P_C(p_1, \delta)]. \tag{11.24}$$

Therefore, the difference in unconditional power between MSP and MPP is dependent on the distribution of $p_1$, and consequently, dependent on the true difference $\delta$ and the stopping boundaries at the first stage $(\alpha_1, \beta_1)$.

Note that in Bauer and Kohne's method using Fisher's combination (Bauer and Kohne, 1994), which leads to the equation

$$\alpha_1 + \ln(\beta_1 / \alpha_1)e^{-(1/2)\chi_{4,1-\alpha}^2} = \alpha,$$

it is obvious that the determination of $\beta_1$ leads to a unique $\alpha_1$, and consequently $\alpha_2$. This is a non-flexible approach. However, it can be verified that the method can be generalized to $\alpha_1 + \alpha_2 \ln \beta_1 / \alpha_1 = \alpha$, where $\alpha_2$ does not have to be

$$e^{-(1/2)\chi_{4,1-\alpha}^2}.$$

Note that Tsiatis and Mehta (2003) indicated that for any adaptive design with sample size adjustment, there exists a more powerful group sequential design. It, however, should be noted that the efficacy gain by the classic group sequential design is at the price of a cost. For example, as the number of interim analyses increases (e.g., from 3 to 10), the associated cost may increase substantially. Also, the optimal design is under the condition of a pre-specified error-spending function, but adaptive designs do not require in general a fixed error-spending function.

## 11.4 Analysis for Category II Adaptive Designs

Now, consider a Category II two-stage phase II/III seamless adaptive designs, which have the same study objectives but different study endpoints (continuous endpoints). Let $x_i$ be the observed value of the study endpoint (e.g., a biomarker) from the $i$th subject in phase II (stage 1), $i = 1,...,n$ and $y_j$ be the observed value of the study endpoint (i.e., the primary clinical endpoint) from the $j$th subject in phase III (stage 2), $j = 1,...,m$. Suppose that $x_i's$ and $y_j's$ are independently and identically distributed with $E(x_i) = v$ and $Var(x_i) = \tau^2$, and $E(y_j) = \mu$ $Var(y_j) = \sigma^2$, respectively. Chow et al. (2007) proposed obtaining predicted values of the clinical endpoint based on data collected from the biomarker (or surrogate endpoint) under an established relationship between the biomarker and the clinical endpoint. These predicted values are then combined with the data collected at the confirmatory phase (stage 2) to derive a statistical inference on the treatment effect under investigation. For simplicity, suppose that $x$ and $y$ can be correlated in the following straight-line relationship:

$$y = \beta_0 + \beta_1 x + \varepsilon, \tag{11.25}$$

where $\varepsilon$ is the random error with zero mean and variance $\varsigma^2$. $\varepsilon$ is assumed to be independent of $x$. In practice, we assume that this relationship is well established. In other words, the parameters $\beta_0$ and $\beta_1$ are assumed known. Based on equation (11.25), the observations $x_i$ observed in the first stage can then be transformed into $\beta_0 + \beta_1 x_i$ (denoted by $\hat{y}_i$). $\hat{y}_i$ is then considered as the observation of the clinical endpoint and combined with those observations $y_i$ collected in the second stage to estimate the treatment mean $\mu$. Chow et al. (2007) proposed the following weighted-mean estimator:

$$\hat{\mu} = \omega \bar{\hat{y}} + (1 - \omega)\bar{y}, \tag{11.26}$$

where $\bar{\hat{y}} = \dfrac{1}{n}\Sigma_{i=1}^{n}\hat{y}_i$, $\bar{y} = \dfrac{1}{m}\sum_{j=1}^{m} y_j$, and $0 \le \omega \le 1$. It should be noted that $\hat{\mu}$ is the

minimum variance-unbiased estimator among all weighted-mean estimators when the weight is given by

$$\omega = \frac{n/(\beta_1^2 \tau^2)}{n/(\beta_1^2 \tau^2) + m/\sigma^2} \tag{11.27}$$

if $\beta_1, \tau^2$ and $\sigma^2$ are known. In practice, $\tau^2$ and $\sigma^2$ are usually unknown and are commonly estimated by

$$\hat{\omega} = \frac{n/s_1^2}{n/s_1^2 + m/s_2^2}, \tag{11.28}$$

where $s_1^2$ and $s_2^2$ are the sample variances of $\hat{y}_i's$ and $y_j's$, respectively. The corresponding estimator of $\mu$ is denoted by

$$\hat{\mu}_{GD} = \hat{\omega}\bar{\hat{y}} + (1 - \hat{\omega})\bar{y}, \tag{11.29}$$

and is referred to as the Graybill–Deal (GD) estimator of $\mu$. Note that Meier (1953) proposed an approximate unbiased estimator of the variance of the GD estimator, which has bias of order $O(n^{-2} + m^{-2})$. Khatri and Shah (1974) gave an exact expression of the variance of this estimator in the form of an infinite series, which is given as

$$\widehat{\mathrm{Var}}(\hat{\mu}_{\mathrm{GD}}) = \frac{1}{n/S_1^2 + m/S_2^2}\left[1 + 4\hat{\omega}(1-\hat{\omega})\left(\frac{1}{n-1} + \frac{1}{m-1}\right)\right].$$

Based on the GD estimator, the comparison of the two treatments can be made by testing the following hypotheses:

$$H_0: \mu_1 = \mu_2 \text{ versus } H_1: \mu_1 \neq \mu_2. \tag{11.30}$$

Let $\hat{y}_{ij}$ be the predicted value (based on $\beta_0 + \beta_1 x_{ij}$), which is used as the prediction of $y$ for the $j$th subject under the $i$th treatment in phase II (stage 1). From equation (11.29), the GD estimator is given by

$$\hat{\mu}_{\mathrm{GD}i} = \hat{\omega}_i \bar{\hat{y}}_i + (1-\hat{\omega}_i)\bar{y}_i, \tag{11.31}$$

where

$$\bar{\hat{y}}_i = \frac{1}{n_i}\sum_{j=1}^{n_i} \hat{y}_{ij}, \; \bar{y}_i = \frac{1}{m_i}\sum_{j=1}^{m_i} y_{ij}$$

and

$$\hat{\omega}_i = \frac{n_i/S_{1i}^2}{n_i/S_{1i}^2 + m_i/S_{2i}^2},$$

with $S_{1i}^2$ and $S_{2i}^2$ being the sample variances of $(\hat{y}_{i1},\ldots,\hat{y}_{in_i})$ and $(y_{i1},\ldots,y_{im_i})$, respectively. For hypotheses (11.30), consider the following test statistic:

$$\tilde{T}_1 = \frac{\hat{\mu}_{\mathrm{GD}1} - \hat{\mu}_{\mathrm{GD}2}}{\sqrt{\widehat{\mathrm{Var}}(\hat{\mu}_{\mathrm{GD}1}) + \widehat{\mathrm{Var}}(\hat{\mu}_{\mathrm{GD}2})}}, \tag{11.32}$$

where

$$\widehat{\mathrm{Var}}(\hat{\mu}_{\mathrm{GD}i}) = \frac{1}{n_i/S_{1i}^2 + m_i/S_{2i}^2}\left[1 + 4\hat{\omega}_i(1-\hat{\omega}_i)\left(\frac{1}{n_i-1} + \frac{1}{m_i-1}\right)\right]$$

is an estimator of $\text{Var}(\hat{\mu}_{\text{GD}i})$, $i = 1, 2$. Consequently, an approximate $100(1 - \alpha)\%$ confidence interval is given as

$$\left( \hat{\mu}_{\text{GD}1} - \hat{\mu}_{\text{GD}2} - z_{\alpha/2}\sqrt{V_T}, \ \hat{\mu}_{\text{GD}1} - \hat{\mu}_{\text{GD}2} + z_{\alpha/2}\sqrt{V_T} \right), \tag{11.33}$$

where $V_T = \text{Var}(\hat{\mu}_{\text{GD}1}) + \text{Var}(\hat{\mu}_{\text{GD}2})$. As a result, the null hypothesis $H_0$ is rejected if the above confidence interval does not contain 0. Thus, under the local alternative hypothesis that $H_1: \mu_1 - \mu_2 = \delta \neq 0$, the required sample size to achieve a $1 - \beta$ power satisfies

$$-z_{\alpha/2} + |\delta| / \sqrt{\text{Var}(\hat{\mu}_{\text{GD}1}) + \text{Var}(\hat{\mu}_{\text{GD}2})} = z_\beta.$$

Thus, if we let $m_i = \rho n_i$ and $n_2 = \gamma n_1$, then, denoted by $N_T$, the total sample size required for achieving a desired power for detecting a clinically meaningful difference between the two treatments is $(1 + \rho)(1 + \gamma)n_1$, which is given by

$$n_1 = \frac{1}{2}AB\left(1 + \sqrt{1 + 8(1 + \rho)A^{-1}C}\right), \tag{11.34}$$

where

$$A = \frac{(z_{\alpha/2} + z_\beta)^2}{\delta^2},$$

$$B = \frac{\sigma_1^2}{\rho + r_1^{-1}} + \frac{\sigma_2^2}{\gamma(\rho + r_2^{-1})}$$

and

$$C = B^{-2}\left[ \frac{\sigma_1^2}{r_1(\rho + r_1^{-1})^3} + \frac{\sigma_2^2}{\gamma^2 r_2(\rho + r_2^{-1})^3} \right],$$

with $r_i = \beta_1^2 \tau_i^2 / \sigma_i^2$, $i = 1, 2$.

If one wishes to test for the following superiority hypotheses

$$H_1: \mu_1 - \mu_2 = \delta_1 > \delta,$$

then the required sample size for achieving $1 - \beta$ power satisfies

$$-z_\alpha + (\delta_1 - \delta) / \sqrt{\text{Var}(\hat{\mu}_{\text{GD}1}) + \text{Var}(\hat{\mu}_{\text{GD}2})} = z_\beta.$$

This gives

$$n_1 = \frac{1}{2}DB\left(1 + \sqrt{1 + 8(1 + \rho)D^{-1}C}\right),\tag{11.35}$$

where $D = \frac{(z_\alpha + z_\beta)^2}{(\delta_1 - \delta)^2}$. For the case of testing for equivalence with a signifi-cance level $\alpha$, consider the local alternative hypothesis that $H_1: \mu_1 - \mu_2 = \delta_1$ with $|\delta_1| < \delta$. The required sample size to achieve $1 - \beta$ power satisfies

$$-z_\alpha + (\delta - \delta_1)/\sqrt{\mathrm{Var}(\hat{\mu}_{GD1}) + \mathrm{Var}(\hat{\mu}_{GD2})} = z_\beta.$$

Thus, the total sample size for two treatment groups is $(1 + \rho)(1 + \gamma)n_1$ with $n_1$ given

$$n_1 = \frac{1}{2}EB\left(1 + \sqrt{1 + 8(1 + \rho)E^{-1}C}\right),\tag{11.36}$$

where $E = \frac{(z_\alpha + z_{\beta/2})^2}{(\delta - |\delta_1|)^2}$.

Note that formulas for sample size calculation and allocation for testing equality, non-inferiority, superiority, and equivalence for other data types such as binary response and time-to-event endpoints can be similarly obtained.

## 11.5 Analysis for Category III and IV Adaptive Designs

In this section, statistical inference for Category III and IV phase II/III seam-less adaptive designs will be discussed. For a Category III design, the study objectives at different stages are different (e.g., dose selection versus efficacy confirmation), but the study endpoints are the same at different stages. For a Category IV design, both study objectives and endpoints at different stages are different (e.g., dose selection versus efficacy confirmation with surrogate endpoint versus clinical study endpoint).

As indicated earlier, how to control the overall type I error rate at a pre-specified level is one of the major regulatory concerns when adaptive design methods are employed in confirmatory clinical trials. Another concern is how to perform power analysis for sample size calculation/allocation for achieving individual study objectives originally set by the two separate stud-ies (different stages). In addition, also another concern is how to combine data collected from both stages for a combined and valid final analysis. Under a Category III or IV phase II/III seamless adaptive design, in addition, the investigator plans to have an interim analysis at each stage. Thus, if we consider the initiation of the study, first interim analysis, end of stage 1 analysis, second interim analysis, and final analysis as critical milestones, the two-stage adaptive design becomes a four-stage transitional seamless trial design. In what follows, we will focus on the analysis of a four-stage transitional seamless design without (non-adaptive version) and with (adaptive version) adaptations, respectively.

### 11.5.1 Non-Adaptive Version

For a given clinical trial comparing $k$ treatment groups, $E_1, ..., E_k$ with a control group $C$, suppose a surrogate (biomarker) endpoint and a well-established clinical endpoint are available for the assessment of the treatment effect. Denoted by $\theta_i$ and $\psi_i$, $i = 1, ..., k$ the treatment effect comparing $E_i$ with $C$ assessed by the surrogate (biomarker) endpoint and the clinical endpoint, respectively. Under the surrogate and clinical endpoints, the treatment effect can be tested by the following hypotheses:

$$H_{0,2}: \psi_1 = ... = \psi_k, \tag{11.37}$$

which is for the clinical endpoint, while the hypothesis

$$H_{0,1}: \theta_1 = ... \theta_k, \tag{11.38}$$

is for the surrogate (biomarker) endpoint. Chow and Lin (2015) assumed that $\psi_i$ is a monotone increasing function of the corresponding $\theta_i$ and proposed to test the hypotheses (11.37) and (11.38) at three stages (i.e., stage 1, stage 2a, stage 2b, and stage 3) based on accrued data at four interim analyses. Their proposed tests are briefly described below. For simplicity, the variances of the surrogate (biomarker) endpoint and the clinical outcome are denoted by $\sigma^2$ and $\tau^2$, which are assumed known.

*Stage 1.* At this stage, $(k + 1)n_1$ subjects are randomly assigned to receive either one of the $k$ treatments or the control at a 1:1 ratio. In this case, we have $n_1$ subjects in each group. At the first interim analysis, the most effective treatment will be selected based on the surrogate (biomarker) endpoint and proceed to subsequent stages. For pairwise comparison, consider test statistics $\hat{\theta}_{i,1}$, $i = 1, ..., k$ and $S = \text{argmax}_{1 \leq j \leq k} \hat{\theta}_{i,1}$. Thus, if $\hat{\theta}_{S,1} \leq c_1$ for some pre-specified critical value $c_1$, then the trial is stopped and we are in favor of $H_{0,1}$. On the other hand, if $\hat{\theta}_{S,1} > c_{1,1}$, then we conclude that the treatment $E_S$ is considered the most promising treatment and proceed to subsequent stages. Subjects who receive either the promising treatment or the control will be followed for the clinical endpoint. Treatment assessment for all other subjects will be terminated but will undergo necessary safety monitoring.

*Stage 2a.* At stage 2a, $2n_2$ additional subjects will be equally randomized to receive either the treatment $E_S$ or the control $C$. The second interim analysis is scheduled when the short-term surrogate measures from these $2n_2$ stage 2 subjects and the primary endpoint measures from those $2n_1$ stage 1 subjects who receive either the treatment $E_S$ or the control $C$ become available. Let $T_{1,1} = \hat{\theta}_{S,1}$ and $T_{1,2} = \hat{\psi}_{S,1}$ be the pairwise test statistics from stage 1 based on the surrogate endpoint and the primary endpoint, respectively, and $\hat{\theta}_{S,2}$ be the statistic from stage 2 based on the surrogate. If

$$T_{2,1} = \sqrt{\frac{n_1}{n_1 + n_2}}\hat{\theta}_{S,1} + \sqrt{\frac{n_2}{n_1 + n_2}}\hat{\theta}_{S,2} \leq c_{2,1},$$

then stop the trial and accept $H_{0,1}$. If $T_{2,1} > c_{2,1}$ and $T_{1,2} > c_{1,2}$, then stop the trial and reject both $H_{0,1}$ and $H_{0,2}$. Otherwise, if $T_{2,1} > c_{2,1}$ but $T_{1,2} \leq c_{1,2}$, then we will move on to stage 2b.

*Stage 2b.* At stage 2b, no additional subjects will be recruited. The third interim analysis will be performed when the subjects in stage 2a complete their primary endpoints. Let

$$T_{2,2} = \sqrt{\frac{n_1}{n_1 + n_2}}\hat{\psi}_{S,1} + \sqrt{\frac{n_2}{n_1 + n_2}}\hat{\psi}_{S,2},$$

where $\hat{\psi}_{S,2}$ is the pairwise test statistic from stage 2b. If $T_{2,2} > c_{2,2}$, then stop the trial and reject. Otherwise, we move on to stage 3.

*Stage 3.* At stage 3, the final stage, $2n_3$ additional subjects will be recruited and followed till their primary endpoints. At the fourth interim analysis, define

$$T_3 = \sqrt{\frac{n_1}{n_1 + n_2 + n_3}}\hat{\psi}_{S,1} + \sqrt{\frac{n_2}{n_1 + n_2 + n_3}}\hat{\psi}_{S,2} + \sqrt{\frac{n_1}{n_1 + n_2 + n_3}}\hat{\psi}_{S,3},$$

where $\hat{\psi}_{S,3}$, is the pairwise test statistic from stage 3. If $T_3 > C_3$, then stop the trial and reject; $H_{0,2}$ otherwise, accept $H_{0,2}$. The parameters in the above designs, $n_1, n_2, n_3, c_{1,1}, c_{1,2}, c_{2,1}, c_{2,2}$, and $c_3$ are determined such that the procedure will have a controlled type I error rate of $\alpha$ and a target power of $1 - \beta$.

In the above design, the surrogate data in the first stage are used to select the most promising treatment rather than assessing $H_{0,1}$. This means that upon completion of stage 1, a dose does not need to be significant in order to be used in subsequent stages. In practice, it is recommended that the selection criterion be based on precision analysis (desired precision or maximum error allowed) rather than power analysis (desired power). This property is attractive to the investigator since it does not suffer from any lack of power because of limited sample sizes.

As discussed above, under the four-stage transitional seamless design, two sets of hypotheses, namely $H_{0,1}$ and $H_{0,2}$ are to be tested. Since the rejection of $H_{0,2}$ leads to the claim of efficacy, these hypotheses are of primary interest. However, the interest of controlling the overall type I error rate at a pre-specified level of significance, $H_{0,1}$ need to be tested following the principle of closed testing procedure to avoid any statistical penalties.

In summary, the two-stage phase II/III seamless adaptive design is attractive due to its efficiency (i.e., potentially reducing the lead time between studies – a phase II trial and a phase III study), and flexibility (i.e., making an early decision and taking appropriate actions – e.g., stop the trial early or delete/add dose groups).

## 11.5.2 Adaptive Version

The approach for trial design with non-adaptive version discussed in the previous section is basically a group sequential procedure with treatment selection at interim. There are no additional adaptations involved. With additional adaptations (adaptive version), Tsiatis and Metha (2003) and Jennison and Turnbull (2006) argue that adaptive designs typically suffer from the loss of efficiency and hence are typically not recommended in regular practice. Proschan et al. (2006), however, also indicated that in some scenarios, particularly when there is not enough primary outcome information available, it is appealing to use an adaptive procedure as long as it is statistically valid and justified. The transitional feature of the multiple-stage design enables us not only to verify whether the surrogate (biomarker) endpoint is predictive of the clinical outcome, but also to modify the design adaptively after the review of interim data. A possible modification is to adjust the treatment effect of the clinical outcome while validating the relationship between the surrogate (e.g., biomarker) endpoint and the clinical outcome. In practice, it is often assumed that there exists a local linear relationship between $\psi$ and $\theta$, which is a reasonable assumption if we focus only on the values at a neighborhood of the most promising treatment $E_S$. Thus, at the end of stage 2a, we can re-estimate the treatment effect of the primary endpoint using

$$\hat{\delta}_S = \frac{\hat{\psi}_{s,1}}{\hat{\theta}_{s,1}} T_{2,1}.$$

Consequently, sample size can be re-assessed at Stage 3 based on a modified treatment effect of the primary endpoint $\delta = \max\{\delta_S, \delta_0\}$, where $\delta_0$ is a minimally clinically relevant treatment effect. Suppose $m$ is the re-estimated

Stage 3 sample size based on $\delta$. Then, there is no modification for the procedure if $m \leq n_3$. On the other hand, if $m > n_3$, then $m$ (instead of $n_3$ as originally planned) subjects per arm will be recruited at Stage 3. The detailed justification of the above adaptation can be found in Chow (2011).

### 11.5.3 A Case Study – Hepatitis C Infection

A pharmaceutical company is interested in conducting a clinical trial for the evaluation of safety, tolerability, and efficacy of a test treatment for patients with hepatitis C virus infection. For this purpose, a two-stage seamless adaptive design is considered. The proposed trial design is to combine two independent studies (one phase Ib is the study for treatment selection and one phase III is the study for efficacy confirmation) into a single study. Thus, the study consists of two stages: treatment selection (stage 1) and efficacy confirmation (stage 2). The study objective at stage 1 is for treatment selection, while the study objective at stage 2 is to establish the non-inferiority of the treatment selected from stage 1 as compared to the standard of care (SOC). Thus, this is a typical Category IV design (a two-stage adaptive design with different study objectives at different stages).

For genotype 1 HCV patients, the treatment duration is usually 48 weeks of treatment followed by a 24-week follow-up. The well-established clinical endpoint is the sustained virologic response (SVR) at week 72. The SVR is defined as an undetectable HCV RNA level (<10 IU/mL) at week 72. Thus, it will take a long time to observe a response. The pharmaceutical company is interested in considering a biomarker or a surrogate endpoint such as a regular clinical endpoint with short duration to make early decision for the treatment selection of four active treatments under study at the end of stage 1. As a result, the clinical endpoint of early virologic response (EVR) at week 12 is considered as a surrogate endpoint for treatment selection at stage 1. At this point, the trial design has become a typical Category IV adaptive trial design (i.e., a two-stage adaptive design with different study endpoints and different study objectives at different stages). The resultant Category IV adaptive design is briefly outlined in Figure 11.1.

*Stage 1.* At this stage, the design begins with five arms (four active treatment arms and one control arm). Qualified subjects are randomly assigned to receive one of the five treatment arms at a 1:1:1:1:1 ratio. After all stage 1 subjects have completed week 12 of the study, an interim analysis will be performed based on EVR at week 12 for treatment selection. Treatment selection will be made under the assumption that the 12-week EVR is predictive of 72-week SVR. Under this assumption, the most promising treatment arm will be selected using precision analysis under some pre-specified selection criteria. In other words, the treatment arm with highest confidence level for achieving statistical significance (i.e., the observed difference as compared to the control is not by chance alone) will be selected. Stage 1 subjects who have not yet completed the study protocol will

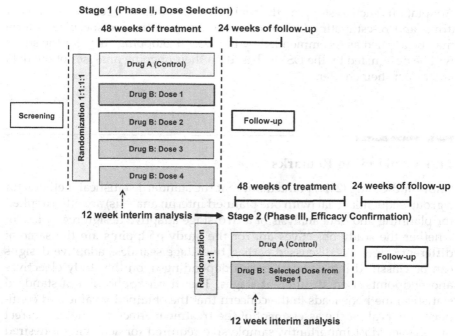

**Stage 1 (Phase II, Dose Selection)**

FIGURE 11.1
A diagram of four-stage transitional seamless trial design.

continue with their assigned therapies for the remainder of the planned 48 weeks, with a final follow-up at week 72. The selected treatment arm will then proceed to stage 2.

*Stage 2*. At stage 2, the selected treatment arm from stage 1 will be tested for non-inferiority against the control (SOC). A separate cohort of subjects will be randomized to receive either the selected treatment from stage 1 or the control (SOC) at a 1:1 ratio. A second interim analysis will be performed when all stage 2 subjects have completed week 12 and 50% of the subjects (stage 1 and stage 2 combined) have completed 48 weeks of treatment and follow-up of 24 weeks. The purpose of this interim analysis is twofold: First, it is to validate the assumption that EVR at week 12 is predictive of SVR at week 72, and second, it is to perform sample size re-estimation in order to determine whether the trial will achieve study objective (establishing non-inferiority) with the desired power if the observed treatment preserves till the end of the study.

Statistical tests as described in the previous section will be used to test non-inferiority hypotheses at interim analyses and at the end of stage analyses. For the two planned interim analyses, the incidence of EVR at week 12 as well as safety data will be reviewed by an independent data safety monitoring board (DSMB). The commonly used O'Brien–Fleming type of conservative boundaries will be applied for controlling the overall type I error rate at 5% [27].

Adaptations such as stopping the trial early, discontinuing selected treatment arms, and re-estimating the sample size based on the pre-specified criteria may be applied as recommended by the DSMB. Stopping rules for the study will be designated by the DSMB, based on their ongoing analyses of the data and as per their charter.

## 11.6  Concluding Remarks

Chow and Chang (2011) pointed out that the standard statistical methods for a group sequential trial (with one planned interim analysis) are often applied for planning and data analysis of a two-stage adaptive design regardless of whether the study objectives and/or the study endpoints are the same at different stages. As discussed earlier, two-stage seamless adaptive designs can be classified into four categories depending upon the study objectives and endpoints used at different stages. The direct application of standard statistical methods leads to the concern that the obtained $p$-value and confidence interval for the assessment of the treatment effect may not be correct or reliable. Most importantly, sample size required for achieving a desired power obtained under a standard group sequential trial design may not be sufficient for achieving the study objectives under the two-stage seamless adaptive trial design, especially when the study objectives and/or study endpoints at different stages are different.

As indicated in the 2010 FDA draft guidance on adaptive clinical trial design, adaptive designs were classified as either well-understood designs or less well-understood designs depending upon the availability of the well-established statistical methods of specific designs (FDA, 2010, 2019c). In practice, most of the adaptive designs (including the two-stage seamless adaptive designs discussed in this chapter) are considered less well-understood designs. Thus, the major challenge is not only the development of valid statistical methods for those less well-understood designs, but also the development of a set of criteria for choosing an appropriate design among these less well-understood designs for valid and reliable assessment of test treatment under investigation.

# 12

## Master Protocol – Platform Trial Design

## 12.1 Introduction

In clinical trials, a master protocol is defined as a protocol designed with multiple sub-studies, which may have different objectives and involves coordinated efforts to evaluate one or more investigational drugs in one or more disease sub-types within the overall trial structure (Woodcock and LaVange, 2017). Clinical trials utilizing the concept of master protocol are usually referred to as platform trials, which are to test multiple therapies in one indication, one therapy for multiple indications, or both. Woodcock and LaVange (2017) indicated that a master protocol may be used to conduct the trials for exploratory purposes or to support a marketing application, and can be structured to evaluate different drugs compared to their respective controls or to a single common control. The sponsor can design the master protocol with a fixed or adaptive design with the intent to modify the protocol in order to incorporate or terminate individual sub-studies within the master protocol. Individual drug sub-studies under the master protocol can incorporate an initial dose-finding phase, e.g., in pediatric patients when sufficient adult data are available to inform a starting dose and the investigational drug provides the prospect of direct clinical benefit to pediatric patients.

Master protocols may involve one or more interventions in multiple diseases or a single disease, as defined by the current disease classification, with multiple interventions, each targeting a particular biomarker-defined population or disease subtype. Under this broad definition, a master protocol consists of three distinct entities: umbrella, basket, and platform trials, which are illustrated in Table 12.1, Figures 12.1 and 12.2, respectively.

A master protocol may involve direct comparisons of competing therapies or be structured to evaluate, in parallel, different therapies relative to their respective controls. Some take advantage of the existing infrastructure to capitalize on similarities among trials, whereas others involve setting up a new trial network specific to the master protocol. All require intensive pre-trial discussion among sponsors contributing therapies for evaluation and parties involved in the conduct and governance of the trials to ensure that

**TABLE 12.1**

Types of Master Protocols

| Type of Trial | Objective |
|---|---|
| Umbrella | To study multiple targeted therapies in the context of a single disease |
| Basket | To study a single targeted therapy in the context of multiple diseases or disease subtypes |
| Platform | To study multiple targeted therapies in the context of a single disease in a perpetual manner, with therapies allowed to enter or leave the platform on the basis of a decision algorithm |

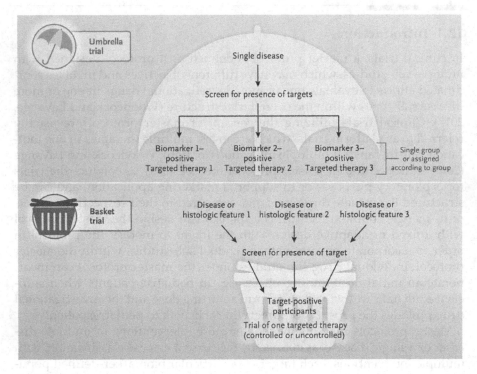

**FIGURE 12.1**
Umbrella trial and basket trial. (Woodcock and LaVange, 2017)

issues surrounding data use, publication rights, and the timing of regulatory submissions are addressed and resolved before the start of the trial.

A platform trial is an exploratory multi-arm clinical trial evaluating one or more treatments on one or more cohorts (or populations) with an objective to screen and identify promising treatments in connection with some cohorts for further investigation. Thus, platform trial designs have the potential to dramatically increase the cost-effectiveness of drug development, leading to more life-altering medicines for patients suffering from serious or

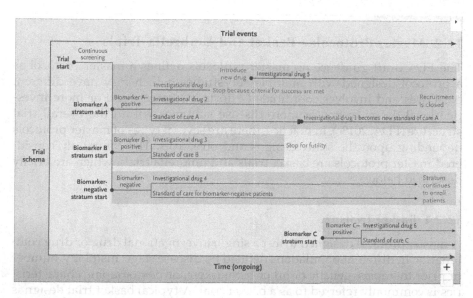

**FIGURE 12.2**
Potential design of a platform trial involving a single disease. (Woodcock and LaVange, 2017)

life-threatening illnesses. Platform trial designs are extremely useful in areas of rare diseases, in which it may be quite difficult to accrue sufficient patients for traditional trial designs, and oncology, where there is increasing molecular understanding with small biomarker-defined subgroups from which it is also a challenge to accrue for reasonable testable clinical trial hypotheses. In practice, a platform trial is typically followed by confirmative studies further investigating potential arms identified by the screening outcomes.

As indicated by Saville and Berry (2016), a platform trial is an efficient trial designed to evaluate multiple treatments and combinations of treatments, in heterogeneous patient populations, with the capability to add new treatments in the future and eliminate investigational treatments lacking efficacy.

The purpose of this chapter is not only to provide a review of platform trial design with a master protocol but also to outline potential challenges from multiple perspectives. In the next section, types of master protocols (platform trials), including basket and umbrella trials, are briefly introduced. Regulatory perspectives, including the potential use of real-world data and real-world evidence to inform platform trial design, are discussed in Section 12.3. The challenges regarding statistical validity (potential operational bias) and efficiency from statistical perspectives are provided in Section 12.4. Some applications of platform trial design in rare diseases and oncology drug development are discussed in Section 12.5. Concluding remarks are given in the last section of this chapter.

## 12.2 Master Protocols - Basket and Umbrella Trials

The FDA draft guidance on master protocols defines a master protocol as a protocol designed with multiple sub-studies, which may have different objectives and involves coordinated efforts to evaluate one or more investigational drugs in one or more disease subtypes within the overall trial structure (FDA, 2018c). In practice, there are several types of master protocols depending upon the study designs and objectives. The commonly considered master protocols are basket trials and umbrella trials, which are briefly described below.

### 12.2.1 Basket Trials

A master protocol designed to test a single investigational drug or drug combination in different populations defined by disease stage, histology, number of prior therapies, genetic or other biomarkers, or demographic characteristics is commonly referred to as a basket trial. A typical basket trial design is illustrated in Figure 12.3.

A basket trial involves multiple diseases or histologic features (i.e., in cancer). After participants are screened for the presence of a target, target-positive participants are entered into the trial; as a result, the trial may involve many different diseases or histologic features. A master protocol for a basket trial could contain multiple strata that test various biomarker–drug pairs.

As it can be seen from Figure 12.3, the sub-studies within the basket trials are usually designed as single-arm activity-estimating trials with overall response rate (ORR) as the primary endpoint. A strong response signal seen in a sub-study may allow for expansion of the sub-study to generate data that could potentially support a marketing approval. Each sub-study should include specific objectives, the scientific rationale for inclusion of each population, and a detailed statistical analysis plan (SAP) that includes sample size justification and stopping rules for futility.

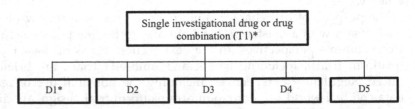

\* T = investigational drug; D = protocol defined subpopulation in multiple disease subtypes.

**FIGURE 12.3**
Schematic representation of a master protocol with basket trial design. \*T = investigational drug; D = protocol-defined subpopulation in multiple disease subtypes.

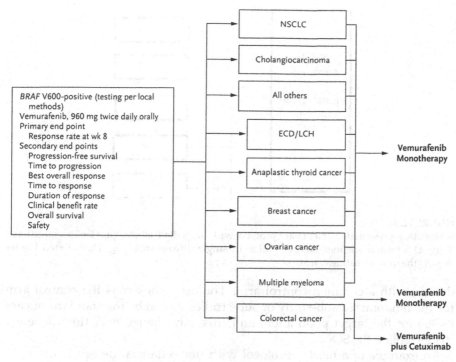

**FIGURE 12.4**
Vemurafenib in nonmelanoma cancers harboring BRAF V600 mutations. *NSCLC = non-small-cell lung cancer; ECD = Erdheim–Chester disease; LCH = Langerhans cell histiocytosis. (Hyman et al., 2015.)

An example of a master protocol with basket trial design is the phase II trial evaluating vemurafenib in multiple nonmelanoma cancers with BRAF V600 mutations (see Figure 12.4).

## 12.2.2 Umbrella Trials

A master protocol designed to evaluate multiple investigational drugs administered as single drugs or as drug combinations in a single disease population is commonly referred to as umbrella trial. A typical umbrella trial is illustrated in Figure 12.5.

An umbrella trial evaluates various (often biomarker-defined) subgroups within a conventionally defined disease. Patients with the disease are screened for the presence of a biomarker or other characteristics, and then assigned to a stratum on the basis of the results. Multiple drugs are studied in the various strata, and the design may be randomized or use external controls depending on the disease (Woodcock and LaVange, 2017).

As indicated in the FDA draft guidance, umbrella trials can employ randomized controlled designs to compare the activity of the investigational

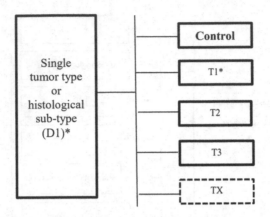

**FIGURE 12.5**
Schematic representation of a master protocol with umbrella trial design. *T = investigational drug; D = protocol-defined subpopulation in single disease subtypes; TX = dotted border depicts the future treatment arm.

drug(s) with a common control arm. The drug chosen as the control arm for the randomized sub-study or sub-studies should be the standard of care (SOC) for the target population, and this may change over time if newer drugs replace the SOC.

An example of a master protocol with umbrella trial design is the original version of the LUNG-MAP trial (Herbst et al., 2015), a multidrug, multi-sub-study, biomarker-driven trial in patients with advanced/metastatic squamous cell carcinoma of the lung. Eligible patients were assigned to sub-studies based on their biomarkers or to a nonmatched therapy sub-study for patients not eligible for the biomarker-specific sub-studies. Within the sub-studies, patients were randomized to a biomarker-driven target or SOC therapy (see Figure 12.6).

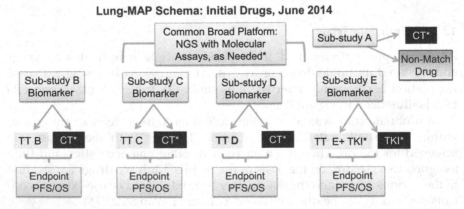

**FIGURE 12.6**
LUNG-MAP trial in patients with squamous cell carcinoma of the lung. (Herbst et al., 2015.)

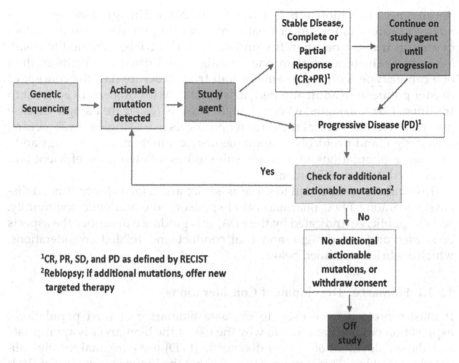

**FIGURE 12.7**
National Cancer Institute Match Trial Scheme. (Abrams et al., 2014)

### 12.2.3 Other Trial Designs

Master protocol designs may also incorporate design features common to both basket and umbrella trials, and may evaluate multiple investigational drugs and/or drug combination regimens across multiple tumor types. A typical example of a master protocol with a complex trial design is the NCI-MATCH trial, which aims to establish whether patients with one or more tumor mutations, amplifications, or translocations in a genetic pathway of interest identified in solid tumors or hematologic malignancies derive clinical benefits if treated with drugs targeting that specific pathway in a single-arm design (see Figure 12.7).

## 12.3 Regulatory Perspectives and Statistical Considerations

In 2018, FDA published a draft guidance for industry entitled *Master Protocols: Efficient Clinical Trial Design Strategies to Expedite Development of Oncology Drugs and Biologics* to assist the sponsors in oncology drug development in a more efficient way (FDA, 2018). This guidance provides advice to sponsors for cancer treatment regarding the design and conduct of clinical trials, other than first-in-human (FIH) trials, intended to *simultaneously* evaluate more than one investigational drug and/or more than one cancer type within the same overall trial structure (i.e., the concept of master protocol) in adult and pediatric patients with cancers. In contrast to traditional trial designs, where a single drug is tested in a single disease population in one clinical trial, master protocols use a single infrastructure, trial design, and protocol to simultaneously evaluate multiple drugs and/ or disease populations in multiple sub-studies, allowing for efficient and accelerated drug development.

This guidance is intended to serve as advice and a focus for continued discussions among FDA, pharmaceutical sponsors, the academic community, and the public. As indicated by the FDA, this guidance describes the aspects of master protocol designs and trial conduct and related considerations, which are briefly outlined below.

### 12.3.1 Biomarker Development Considerations

If master protocols are used to evaluate biomarker-defined populations, explanation or justification as to why the use of the biomarker is appropriate and the employment of *in vitro* diagnostic (IVD) tests are analytically validated is provided. This is mainly because that the inappropriate use of IVDs may result in inaccurate and/or unreliable evaluation of the drug product under investigation. Moreover, clinical trials using IVD tests that are not fully validated is considered violation of good clinical practice (GCP) and consequently can be placed on clinical hold for deficiencies in meeting the study objectives. Thus, the sponsor may be required to submit data for analytical/scientific validation of the IVD tests in order for FDA to determine whether the clinical results are reliable and interpretable.

In practice, it is then suggested that sponsors should establish procedures for sample acquisition, handling, and the testing and analysis plans as early as possible in the biomarker development program. Furthermore, when the trial uses an investigational IVD, it is also suggested that the sponsor and institutional review boards (IRBs) should assess what investigational device application requirements apply using the criteria as described in 21 CFR 812.2, which address level of risk (e.g., significant risk, nonsignificant risk) that the device presents to trial subjects. FDA encourages the sponsors contacting the appropriate center at FDA, e.g., the Center for Devices and Radiological Health (CDRH) or the Center for Biologics Evaluation and Research (CBER) for the device. In addition, FDA also encourages the sponsors submitting all information regarding the oncology co-development program, including IVD information in the IND (investigational new drug application) submitted to the Center for Drug Evaluation and Research (CDER) or CBER for risk assessment of the intended trials. A sponsor interested in the development of a specific biomarker test for marketing (as a device) should consult the appropriate center at FDA such as CDRH or CBER responsible for review of the IVD.

## 12.3.2 Safety Consideration

The other consideration when the master protocols are used is safety concern and monitoring. For safety assessment, the 2018 FDA guidance recommends that the following should be considered: (i) informed consent document, (ii) safety monitoring and reporting plans, (iii) independent safety assessment committee, and (iv) IRB/independent ethics committee (IEC) (FDA, 2018), which are briefly described below.

**Informed Consent Document** – For clinical trials, informed consent documents are necessarily submitted to the IRB for review for patient protection. In addition to submitting informed consent documents to the IRB for review, the sponsor may need to submit the original and all updated informed consent documents to the IND to allow FDA in order to assess that patients have the information to make informed decisions regarding participation in the trial.

In addition to new safety information, updates to the informed consent document should include all clinically important protocol modifications. Protocol amendments submitted under 21 CFR 312.30 should be accompanied by the revised informed consent documents unless immediate modifications are needed for patient safety in which case the sponsor should submit the revised informed consent document as soon as possible.

**Safety Monitoring and Reporting Plans** – The sponsor is required to ensure proper monitoring of the investigations and to ensure that the investigations are conducted in accordance with the general investigational plan and protocols contained in the IND. The sponsor should establish a systematic approach that ensures a rapid communication of serious safety issues to clinical investigators and regulatory authorities under IND safety reporting regulations. In addition, the approach should describe the process for the rapid implementation of protocol amendments to address serious safety issues. The 2018 FDA guidance also indicated that original IND should contain a proposed plan for periodic submissions of a cumulative summary of safety, as described under 21 CFR 312.32(c)(3), which is more frequent than annually. The summary of safety should include information on any action taken for safety reasons for each investigational drug during that reporting period across the clinical development program for the investigational drug. The sponsor should reference the most recent cumulative safety report in support of protocol amendments proposing modification of the existing or new sub-studies. Given the complexity of and the generally rapid accrual to these trials, resulting in increased risks to patients of failure to promptly identify adverse events, sponsors should select medical monitors who have training and experience in cancer research and in the conduct of clinical trials, so that safety information can be promptly assessed.

**Independent Safety Assessment Committee** – For all master protocols, the sponsor should institute an independent safety assessment committee (ISAC) or an independent data monitoring committee (IDMC) structured to assess safety in addition to efficacy. The sponsor should describe in the IND the constitution of this committee and the definition of its responsibilities. The committee should complete the real-time review of all serious adverse events as defined in FDA regulations and periodically assess the totality of safety information in the development program. The ISAC or IDMC should have the responsibility for conducting pre-specified and ad hoc assessments of safety to recommend protocol modifications or other actions, including, but not limited, to the following:

i. Discontinuing or modifying a sub-study based on safety information obtained from the protocol or from information external to the trial;

ii. Changing the eligibility criteria if the risks of the intervention appear to be higher in a particular subgroup;

iii. Altering the drug dosage and/or schedule if the adverse events observed appear likely to be mitigated by such changes;

iv. Instituting screening procedures that could identify those subjects at an increased risk of a particular adverse event;

v. Identifying information needed to inform current and future trial subjects of newly identified risks via changes in the informed consent document and, if appropriate, recommending re-consent of current subjects to continue trial participation.

**Institutional Review Board/Independent Ethics Committee** – A sponsor must not initiate a clinical trial until an IRB or IEC has reviewed and approved the protocol, and the trial remains subject to continuing review by an IRB. Once approved, the investigator should provide cumulative safety information to the IRB along with other information required by the IRB to allow the IRB to meet its requirements. Because of the complexity of master protocols, in general, the sponsor is expected to make an assessment of safety more frequently than on an annual basis and to provide this information to the investigator. Sponsors are required to "keep each participating investigator informed of new observations discovered by or reported to the sponsor on the drug, particularly with respect to adverse effects and safe use."

The investigator must convey this information to the IRB during the time of the IRB's continuing review, or sooner, if the information is an unanticipated problem involving risk to human subjects or others. This information can include a description of the detailed plan for timely, periodic communication of trial progress, cumulative safety information, and other reports from the ISAC or IDMC. This information is necessary to allow the IRB to evaluate, e.g., the risks to patients of the ongoing investigation and the adequacy of the informed consent document. To facilitate IRB review of master protocols, FDA recommends the use of a central IRB. The central IRB should have adequate resources and appropriate expertise to review master protocols in a timely and thorough manner. When necessary, an IRB can invite individuals with competence in special areas (i.e., consultants) to assist in the review of complex issues that require expertise beyond or in addition to that available on the IRB.

Given the rapid accumulation of safety data and the complexity of the trial design, IRBs should consider convening additional meetings (i.e., ad hoc meetings of an existing IRB) to review the evolving safety information, provided regulatory requirements in 21 CFR part 56, such as quorum, can be met. Alternatively, a separate, duly constituted specialty IRB can be established and specifically charged with meeting on short notice to review new information and/or modifications to trials with master protocols. Such an IRB would need to satisfy the same requirements of any IRB (i.e., 21 CFR part 56); however, it could be designed to facilitate a quorum by keeping membership to a minimum (i.e., 21 CFR 56.107 requires that each IRB has at least five members) and being composed of experienced members who are capable of meeting and reviewing trial-related materials on short notice. Ad hoc meetings of an existing IRB or the establishment of a separate specialty IRB designed to facilitate the review of trials with master protocols are acceptable approaches that, if appropriately constituted and operated, can satisfy the regulatory requirement for IRB oversight. Irrespective of the type of IRB that is used, if the master protocol includes plans to enroll pediatric patients in the trial, we recommend that the IRB include (either as a member or an invited nonvoting expert) an individual or individuals who have expertise in the management of pediatric oncology patients and experience with the regulatory requirements, including parental permission and assent requirements, for the enrollment of pediatric patients in clinical investigations.

### 12.3.3 Other Regulatory Considerations

**Nonrandomized, Activity-Estimating Design** – In nonrandomized protocols, where the primary endpoint is ORR, the planned sample size should be sufficient to rule out a clinically unimportant response rate based on the lower bound of the 95% confidence interval around the observed response rate. The analysis plan should describe the futility analyses to be conducted. FDA recommends designs like the Simon two-stage design that limit exposure to an ineffective drug. If a sponsor anticipates that the results would form the primary basis of an efficacy claim in a marketing application, the clinical protocol and SAP should ensure that collected data are of adequate quality for this purpose.

Additionally, the SAP should pre-specify the timing of the final analysis, ensure adequate data collection and follow-up on all patients for efficacy and safety, and describe the plan for independent review of confirmed ORR in solid tumors for each sub-study. If preliminary results from a sub-study or sub-studies suggest a major advance over the available therapy, the sponsor should meet with the review division to discuss modifications to the protocol.

**Randomized Designs** – If a sponsor incorporates randomization into an umbrella trial design, FDA strongly recommends the use of a common control arm when possible.

**Master Protocols Employing Adaptive/Bayesian Design** – In master protocols that incorporate adaptive designs, the SAP should provide all information described in the guidance for industry Adaptive Design Clinical Trials for Drugs and Biologics and the draft guidance for industry *Enrichment Strategies for Clinical Trials to Support Approval of Human Drugs and Biological Products*, and describe plans for futility analyses. Master protocols can use a Bayesian statistical method or other methods for planning or modifying the sample size, dropping an arm, or other adaptive strategies. The SAP should include the details on implementation of Bayesian or other methods.

**Master Protocols with Biomarker-Defined Subgroups** – In master protocols with basket or complex design, where patient assignment to a treatment arm is based on the presence of a specific biomarker of interest, the protocol should clearly specify how patients with more than one biomarker of interest will be assigned to sub-studies. There are two approaches to making such assignments that FDA considers acceptable from a clinical trial design perspective, but other approaches may also be appropriate. One approach is to prioritize biomarkers or treatments. For example, in the BATTLE-1 trial, investigators ranked the biomarker groups based on their predictive values and assigned patients with multiple biomarkers to the group for one of their biomarkers that has the highest predictive value. The other approach is based on a pre-specified randomization ratio. For example, the LUNG-MAP trial uses a reverse ratio of prevalence rates. Using reverse prevalence ratios, patients in the trial with tumors that have biomarkers with low prevalence have a greater likelihood to be assigned to a sub-study for the lower prevalence population.

## 12.3.4 Statistical Considerations

In addition to regulatory considerations of master protocol designs and trial conduct, this section will focus on statistical aspects of master protocol designs and trial conduct. These statistical aspects include, but are not limited to, (i) randomization (e.g., response-adaptive randomization), (ii) adaptive dose finding (e.g., selecting, deleting, modifying, and/or adding treatment arms), (iii) selection criteria, (iv) control of type I error, and (v) power analysis for sample size. These aspects are briefly outlined below.

**Randomization** – For clinical studies utilizing the concept of master protocol design, including basket design, umbrella design, and platform trial design, adaptive randomization like response-adaptive randomization is often considered to increase the probability of success. The use of response-adaptive randomization has the following advantages: (i) It assigns more patients to more promising (or effective) treatment arms, (ii) it reduces patient failures, (iii) it helps in obtaining more safety and efficacy information for effective arms, and (iv) most importantly, it shortens the evaluation time for effective arms. However, the use of response-adaptive randomization also suffers the following limitations: (i) It may cause treatment imbalance in certain covariates, (ii) it may produce biased estimates due to dependent samples, (iii) it is much more complicated in practice, (iv) it may cause unexpected delays between enrollment and response, and (v) most importantly, statistical methods for specific master protocol designs may not be fully developed, which may post challenging statistical issues depending upon its complexity.

**Adaptive Dose Selection** – In cancer clinical research, complex innovative designs like a two-stage seamless adaptive trial design using master protocol are often considered in dose-finding or treatment selection studies. A two-stage adaptive seamless design consists of two stages: The first stage is for dose-finding or treatment selection, and the second stage is for efficacy confirmation. At first stage, the most promising dose is usually selected after the review of interim analysis results. At interim, there are only limited number of subjects (data) available for an accurate and reliable assessment of treatment effects of each treatment arm in the study. Thus, it is a concern as if a wrong decision for choosing the promising treatment arm is made especially based on limited data available at interim.

In practice, under a two-stage adaptive seamless design for dose finding or treatment selection, two important issues, namely, multiplicity and criteria for dose or treatment selection, are evitably raised. Multiplicity is related to (i) early stopping/suspending rules or selection criteria based on either safety or efficacy and (ii) control of the overall type I error rate. Zheng and Chow (2019) compared the following criteria for dose finding or treatment selection: (i) precision analysis, (ii) power analysis (conditional power), (iii) predictive probability of success, and (iv) probability of being the best dose or treatment under various scenarios (with a limited number of subjects available at interim).

The issues of multiplicity and criteria for dose selection could be more problematic if we have the options of (i) adding new arms during the trial and (ii) prioritizing treatment arms to start the trial and add in arms during the trial. The adaptive dose-finding design could be more complex if using active control and/or concurrent control versus all control.

**Control of Type I Error Rate** – For platform trial design, the type I error rate can be controlled at two different levels: at individual arm level and at the trial level. For the control of arm level type I error, the probability of incorrectly identifying an ineffective treatment as effective should be carefully evaluated. For the control of trial-level (family-wise) type I error, the probability of incorrectly selecting at least one ineffective treatment as promising from a study with several ineffective treatments should be assessed.

It should be noted that the control of the trial-level type I error rate depends upon the study objectives (the union–intersection concept versus the intersection–union concept). Under the null hypotheses, several statistical methods such as Bonferroni correction (Bonferroni, 1936), Holm method (Holm, 1979), Hochberg approach (Hochberg, 1988), and Hommel's procedure (Hommel, 2001) can be used to assess multiplicity for the control of type I error rate.

**Power Analysis for Sample Size** – For a platform trial comparing multiple treatment arms, power analysis for sample size calculation is often performed based on the arm-level power (controlling arm-level type I error rate) or trial-level power (controlling trial-level type I error rate). Arm-level power is the probability of correctly identifying an effective treatment as effective, while the definition of the trial-level power depends on the study objective. When the objective is to find one effective arm, there is the probability of correctly selecting at least one effective treatment as promising. When the objective is to find the best effective arm, there is the probability of correctly selecting one of the most effective treatments as promising. Sample size requirement for achieving the given power can then be performed under the alternative hypothesis.

In practice, we may consider the maximum sample size in total (i.e., the total of maximum sample size per arm). It, however, should be noted that the resultant sample size may not be feasible especially for rare disease drug development where there are only limited sample sizes (per arm) available.

Power calculation is made under the alternative hypothesis. Due to the complexity of the platform trial design, clinical trial simulation is often considered for the evaluation of the operating characteristics, including power analysis of multiple comparisons or tests.

**Statistical Methods** – Specific statistical methods may not be fully developed due to the complexity of specific platform trial design. In practice, it is suggested that the analysis of a full model that incorporates all treatment arms should be considered. Under the full model, data collected from individual arm will be combined for a more accurate and reliable assessment of the treatment effects. The advantage for the analysis of a full model is that we can fully utilize data collected from all treatment arms. However, one of the complications of the full model analysis is that the population may be heterogeneous across different treatment arms. Moreover, there may be treatment imbalance. In this case, how to control the trial-level type I error rate is a great concern.

Alternatively, in the interest of treatment balance, one may consider applying Bayesian analysis method to borrow information across the relevant treatment arms. In clinical trials, it is well recognized that more information can increase the precision and reliability of the estimates. The problem is not only how to obtain (borrow) more information and what information can be borrowed. An intuitive approach is to borrow information from other similar studies, including different drugs on the same population, the same drug on different populations, or combinations of a certain drug with different SOCs on the same population. In practice, however, it is a concern that one may borrow inadequate information due to possible interactions among studies, drugs (treatments), and populations. Borrowing information may impact the type I error rate and power. For example, when all arms are actually ineffective, we may lower the type I error. On the other hand, when all arms are actually effective, we may increase the power. When ineffective and effective arms are mixed, we may reduce the power for effective arms, but increase the type I error for ineffective arms. In practice, although Bayesian approach for borrowing information is useful, it should be performed with caution.

## 12.4 Applications of Platform Trial Design

### 12.4.1 I-SPY2 Program

A typical example for the application of master protocol is the I-SPY (investigation of serial studies to predict your therapeutic response with imagine and molecular analysis) program. The mission of the I-SPY2 program is to accelerate the development of improved y = treatment for I-SPY program, which was designed to integrate and link phase I (I-SPY phase 1), phase 2 (I-SPY2), and phase III (I-SPY3) evaluations to establish a dynamic pipeline of novel agents. The centerpiece of the program is the phase 2 I-SPY2 adaptive randomized study (*I-SPY2 Trial: Neoadjuvant and Personalized Adaptive Novel Agents to Treat Breast Cancer*) that utilizes a master protocol comparing multiple arms with a common control (Rugo et al., 2016; Park et al., 2016; Esserman et al., 2019). The key feature of the I-SPY2 study is the use of a master protocol that allows multiple drugs from multiple companies to be in the trial for evaluation simultaneously.

I-SPY2 trial was a biomarker-driven study. Thus, all participants were subjected to baseline assessments for hormone receptor (HR) status, HER2 status, and a 70-gene assay (Cardoso et al., 2016) that correlates with the risk of recurrence. Each subject was then classified into one of eight pre-defined disease sub-types. I-SPY2 was designed to increase the success rate of phase 3 trials, assess which subset of disease has the superior benefit for specific agents, and shorten the development process of phase 3 trials (see Table 12.2). In I-SPY2 trial, a Bayesian adaptive randomization procedure was used to assign subjects to competing drug regimens. In other words, Bayesian adaptive randomization assigns women to arms that have a higher likelihood of achieving a good outcome in their tumor sub-type. Thus, unlike traditional randomized clinical trials, the number of patients being assigned to each treatment arm (agent) in I-SPY2 is influenced by the trial results.

The primary endpoint of the I-SPY2 trial was the pathologic complete response (pCR) assessed histologically at the time of surgery. For the

**TABLE 12.2**

I-SPY2 Study Design

| Principle | Solution |
|---|---|
| Test agents where they matter most | • Neoadjuvant setting, poor prognosis cancers<br>• Integrate advocates into trial planning |
| Rapidly learn to tailor agents | • Adaptive design<br>• Neoadjuvant therapy<br>• Integration of biomarkers, imaging |
| Optimize phase 3 trials | • Graduate drugs with predicted probability of success in phase 3 trials for given biomarker profile |
| Drive organizational efficiency | • Adaptive design<br>• Master IND and master CTA<br>• Test drugs by class, across many companies<br>• Shared cost of profiling<br>• Financial support separated from drug supply<br>• Shared IT infrastructure<br>• Protocol and ICF structure to minimize delays |
| Use team approach | • Democratize access to data<br>• Share credit and opportunity<br>• Collaborative process for development |

*Source:* Esserman et al. (2019), Table 1.1, p. 7.

purpose of assessing the efficacy of each arm/agent, ten clinically relevant signatures were defined prospectively. These are similar to the sub-types used in randomization. As the trial progresses, each arm/agent is continually assessed against pre-defined criteria for success or failure within each of the ten signatures. Based on Bayesian adaptive approach, I-SPY2 trial results were reported in terms of probabilities for each of the ten signatures: (i) the estimated pCR rate in the agent/control arms for the given signature, (ii) the Bayesian probability that the agent is superior to control, and (iii) the Bayesian probability of success in the hypothetical 300-patient phase 3 trial (Rugo et al., 2016) (Figure 12.8).

The red curves represent the patients treated with V/C plus paclitaxel followed by AC, and the blue curves represent the concurrent controls. The corresponding 95% probability distributions (represented by the width of the curve) are shown for each. The mean of each distribution is the estimated pCR rate.

## 12.4.2 Ebola Platform Trial Design

Another typical example would be the use of platform trial design for the evaluation of Ebola. Based on recent outbreaks of the deadly Ebola virus, academics and industry have partnered to preemptively plan a platform trial design that can be sued in the next epidemic (Berry et al., 2016). Ebola is unique in that the outcome of interest is relatively short (14-day mortality) and there is no known effective therapy. For a number of reasons unique to the disease and third-world location, it is impractical to conduct traditional

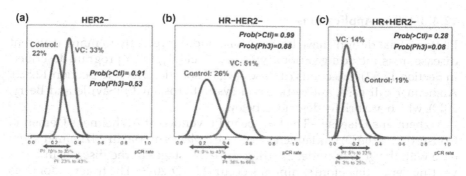

**FIGURE 12.8**
Estimated pCR rate for the signatures evaluated for V/C versus concurrent HER2-negative control. (a) Probability distribution for all patients with HER2-negative disease; (b) probability distribution for patients with TNBC (HR-negative/HER2-negative); (c) probability distribution for patients with HR-positive, HER2-negative disease. ( Rugo et al., 2016.)

clinical trials with one or two experimental treatments per trial. Rather, a platform trial that focuses on the best treatment of the disease is ideal in this setting. Response-adaptive randomization is critical in this context, as we want to quickly move away from ineffective therapies and allocate patients to the more promising arms. The trial has primary and secondary agents, as well as the combinations of primary agents plus secondary agents. Controversially, a SOC arm can also be included, which receives a minimum of 20% patients until it is replaced by a more effective agent. 50% of the remaining patients are allocated to single-agent arms and the other 50% are allocated to combination arms. However, the master protocol allows the exclusion of a control arm depending upon the available therapies at the time of the epidemic. The design incorporates Bayesian response-adaptive randomization and modeling for population drift.

A notable feature of this design is that the treatment labels in the master protocol are completely generic, and when the time comes to implement this trial, these specific drugs/agents are simply inserted into the trial based on the master protocol, with simulations updated depending upon the number of available treatments, and the trial design is ready for implementation. Speed is of the essence in this setting, as the trial must begin enrolling patients before the epidemic peaks (Figure 12.9).

| Regimens | | Treatments | | | | | |
|---|---|---|---|---|---|---|---|
| | | P1 | P2 | P3 | P4 | S1 | S2 |
| Treatments | P1 | | | | | | |
| | P2 | | | | | | |
| | P3 | | | | | | |
| | P4 | | | | | | |

**FIGURE 12.9**
Ebola platform trial design.

### 12.4.3 Other Applications

Platform trial designs have become very popular recently in many different disease areas, including various types of cancer (e.g., I-SPY program described in Section 12.5.1), infectious diseases (e.g., Ebola described in Section 12.5.2), Alzheimer's disease, antibiotics, and cystic fibrosis (CF) (Saville and Berry, 2019), which are briefly described below.

**Alzheimer's Disease** – The European Prevention of Alzheimer's Dementia Consortium (EPAD) has launched a phase 2 platform trial in Alzheimer's disease with the goal of understanding the early stages of the disease and preventing dementia before symptoms occur (EPAD, 2018). The trial randomizes patients to multiple interventions within multiple groups' biomarker-driven subgroups. The primary analysis uses an innovative Bayesian model to compare the rate of disease progression for a given treatment versus placebo. The trial design includes regulatory master protocol evolution analysis and monitors interventions for early futility or success based on the measured cognitive effect.

**Antibiotics** – As indicated by Saville and Berry (2019), ADAPT is a platform trial design in antibiotics with the goal of providing an efficient pathway to evaluate multiple potential novel agents in multiple body sites, in a unique context where a successful treatment for a resistant pathogen should be used minimally to preserve its effectiveness (APD, 2017). The trial design includes the sharing of a control arm across multiple agents, adaptive starting and stopping of novel agents with a potentially perpetual timeframe, and hierarchical borrowing of information across body sites. This study design, however, does not use response-adaptive randomization because finding the second and third best drugs are highly valued objectives, as opposed to single goal of finding the best drug.

**Cystic Fibrosis** – A platform trial design is also currently being planned in Australia to investigate the treatment of patients with CF, which is a genetic disorder characterized by persistent respiratory infections. The trial design incorporates multiple domains for the treatment of CF-induced pulmonary exacerbations, such as primary antibiotic, adjunct antibiotic, muco-active therapy, immunomodulation therapy, and airway clearance, in which each domain has multiple factors of treatment.

## 12.5 Concluding Remarks

A platform trial design is a cost-effective and attractive study design in studying multi-arm clinical trial evaluating one or more treatments on one or more cohorts (or populations) with an objective to screen and identify promising treatments in connection with some cohorts. Platform trial design is useful in rare disease drug development especially when the study objective is to identify promising doses (among a number of doses of a drug product) or promising treatments of single disease like Ebola virus within a relatively short period of time with a limited number of subjects available. The use of complex innovative trial design like adaptive trial design using master protocol (either basket, umbrella, or platform trial) in conjunction with real-world data/real-world evidence is not only cost-effective but also can shorten the development process.

Although adaptive platform trial designs discussed in this chapter provide a path for evaluating multiple treatments of critical diseases such as cancer and rare diseases, under the framework of master protocol, standard methods may not be appropriate for the assessment of treatment effect under study. However, under certain complex innovative design such as a two-stage (e.g., phase I/II or phase II/III) seamless adaptive design) statistical methods that have been developed (see, e.g., Chow and Lin, 2015), statistical methods for complex adaptive trial design with master protocol for basket, umbrella, and/or platform trial have not yet been fully developed. Further regulatory guidance and/or statistical methodology research is needed.

# 13

## Gene Therapy for Rare Diseases

### 13.1 What Is Gene Therapy?

Gene therapy is a technique that modifies a person's genes to treat or cure disease. Gene therapies can work by several mechanisms, namely, (i) replacing a disease-causing gene with a healthy copy of the gene, (ii) inactivating a disease-causing gene that is not functioning properly, and (iii) introducing a new or modified gene into the body to help treat a disease. As indicated in the 2018 FDA guidance on *Long Term Follow-Up After Administration of Human Gene Therapy Products*, human gene therapy seeks to modify or manipulate the expression of a gene or to alter the biological properties of living cells for therapeutic uses (FDA 2018a,b). Gene therapy products are being studied to treat diseases, including cancer, genetic diseases, and infectious diseases.

There are a variety of types of gene therapy products, including (i) plasmid DNA, (ii) viral vectors, (iii) bacterial vectors, (iv) human gene editing technology, and (v) patient-derived cellular gene therapy products. For plasmid DNA, circular DNA molecules can be genetically engineered to carry therapeutic genes into human cells. For viral vectors, viruses have a natural ability to deliver genetic material into cells, and therefore, some gene therapy products are derived from viruses. Once viruses have been modified to remove their ability to cause infectious disease, these modified viruses can be used as vectors (vehicles) to carry therapeutic genes into human cells. For bacterial vectors, bacteria can be modified to prevent them from causing infectious disease and then used as vectors (vehicles) to carry therapeutic genes into human tissues. For human gene editing technology, the goals are to disrupt harmful genes or to repair mutated genes. For patient-derived cellular gene therapy products, cells are removed from the patient, genetically modified (often using a viral vector), and then returned to the patient.

As indicated by the National Institutes of Health (NIH) reports, nearly 7,000 rare diseases affect more than 25 million Americans. Approximately 80% of rare diseases are caused by a single-gene defect, and about half of all rare diseases affect children. Since most rare diseases have no approved therapies, there is a significant unmet need for effective treatments, and many rare diseases are serious or life-threatening conditions. Additionally, many rare diseases exhibit a number of variations or sub-types. As a result, the development of gene therapy products in the area of rare diseases has become very popular lately.

The remaining of this chapter is organized as follows. In the next section, FDA's guidance on gene therapy for rare diseases is reviewed. Some statistical considerations are given in Section 13.3. A detailed discussion of a recent FDA-approved gene therapy regulatory submission is provided in Section 13.4. Some concluding remarks are given in the last section of this chapter.

## 13.2 FDA Guidance on Gene Therapy for Rare Diseases

To assist the sponsors for developing a human gene therapy product for rare diseases, FDA published a guidance on *Human Gene Therapy for Rare Diseases* in 2018 (FDA, 2018b). The guidance is intended to provide recommendations to the sponsors developing human gene therapy products intended to treat a rare disease in adult and/or pediatric patients regarding the manufacturing, preclinical, and clinical trial design issues for all phases of the clinical development program (FDA, 2018a,b).

Many rare disorders are serious, with no approved treatments, and represent substantial unmet medical needs for patients. Because of phenotypic heterogeneity, disease manifestations are likely to vary in onset and severity. Natural history study can potentially provide critical information to guide every stage of drug development from drug discovery to determining effectiveness and safety of the drug in treating a disease (FDA, 2015b). However, there may be insufficient information on the natural history of the disease to inform the selection of a historical comparator or to inform the clinical endpoint selection in the future clinical trials.

In a majority of these disorders, clinical manifestations appear early in life, and there are ethical and regulatory considerations regarding the enrollment of children in clinical trials. These considerations should factor into the design of both early- and late-phase clinical trials. Further details of general considerations for gene therapy clinical trials are available in a separate guidance document (FDA, 2015c). FDA recommends that the following important elements should be considered during clinical development of investigational gene therapy products intended for the treatment of rare diseases.

**Study Population** – The study population should be selected based on the existing preclinical or clinical data to determine the potential risks and benefits for the study subjects. In addition, sponsors should consider whether the proposed study population is likely to provide informative safety and/or efficacy data (FDA, 2015b,c). When conducting gene therapy clinical trials for rare diseases, the following general principles follow:

i. As stated in the FDA guidance, if the disease is caused by a genetic defect, the sponsor should perform genetic tests for the specific defects in all clinical trial subjects (FDA, 2015b). This information is important and helpful to ensure correct diagnosis of the disorder of interest. Since many of the disorders may involve either deletions or functional mutations at any of several loci within a specific gene, safety and effectiveness may be linked to specific genotype in unpredictable ways. Thus, early understanding of such associations may help in planning future clinical trials;

ii. As preexisting antibody to the gene therapy product under study may limit its potential for therapeutic effect, FDA suggested that sponsors may choose to exclude patients with preexisting antibodies to the product. In such cases, it is strongly suggested that the sponsor should develop a companion diagnostic to detect antibodies to the product. If an *in vitro* companion diagnostic is needed to appropriately select patients for study, then submission of the marketing application for the companion diagnostic and submission of the BLA (biologics license application) for the product should be coordinated to support contemporaneous marketing authorizations (FDA, 2015b);

iii. At the designing phase of gene therapy clinical trials, severity of the disease under study and the anticipated risk and potential benefits to subjects should be considered (FDA, 2015b). In addition, subjects with severe or advanced disease might experience confounding adverse events related to the underlying disease that should also be considered;

iv. Since most rare diseases are pediatric diseases or have onset of manifestations in childhood, pediatric studies are a critical part of drug development and need to be taken into consideration. Treatment in pediatric patients, however, cannot proceed without addressing ethical considerations. As indicated in 21 CFR 50.53, the administration of an investigational drug in children must offer a prospect of direct clinical benefit to individually enrolled patients, and the risk must be justified by the anticipated benefit unless the risks of an investigational drug are no more than a minor increase over the minimal risk. In addition, the anticipated risk–benefit profile must be at least as favorable as that presented by accepted alternative treatments (21 CFR 50.52). Note that adequate provisions must be made to obtain the permission of the parents and the assent of the child as per 21 CFR 50.55. 221;

v. The risks of most gene therapy products include the possibility of unintended effects that may be permanent, along with adverse effects due to invasive procedures that may be necessary for product administration. Because of these risks, it is generally not acceptable to enroll normal, healthy volunteers into gene therapy studies. A well-written informed consent document is also essential.

**Study Design** – For rare diseases drug development, there are often a limited number of qualified patients available for intended clinical studies. Thus, it is not feasible to enroll a number of subjects for all studies conducted under different phases of the clinical development. Limitation in the number of prospective subjects warrants the collection of as much pertinent data (e.g., adverse events, efficacy outcomes, biomarkers) as possible from every subject, starting from the first-in-human study. All such data may be valuable to inform the design of subsequent studies (e.g., selection of study populations and endpoints). As a result, it is suggested that the sponsor should follow the following general principles for the selection of an appropriate study design when developing gene therapy products for rare diseases (FDA, 2015b):

  i. In practice, a randomized, concurrent controlled trial is generally recommended, which is considered gold standard for establishing effectiveness and provides treatment-related safety data. FDA notes that randomization in early stages of development is strongly encouraged when feasible;

 ii. Sponsors should consider designing their first-in-human study to be an adequate and well-controlled investigation that has the potential to provide substantial evidence of effectiveness and safety of the test treatment under study in support of regulatory submission and marketing application;

iii. In case where the performance of the test treatment under study depends upon subjects with different disease stages or severities, sponsors are encouraged to consider stratified randomization based on disease stage/severity;

 iv. For some gene therapies with different indications (e.g., a genetic skin disease), an intra-subject control design (e.g., a crossover design like *n*-of-1 trial design) may be useful. Comparisons of local therapeutic effects can be facilitated by the elimination of variability among subjects in inter-subject designs;

  v. In many cases, a single-arm trial using historical controls, sometimes including an initial observation period, may be considered if there are feasibility issues while conducting a randomized, controlled trial;

 vi. If a single-arm trial design with a historical control is used, then knowledge of the natural history of disease is critical. Natural history data may provide the basis of a historical control, but only if the control and treatment populations are adequately matched, in terms of demographics, concurrent treatment, disease state, and other relevant factors (see, e.g., FDA, 2015b,c, 2019a). In case where randomized, concurrent controlled trials cannot be conducted and the natural history is well characterized, sponsors may consider the clinical performance of available therapies (if there are any) when setting the performance goal or criteria against which the product effect will be tested;

vii. As the study power for detecting treatment effect is likely to diminish due to small sample size and potentially high inter-subject variability, FDA suggested that alternative trial designs and statistical techniques that maximize data from a small and potentially heterogeneous group of subjects should be considered. FDA also noted that ideally, utilizing as an endpoint a treatment outcome that virtually never occurs in the natural course of the disease would greatly facilitate the design and cogency of small trials;

viii. In addition, FDA suggested that adequate measures to minimize bias should be undertaken. The preferred approach to minimize bias is to use a study design that includes blinding.

**Dose Selection** – Dose selection plays an important role in gene therapy drug development for rare diseases. As indicated in the FDA guidance, the following general principles should be considered (FDA, 2015b):

i. Dose selection should be informed by all available sources of clinical information (e.g., publications, experience with similar products, experience in related patient populations);

ii. Leveraging non-human data obtained in animal models of disease and in vitro data may be, in some cases, the only way to estimate a starting human dose that is anticipated to provide benefit. Additional dosing information can be obtained from predictive models based on current understanding of in vitro enzyme kinetics (including characterizing the enzyme kinetics in relevant cell lines) and allometric scaling;

iii. For early-phase studies, clinical development of gene therapy products should include the evaluation of two or more dose levels to help identify the potentially therapeutic dose(s). Ideally, placebo controls should be added to each dose cohort;

iv. In practice, some gene therapy products may have an extended duration of activity, so that repeated dosing may not be an acceptable risk until there is a preliminary understanding of the product's toxicity and duration of activity.

**Biomarker Consideration** – In the gene therapy product development program, FDA also encourages sponsors to identify and validate biomarkers early and to leverage all available information from published investigations for the disease or related diseases (FDA, 2015b). Some biomarkers or endpoints are very closely linked to (or predictive of) the underlying pathophysiology of the disease (e.g., a missing metabolite in a critical biosynthetic pathway).

In this case, total or substantial restoration of the biosynthetic metabolic pathway may generally be expected to confer clinical benefits. Changes in such biomarkers could be used during drug development for dose selection, or even as an early demonstration of drug activity.

**Safety Considerations** – For safety, FDA indicated that gene therapy clinical trials should include a monitoring plan that is adequate to protect the safety of clinical trial subjects. The elements and procedures of the monitoring plan should be based upon what is known about the gene therapy product, including preclinical toxicology, CMC (chemical, manufacturing, and control) information, and previous human experience (if available) with the proposed product or related products (FDA, 2015b). When there is limited previous human experience with a specific gene therapy product, administration to several subjects concurrently may expose those subjects to unacceptable risk. In practice, most first-in-human trials of gene therapy products should stagger administration to consecutively enrolled subjects, for at least an initial group of subjects, followed by staggering between dose cohorts. This approach limits the number of subjects who might be exposed to an unanticipated safety risk (FDA, 2015b).

Regarding immunogenicity, FDA also indicated that innate and adaptive immune responses directed against one or more components of gene therapy products (e.g., against the vector and/or transgene) may impact the product safety and efficacy. Early development of appropriate assays to measure the product-directed immune responses may be critical to program success. Development of neutralizing and non-neutralizing immune responses that are directed against the product should be monitored throughout the clinical trial (FDA, 2014b). In addition, because of the unique nature of the mechanism of action involving genetic manipulation, a potential exists for serious long-term effects that may not be apparent during development or even at the time of an initial licensure. The long-term safety of gene therapy products is currently unknown. The appropriate duration of the long-term follow-up depends on the results of preclinical studies with this product, knowledge of the disease process, and other scientific information (FDA, 2018a).

FDA also noted that early-phase gene therapy clinical trial protocols should generally include study stopping rules, which are criteria for halting the study based on the observed incidence of particular adverse events (FDA, 2015c). The objective of study stopping rules is to limit subject exposure to risk in the event that safety concerns arise. Well-designed stopping rules may allow sponsors to assess and address risks identified as the trial proceeds, and to amend the protocol to mitigate such risks or to assure that human subjects are not exposed to unreasonable and significant risk of illness or injury. The potential for viral shedding should be addressed early in product development (FDA, 2015c).

**Efficacy Endpoints** – As indicated by the FDA, for the evaluation of efficacy of gene therapy products, the well-established, disease-specific efficacy endpoints are not available in many rare diseases (FDA, 1998). In general, demonstration of clinical benefit of a gene therapy product follows the same principles as for any other drug products. However, in some cases, there may be unique characteristics of gene therapy products like a protein that is expressed by a gene therapy product which may have different bioactivity than the standard enzyme replacement therapy that warrants additional considerations in both pre-approval and post-marketing. Thus, it is suggested that prior to commencing clinical trials of gene therapy products for rare diseases, it is critically important to have a discussion with FDA about the primary efficacy endpoints. FDA suggested that endpoint selection for a clinical trial of a gene therapy product for a rare disease should consider the following:

i. Sponsors should utilize the pathophysiology and natural history of a disease as fully as possible at the outset of product development. Full understanding of mechanism of product action is not required for product approval; however, understanding of pathophysiology is important for the selection of endpoints in planning clinical trials;

ii. For sponsors that are considering seeking accelerated approval of a gene therapy product for a rare disease pursuant to section 506(c) of the Federal Food, Drug, and Cosmetic Act (FD&C Act) based on a surrogate endpoint, it is particularly important to understand the pathophysiology and natural history of the disease in order to help identify potential surrogate endpoints that are reasonably likely to predict clinical benefit;

iii. Sponsors should identify the specific aspects of the disease that are meaningful to the patient and might also be affected by the gene therapy product's activity;

iv. Considerable information can be gained by collecting clinical measurements repeatedly over time. Such longitudinal profile allows the assessments of effect, largely based on within patient changes, that otherwise could not be studied.

**Patient Experience** – Since patient experience data may provide important additional information about the clinical benefit of a gene therapy product, FDA encourages sponsors to collect patient experience data during product development, and to submit such data in the marketing application.

## 13.3 Statistical/Scientific Considerations

In addition to the above important elements to be considered, the following statistical considerations are necessarily implemented during the conduct of gene therapy rare diseases clinical trials.

**Small sample size** – In clinical research, power analysis is often performed for the sample size calculation. The purpose is to achieve a desired power of correctly detecting a clinically meaningful difference at a pre-specified level of significance if such a difference truly exists. However, in some situations such as (i) clinical trials with extremely low incidence rates and (ii) for rare disease drug development clinical trials, power analysis for sample size calculation may not be feasible because (i) it may require a huge sample size for detecting a relatively small difference and (ii) eligible patients may not be available for a small target patient population. In these cases, other procedures for the sample size determination with certain statistical assurance are needed. In this chapter, an innovative method based on a probability monitoring procedure is proposed for the sample size determination. The concept is to select an appropriate sample size for controlling the probability of crossing safety and/or efficacy boundaries. For rare disease clinical development, an adaptive probability monitoring procedure may be applied if a multiple-stage adaptive trial design is used.

**Endpoint selection** – In clinical trials, the selection of appropriate study endpoints is critical for an accurate and reliable evaluation of safety and effectiveness of a test treatment under investigation. For a given study endpoint, it is expected that the study endpoint can achieve the study objective of clinical importance with statistical significance. In practice, however, there are usually multiple endpoints available for the measurement of disease status and/ or therapeutic effect of the test treatment under study. For example, in cancer clinical trials, overall survival, response rate, and/or time-to-disease progression are usually considered as the primary clinical endpoints for the evaluation of safety and effectiveness of the test treatment under investigation. Once the study endpoints have been selected, the sample size required for achieving a desired power is then determined. It, however, should be noted that different study endpoints may result in different sample sizes. In practice, it is usually not clear which study endpoint can best inform the disease status and measure the treatment effect. Moreover, different study endpoints may not translate one another although they may be highly correlated with one another. In this chapter, we intend to develop an innovative endpoint, namely, therapeutic index, based on a utility function to combine and utilize information collected from all study endpoints. More details regarding the development of therapeutic index can be found in Chapter 4.

**The Use of Biomarkers** – As compared to a hard (or gold standard) endpoint like survival, a biomarker often has the following characteristics: (i) It can be measured earlier, easier, and more frequently; (ii) there are less subjects to competing risks; (iii) it is less affected by other treatment modalities; (iv) it can detect a larger effect size (i.e., a smaller sample size is required); and (v) it is predictive of clinical endpoint. The use of biomarker has the following advantages: (i) It can lead to better target population, (ii) it can detect a larger effect size (clinical benefit) with a smaller sample size, and (iii) it allows an early and faster decision-making. Thus, under the assumption that there is a well-established relationship between a biomarker and clinical outcomes, the use of biomarker in rare disease clinical trials can not only allow screening for possible responders at enrichment phase, but also provides the opportunity to detect the signal of potential safety concerns early and provides supportive evidence of efficacy with a small number of patients available.

**Innovative Trial Design** – As indicated in Chapter 2, small patient population is a challenge to rare disease clinical trials. Thus, there is a need for innovative trial designs in order to obtain substantial evidence with a small number of subjects available for achieving the same standard for regulatory approval. These innovative trial designs include, but are not limited to, the $n$-of-1 trial design, an adaptive trial design, master protocols, and a Bayesian design. The use of innovative trial design is not only to fix the dilemma of sample available (e.g., $n$-of-1 trial design and master protocol) but also to allow flexibility to effectively, accurately, and reliably assess the treatment effect and increase the probability of success of the intended clinical trials (e.g., adaptive trial design in conjunction with Bayesian approach for borrowing RWD). More details regarding these innovative trial designs for potential use in rare diseases clinical development are discussed in Chapters 10–12.

**Demonstrating Not-Ineffectiveness** – With only small number of patients available in rare diseases drug development, Chow and Huang (2019) proposed to first demonstrate not-ineffectiveness (or non-inferiority) of the test treatment under investigation with a limited number of patients. Once the not-ineffectiveness of the test treatment has been established, a statistical test was performed to rule out the probability of inconclusiveness and then conclude the effectiveness of the test treatment under investigation. Along this line, Chow (2020) proposed an innovative approach for rare diseases drug development. The idea is implemented by a two-stage trial design. At the first stage, a limited number of patients from a randomized clinical trial (RCT) will be used to establish not-ineffectiveness of the test treatment under investigation once the not-ineffectiveness of the test treatment has been demonstrated. At the second stage, Chow (2020) proposed to combine clinical data from the intended clinical trial and data from the RWD to rule out the probability of inconclusiveness of the test treatment under investigation. If the test result indicates that the probability of inconclusiveness of the test treatment under investigation is negligible, the effectiveness of the test treatment under investigation is claimed.

**The Use of RWE for Regulatory Approval** – The 21st-century Cures Act passed by the United States Congress in December 2016 requires that the FDA shall establish a program to evaluate the potential use of real-world evidence (RWE), which is derived from RWD to (i) support the approval of new indication for a drug approved under section 505 (c) and (ii) satisfy post-approval study requirements. Although RWE offers the opportunities to develop robust evidence using high-quality data and sophisticated methods for producing causal-effect estimates, randomization is feasible. In this chapter, we have demonstrated that the assessment of treatment effect (RWE) based on RWD could be biased due to potential selection and information biases of RWD. Although fit-for-purpose RWE may meet regulatory standards under certain assumptions, it is not the same as substantial evidence (current regulatory standard). In practice, it is then suggested that when there are gaps between fit-for-purpose RWE and substantial evidence, we should make efforts to fill the gaps for an accurate and reliable assessment of the treatment effect.

**Reproducibility of Rare Disease Clinical Trial Results** – Although there will not be sufficient power due to the small sample size available in rare diseases clinical trials, alternatively, we may consider empirical power based on the observed treatment effect and the variability associated with the observed difference adjusted for the sample size required for achieving the desired power (i.e., $N$). The empirical power is also known as the reproducibility probability of the clinical results for future studies if the studies shall be conducted under similar experimental conditions. Shao and Chow (2002) studied how to evaluate the reproducibility probability using this approach under several study designs for comparing means with both equal and unequal variances. When the reproducibility probability is used to provide substantial evidence of the effectiveness of a drug product, the estimated power approach may produce an optimistic result. Alternatively, Shao and Chow (2002) suggested that the reproducibility probability be defined as a lower confidence bound of the power of the second trial. The reproducibility probability can be used to determine the clinical results observed from the rare disease clinical trial that has provided substantial evidence for the evaluation of safety and efficacy of the test treatment under investigation.

## 13.4 Case Study – Approval of Luxturna

In this section, we will discuss a case study regarding gene therapy product Luxturna (voretigene neparvovec-rzyl) recently approved by the FDA on December 19, 2017. Luxturna is a new gene therapy to treat children and adult patients with an inherited form of vision loss that may result in blindness. Luxturna is the first directly administered gene therapy approved in the United States that targets a disease caused by mutations in a specific gene.

### 13.4.1 Background

Biallelic RPE65 mutation-associated retinal dystrophy is a serious and sight-threatening autosomal recessive genetic disorder. The visual function, including visual acuity (VA) and visual field, of the affected individuals declines with age, leading to total blindness in young adulthood. Patients with biallelic RPE65 mutation-associated retinal dystrophy may be given a variety of clinical diagnoses due to the variability in clinical manifestations and ophthalmological examinations. Two common clinical diagnoses that are caused by biallelic mutations in the RPE65 gene are Leber congenital amaurosis type 2 (LCA2) and retinitis pigmentosa type 20 (RP20). There are approximately 1,000–3,000 patients with biallelic RPE65 mutation-associated retinal dystrophy in the United States.

For the treatment of biallelic RPE65 mutation-associated retinal dystrophy, the Argus II Retinal Prosthesis System, an implanted device, is approved in the United States under a Humanitarian Device Exemption (HDE). The device is indicated for use in patients aged 25 and older with severe to profound retinitis pigmentosa (bare light or no light perception in both eyes) by providing electrical stimulation of the retina to induce visual perception. Thus, there is no approved pharmacological treatment for biallelic RPE65 mutation-associated retinal dystrophy.

Luxturna is an adeno-associated virus vector-based gene therapy, which is approved for the treatment of patients with confirmed biallelic RPE65 mutation-associated retinal dystrophy that leads to vision loss and may cause complete blindness in certain patients. Luxturna works by delivering a normal copy of the RPE65 gene directly to the retinal cells. These retinal cells then produce the normal protein that converts light to an electrical signal in the retina in order to restore patient's vision loss. Luxturna uses a naturally occurring adeno-associated virus, which has been modified using recombinant DNA techniques, as a vehicle to deliver the normal human RPE65 gene to the retinal cells in order to restore vision.

Hereditary retinal dystrophies are a broad group of genetic retinal disorders that are associated with progressive visual dysfunction and are caused by mutations in any one of more than 220 different genes. Biallelic RPE65 mutation-associated retinal dystrophy affects approximately 1,000 to 2,000 patients in the United States. Biallelic mutation carriers have a mutation (not necessarily the same mutation) in both copies of a particular gene (a paternal and a maternal mutation). The RPE65 gene provides instructions for making an enzyme (a protein that facilitates chemical reactions) that is essential for normal vision. Mutations in the RPE65 gene lead to reduced or absent levels of RPE65 activity, blocking the visual cycle and resulting in impaired vision. Individuals with biallelic RPE65 mutation-associated retinal dystrophy experience progressive deterioration of vision over time. This loss of vision, often during childhood or adolescence, ultimately progresses to complete blindness.

### 13.4.2 Regulatory Review/Approval Process

The sponsor is receiving a *Rare Pediatric Disease Priority Review Voucher* under a program intended to encourage the development of new drugs and biologics for the prevention and treatment of rare pediatric diseases. A voucher can be redeemed by a sponsor at a later date to receive *Priority Review* of a subsequent marketing application for a different product. Luxturna was the 13th rare pediatric disease priority review voucher issued by the FDA since the program began. The FDA granted the BLA of Luxturna *Priority Review* and *Breakthrough Therapy* designations. Luxturna also received *Orphan Drug* designation, which provides incentives to assist and encourage the development of drugs for rare diseases (see also Section 1.2.1 of Chapter 1 for more details).

### 13.4.3 FDA Review of Clinical Safety and Efficacy

This section discusses the basis for regulatory approval for Luxturna based on a phase I clinical study and a phase III clinical study, which provided the primary evidence of safety and effectiveness for the BLA submission.

**Phase I and Phase III Studies** – The phase 1 study was an open-label, dose-escalation safety study in a total of 12 patients with confirmed biallelic RPE65 mutation-associated retinal dystrophy. Eleven of the 12 patients received subretinal injection of Luxturna each eye with an injection interval ranging from 1.7 to 4.6 years. One patient received subretinal injection in only one eye. Three doses were evaluated in the phase I study. There was no clear dose effect for safety, bioactivity, or preliminary efficacy. The high dose ($1.5 \times 1,011$ vector genome in an injection volume of 0.3 mL) was chosen for the phase III study.

The phase III study was an open-label, randomized, controlled, crossover trial. It was designed to evaluate the efficacy and safety of sequential subretinal injection of Luxturna to each eye. Patients were randomized in a 2:1 ratio to either the Luxturna treatment group or the observational control group, and were followed for one year for the primary efficacy assessment. Patients in the control group were crossed over to receive Luxturna after 1 year of observation. The phase III study enrolled 31 patients from two sites in the United States. Of the 31 enrolled patients, 21 patients were randomized to the treatment group with discontinuation of one patient at the baseline visit. Ten patients were randomized to the control group with withdrawal of consent of one patient at the screening visit. The nine patients in the control group were crossed over to receive Luxturna after one year of observation. The average age of the 31 randomized patients was 15 years (ranging from 4 to 44 years), including 20 (64%) pediatric patients whose age ranges from 4 to 17 years and 11 adult patients. The subretinal injection interval between the two eyes of each patient ranged between 6 and 18 days.

**Efficacy Evaluation** – The primary efficacy endpoint in the phase III study was a change in multi-luminance mobility testing (MLMT) performance from baseline to one year after Luxturna administration. MLMT was designed to measure functional vision, as assessed by the ability of a patient to navigate a course accurately and at a reasonable pace at different levels of environmental illumination. The MLMT was assessed using both eyes and each eye separately at one or more of seven light levels, ranging from 400 lux (corresponding to a brightly light office) to 1 lux (corresponding to a moon-less summer night). Each light level was assigned a score code ranging from 0 to 6. A higher score indicated that a patient was able to pass the MLMT at a lower light level. A score of -1 was assigned to patients who could not pass the MLMT at a light level of 400 lux. The MLMT of each patient was video-taped and assessed by independent, blinded graders. The MLMT score was determined by the lowest light level at which the patient was able to pass the MLMT. The MLMT score change was defined as the difference between the score at baseline and the score at a follow-up visit. A positive MLMT score change indicated an improvement in the MLMT performance. The primary efficacy analyses were assessed at year 1 in the intention-to-treat (ITT) population, defined as all randomized patients. Additional efficacy endpoints included full-field light sensitivity threshold (FST) and VA. For safety, the more serious risks of Luxturna include endophthalmitis (infection inside of the eye), permanent decline in VA, increased intraocular pressure, retinal abnormalities (e.g., retinal tears or breaks), and cataract development and/or progression. The regulatory review of phase I and phase III studies is summarized below.

Table 13.1 summarizes the primary efficacy endpoint results and the median MLMT score change from baseline to year 1 in the Luxturna treat-ment group as compared to the control group. A median MLMT score change of 2 was observed in the Luxturna treatment group, while a median MLMT score change of 0 was observed in the control group, when using both eyes or the first-treated eye. An MLMT score change of 2 or greater is considered a clinically meaningful benefit for functional vision.

Tables 13.2 and 13.3 illustrate the number and percentage of patients with different magnitudes of MLMT score change using both eyes and individual eyes, respectively, at year 1. If we consider both eyes, 11 patients (out of 21, 52%) in the Luxturna treatment group had an MLMT score change of 2 or greater, while only one patient (out of 10, 10%) in the control group had an MLMT score change of 2 (Table 13.2; see also Figure 13.1).

On the other hand, if we consider the first-treated eye and second-treated eye separately, 15 patients (out of 21, 71%) in the Luxturna treatment group

**TABLE 13.1**

Efficacy Results of the Phase 3 Study at Year 1, Compared to Baseline

| Efficacy Outcomes | Luxturna $n = 21$ | Control $n = 10$ | Differences (Luxturna Minus Control | $p$-Value |
|---|---|---|---|---|
| MLMT score change using both eyes, median (min, max) | 2 (0, 4) | 0 (–1, 2) | 2 | 0.001 |
| MLMT score change using first-treated eye, median (min, max) | 2 (0, 4) | 0 (–1, 1) | 2 | 0.003 |

*Source:* FDA Statistical and Clinical Reviews.

**TABLE 13.2**

Magnitude of MLMT Score Change Using Both Eyes at Year 1

| MLMT Score Change | Luxturna n = 21 | Control n = 10 |
|---|---|---|
| −1 | 0 | 3 (30%) |
| 0 | 2 (10%) | 3 (30%) |
| 1 | 8 (38%) | 3 (30%) |
| 2 | 5 (24%) | 1 (10%) |
| 3 | 5 (24%) | 0 |
| 4 | 1 (4%) | 0 |

*Source:* FDA Statistical and Clinical Reviews.

**TABLE 13.3**

Magnitude of MLMT Score Change Using Individual Eyes at Year 1 (ITT)

| Change Score | First-Treated Eye (n = 21) | Control (n = 10) | Second-Treated Eye (n = 21) | Control (n = 10) |
|---|---|---|---|---|
| −1 | 0 | 1 (10%) | 0 | 2 (20%) |
| 0 | 4 (19%) | 6 (60%) | 2 (10%) | 5 (50%) |
| 1 | 2 (10%) | 3 (30%) | 2 (19%) | 3 (30%) |
| 2 | 8 (38%) | 0 | 8 (38%) | 0 |
| 3 | 6 (28%) | 0 | 5 (23%) | 0 |
| 4 | 1 (5%) | 0 | 1 (5%) | 0 |
| 5 | 0 | 0 | 1 (5%) | 0 |

*Source:* FDA Statistical and Clinical Reviews.

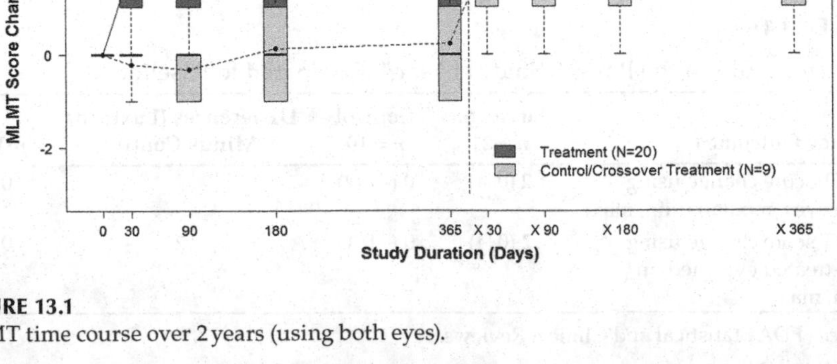

**FIGURE 13.1**
MLMT time course over 2 years (using both eyes).

had an MLMT score change of 2 or greater, while no patient in the control group had a score change of 2 or greater (Table 13.3).

**Safety Assessment** – The safety population included a total of 41 patients (81 eyes) who received subretinal administration of Luxturna (12 patients from phase I and 29 patients from phase III). Twenty-seven (out of 41, 66%) patients had ocular adverse reactions that involved 46 injected eyes (out of 81, 57%). The most common adverse reactions (incidence ≥ 5%) were conjunctival hyperemia, cataracts, increased intraocular pressure, retinal tears, dellen (thinning of corneal stroma), macular hole, eye inflammation, macular breaks, subretinal deposits, eye irritation, eye pain, and maculopathies (wrinkling on the surface of the macula) (Table 13.4). These ocular adverse reactions may have been related to Luxturna, the subretinal injection procedure, the concomitant use of corticosteroids, or a combination of these procedures and products. Most of these ocular adverse reactions were temporary and responded to medical management. There were no deaths in the clinical studies. There were two serious ocular adverse reactions: (i) endophthalmitis (infection inside of the eye) with a series of subsequent complications as a result of the infection and the treatment, and (ii) loss of vision due to fovea thinning as a result of the subretinal injection. Systemic

**TABLE 13.4**

Ocular Adverse Reactions Following Treatment with Luxturna ($n = 41$)

| Adverse Reactions | Patients $n = 41$ | Treated Eyes $n = 81$ |
|---|---|---|
| Any ocular adverse reaction | 27 (66%) | 46 (57%) |
| Conjunctival hyperemia | 9 (22%) | 9 (11%) |
| Cataract | 8 (20%) | 15 (19%) |
| Increased intraocular pressure | 6 (15%) | 8 (5%) |
| Retinal tear | 4 (10%) | 4 (5%) |
| Dellen (thinning of the corneal stroma) | 3 (7%) | 3 (4%) |
| Macular hole | 3 (7%) | 3 (4%) |
| Subretinal deposits[a] | 3 (7%) | 3 (4%) |
| Eye inflammation | 2 (5%) | 4 (5%) |
| Eye irritation | 2 (5%) | 2 (2%) |
| Eye pain | 2 (5%) | 2 (2%) |
| Maculopathy (wrinkling on the surface of the macula) | 2 (5%) | 3 (4%) |
| Foveal thinning and loss of foveal function | 1 (2%) | 2 (2%) |
| Endophthalmitis (infection inside of the eye) | 1 (2%) | 1 (1%) |
| Fovea dehiscence (separation of the retinal layers in the center of the macula) | 1 (2%) | 1 (1%) |
| Retinal hemorrhage | 1 (2%) | 1 (1%) |

*Source:*  Modified from the applicant's BLA submission.
*Note:*
[a]  Transient appearance of a ring-like deposit at the retinal injection site 1–6 days after injection without symptoms.

adverse events included hyperglycemia, nausea, vomiting, and leukocytosis. These systemic events were likely caused by systemic corticosteroid use and reactions to anesthesia.

Of the 41 patients (treated eyes $n = 81$), the following results were observed: (i) 27 (66%) patients showed any ocular adverse reaction; (ii) 46 (57%) subjects experienced conjunctival hyperemia; (iii) 9 (22%) and 9 (11%) had cataract; (iv) 8 (20%) and 15 (19%) have increased intraocular pressure; (v) 6 (15%) and 8 (10%) experienced retinal tear; (vi) 4 (10%) and 4 (5%) experienced dellen (thinning of the corneal stroma); (v) 3 (7%) and 3 (4%) had macular hole; (vi) 3 (7%) and 3 (4%) have subretinal deposits; (vii) 3 (7%) and 3 (4%) experienced eye inflammation; (viii) 2 (5%) and 4 (5%) exhibited eye irritation; (ix) 2 (5%) and 2 (2%) experienced eye pain; (x) 2 (5%) and 2 (2%) showed maculopathy (wrinkling on the surface of the macula); (xi) 2 (5%) and 3 (4%) have foveal, thinning, and loss of foveal function; (xii) 1 (2%) and 2 (2%) have endophthalmitis (infection inside of the eye); (xiii) 1 (2%) and 1 (1%) have fovea dehiscence (separation of the retinal layers in the center of the macula); and (xiv) 1 (2%) and 1 (1%) experienced retinal hemorrhage.

In summary, the safety and efficacy of Luxturna were established in a clinical development program with a total of 41 patients between the ages of 4 and 44 years. All participants had confirmed biallelic RPE65 mutations. The primary evidence of efficacy of Luxturna was based on a phase III study with 31 participants by measuring the change from baseline to year 1 in a subject's ability to navigate an obstacle course at various light levels. The group of patients that received Luxturna demonstrated significant improvements in their ability to complete the obstacle course at low light levels as compared to the control group. The most common adverse reactions due to the treatment with Luxturna included eye redness (conjunctival hyperemia), cataract, increased intraocular pressure, and retinal tear.

As noted by the FDA, Luxturna should be given only to patients who have viable retinal cells as determined by the treating physician(s). Treatment with Luxturna must be done separately in each eye on separate days, with at least six days between surgical procedures. It is administered via subretinal injection by a surgeon experienced in performing intraocular surgery. Patients should be treated with a short course of oral prednisone to limit the potential immune reaction to Luxturna.

**Advisory Committee Recommendations** – A meeting of the Cellular, Tissue, and Gene Therapies Advisory Committee (CTGTAC) was held on October 12, 2017, to provide feedback to FDA regarding clinical efficacy and safety, and an overall benefit–risk assessment of Luxturna. CTGTAC's discussions are summarized below.

   i. A 2-light level improvement in MLMT (i.e., an MLMT score change of 2) is clinically meaningful;

ii. The potential risks associated with subretinal injection of Luxturna and concomitant corticosteroid use are acceptable for the pediatric population, even in the very young population;

iii. The retinal cellular proliferation is not complete until 8–12 months of age, and Luxturna may be diluted or lost during the cellular proliferation process;

iv. Further study may be needed to support the repeated administration of previously treated eyes if the efficacy of Luxturna declines over time.

As a result, the CTGTAC voted 16 (yes) to 0 (no) to the question, *Does voretigene neparvovec-rzyl (Luxturna) have an overall favorable benefit–risk profile for the treatment of patients with vision loss due to confirmed biallelic RPE65 mutation-associated retinal dystrophy?* and in favor of the approval of Luxturna. Note that after the regulatory approval, the sponsor plans to conduct a post-marketing observational study involving patients treated with Luxturna to further evaluate the long-term safety.

## 13.5 Concluding Remarks

In this chapter, a case study regarding Luxturna – which is a rare diseases drug product intended to treat children and adult patients with an inherited form of vision loss that may result in blindness – is provided. The FDA granted Luxturna (i) priority review, (ii) breakthrough therapy, and (iii) orphan drug designations to assist the sponsor in Luxturna development. As discussed in Section 13.4, FDA's recommendation for the regulatory approval of Luxturna was made based on a phase 1 clinical study and a phase 3 clinical study, which involved only 31 patients.

Although the CTGTAC is voted for the approval of Luxturna, several questions and concerns remain. These questions, which are likely to occur due to the small sample size, include, but are not limited to, (i) the treatment effect that may have been contaminated by some covariates and/or interaction/confounding factors; (ii) no scientific/clinical or statistical justification for the selection endpoint and/or clinically meaningful difference; (iii) the study design that can provide an accurate and reliable assessment of the treatment effect, which is not clear; and (iv) statistical methods for the assessment of treatment that may not be adequate (e.g., shift analysis before and after treatment was not done). In addition, one of the major concerns is probably that whether the performance of Luxturna based on a limited number of patients is not by chance alone and hence is reproducibility. Based on a quick assessment, the reproducibility probability was calculated based on the observed responses and the variability associated with the responses of the 31 subjects (see Section 2.4 of Chapter 2). The result indicates that there is a less than 60% probability of reproducibility if the study is conducted under similar experimental conditions in the future.

Thus, it is suggested that statistical considerations as described in Section 13.3 and some innovative thinking regarding study design and analysis should be taken into consideration when conducting rare diseases clinical trials.

# 14

## Clinical Development of NASH Program

### 14.1 Introduction

Nonalcoholic fatty liver disease (NAFLD) is defined by the presence of hepatic accumulation of triglycerides in the hepatocytes in the absence of significant alcohol intake, viral infection, or any other specific etiology of liver disease. It represents a histopathologic spectrum ranging from steatosis alone to nonalcoholic steatohepatitis (NASH), fibrosis, and cirrhosis (Sanyal, 2011).

NASH is defined histologically by the presence of hepatic steatosis with evidence for hepatocyte damage with or without fibrosis. The most important histological feature associated with mortality in NASH is the presence of significant fibrosis. Although recent data suggest that some patients with fatty liver can progress to NASH and clinically significant fibrosis, most of the fibrosis progression seems to occur in patients with NASH. NASH has been recognized as one of the leading causes of cirrhosis in adults, and NASH-related cirrhosis is currently the second indication for liver transplants in the United States (Younossi et al., 2016).

The prevalence of steatohepatitis is increasing worldwide. Patients with obesity, type 2 diabetes (T2DM), and insulin resistance are specifically affected. Patients with a diagnosis of NAFLD have been shown to have a significantly higher risk of cardiovascular disease, and overall and liver-related mortality when compared with an age- and sex-matched general population. NAFLD may also lead to liver failure or hepatocellular carcinoma (Misra et al., 2009).

The diagnosis of NASH requires a liver biopsy with subsequent confirmation of a specific histopathologic pattern. The minimal criteria for the diagnosis of steatohepatitis include the presence of >5% macrovesicular steatosis, inflammation, and liver cell ballooning, typically with a predominantly centrilobular distribution in adults. The NAFLD activity score (NAS) is a validated scoring system that can be used to assess histologic change in studies of both adults and children with NASH (Kleiner et al., 2005). It is an unweighted composite of steatosis, inflammation, and ballooning scores. While a score of 5 or more is associated with a greater likelihood of having NASH, a NAS ≥5 does not necessarily confirm a diagnosis of NASH. Additionally, NAS has not been validated as a marker for likelihood of disease progression (e.g., cirrhosis, mortality) and/or response to therapy (Brunt et al., 2011). The use of NAS is currently limited to clinical trial settings. Additionally, it is recommended that a validated method for the staging of NASH be used for the assessment of changes in disease stage in clinical trials of NASH. The NASH Clinical Research Network (CRN) fibrosis staging system is the most validated system currently available. Total score ranges from 0 to 4 (no fibrosis to cirrhosis) (Brunt et al., 1999).

A major impediment to the development of therapeutics to improve outcomes in NASH is the long natural history of the disease. Several longitudinal cohort studies have provided data about the progression of NASH. However, the rate of progression of fibrosis remains to be quantified and is poorly understood. In a recently published study of 108 NAFLD patients who had serial biopsies, 47% of NASH patients had a progression of fibrosis, and 18% had spontaneous regression of fibrosis over a median follow-up period of 6.6 years (McPherson et al., 2015).

It is required approximately 6–7 years for a one-point progression in fibrosis in patients with NASH (Singh et al., 2015; Sorrentino et al., 2004). The presence of inflammation on the initial biopsy, age, lower platelet count, a low AST/ALT (aspartate transaminase/alanine aminotransferase) ratio, and higher FIB-4 score at the time of the baseline biopsy have been reported to be associated with the development of progressive fibrosis (McPherson et al., 2015). The presence of hypertension was found to be a predictor in some (McPherson et al., 2015; Singh et al., 2015) but not all studies (Argo et al., 2009). This long natural history with some patients spontaneously regressing makes data in clinical trials sometimes difficult to interpret.

In the past decade, adaptive design methods in clinical research have attracted much attention because they offer not only the potential flexibility for identifying clinical benefit of a test treatment under investigation, but also the efficiency for speeding up the development process. The FDA adaptive design draft guidance defines an adaptive design as a clinical study that includes a prospectively planned opportunity for the modification of one or more specified aspects of the study design (FDA, 2010, 2018d). The use of adaptive design methods in clinical trials may allow us to correct assumptions used at the planning stage and select the most promising option early. In addition, adaptive designs make use of cumulative information of the ongoing trial, which provides the investigator an opportunity to react earlier regardless of positive or negative results. Thus, the adaptive design approaches may speed up the drug development process (Chow and Lin, 2015).

## 14.2 Regulatory Perspectives and Registration Pathways

### 14.2.1 Regulatory Guidance

Since currently there are no approved drugs for the treatment of NASH, in December 2018, FDA issued draft guidance on developing drugs to treat patients who have NASH with liver fibrosis to assist the sponsor for the development of NASH with liver fibrosis. As indicated earlier, given the high prevalence of NASH, the associated morbidity, the growing burden of end-stage liver disease, and the limited availability of livers for organ transplantation, the FDA draft guidance provided recommendations for preclinical and clinical development, including trial design and endpoint selection to support the approval of drugs to treat noncirrhotic NASH. The FDA draft guidance meant to assist the sponsor in identifying therapies that will slow the progress or, halt, or reverse NASH and NAFLD will address an unmet medical need. However, FDA indicated that the guidance is not meant to cover the development of drugs to treat cirrhosis caused by NASH or the development of in vitro diagnostics that may be used in developing drugs to treat the disease.

**General Considerations** – As stated in the draft guidance, NAFLD consists of three successive stages: nonalcoholic fatty liver (NAFL), noncirrhotic NASH, and NASH with cirrhosis. The draft guidance provides sponsors a convenient conceptual framework to identify areas of potential future drug development. However, because patients' NAFL can exist for many years and may not progress to NASH, it may be challenging to demonstrate a favorable benefit–risk profile of pharmacological treatments in NAFL patients. Thus, it is suggested that sponsors should consider the following general considerations during drug development for the treatment of noncirrhotic NASH with liver fibrosis (FDA, 2018c):

First, the sponsor should consider using animal models for NASH to screen and identify potential investigational drugs. The sponsor should select a specific animal model based on the mechanism of action of the investigational drug.

Second, if there is a potential for liver toxicity based on animal toxicology studies, the sponsor should institute an appropriate plan to monitor liver safety early in drug development. For such a plan, the sponsor should consider the challenges of effectively recognizing a liver signal in a chronic liver condition like NASH.

As stated in the guidance, until a sponsor can characterize a drug's initial tolerability, preliminary safety, and pharmacokinetics, patients with evidence of abnormal liver synthetic function should be excluded from the early-phase trials (i.e., phase I and early proof-of-concept (POC) clinical trials). In addition, the sponsor should study the effects of hepatic impairment on the drug's pharmacokinetics early during the drug development program in a dedicated hepatic study to support appropriate dosing and dose adjustment across the spectrum of NASH liver diseases.

**Specific Considerations** – In addition to general considerations, the guidance also provided specific considerations regarding phase II development considerations, phase III development considerations, and pediatric considerations, which are summarized below.

For phase II and phase III studies, FDA suggested that sponsors should enroll patients with a histological diagnosis of NASH with liver fibrosis made within 6 months of enrollment, taking into consideration patients' standard of care and background therapy for other chronic conditions. FDA also says that patients' weight should be stable for 3 months prior to enrollment. FDA also indicated that phase III studies for NASH should be double-blind and placebo-controlled with the goal of slowing, halting, or reversing disease progression and improving clinical outcomes.

Because of the slow progression of NASH and the time required to conduct a trial that would evaluate clinical endpoints such as progression to cirrhosis or survival, the FDA recommends sponsors consider several liver histological improvements as endpoints reasonably likely to predict clinical benefit to support accelerated approval under the regulations. FDA-recommended endpoints include (i) resolution of steatohepatitis on the overall histopathological reading and no worsening of liver fibrosis on NASH CRN fibrosis score, (ii) improvement in liver fibrosis greater than or equal to one stage (NASH CRN fibrosis score) and no worsening of steatohepatitis (defined as no increase in NAS for ballooning, inflammation, or steatosis), or (iii) both resolution of steatohepatitis and improvement in fibrosis.

The FDA guidance also provided some caveats for pediatric development. As indicated in the FDA guidance, pediatric NASH appears to have different histological characteristics as well as a different natural history when compared to adult NASH. For reasons that are currently unknown, disease characteristics and progression in pediatric patients may be different.

## 14.2.2 Accelerated/Conditional Marketing Approvals

In practice, FDA has adopted policies to expedite drug development for serious medical conditions where few or no therapies exist. This is an evolving process which began in the 1980s with drugs intended to treat HIV, and continues through the present with multiple regulatory initiatives to facilitate the availability of new drugs to patients – these programs tackle the issue of either accelerating clinical development or accelerating reviews by the regulatory agencies. For example, in 2014, FDA issued a guidance describing "Expedited Programs for Serious Conditions," which consolidated information on fast-track therapy, breakthrough therapy, accelerated approval, and priority review (FDA, 2014a). Similarly, in 2014, the EMA (European Medicines Agency) launched "Adaptive Licensing," which, like the US Accelerated Approval Pathway, is also aimed at accelerating marketing approvals for products throughout the European Union (EMA, 2014). For the development of products to treat NASH, these accelerated pathways offer the potential to use surrogate endpoints to obtain *Accelerated (US) conditional (EU)* marketing approval, with full marketing approval being granted with subsequent confirmatory studies using the well-established and well-defined clinical outcomes. With a registration pathway involving a two-stage process – namely, conditional marketing approval using surrogate endpoints, followed by full marketing approval using the well-defined clinical outcomes – drug development programs to treat NASH seem well suited to continuous adaptive clinical studies. Conditional/ accelerated marketing authorization are not new concepts. For example, in the 1980s, accelerated marketing approvals were granted for a number of drugs to treat HIV based on surrogate endpoints, with post-approval studies being required for full approval. Indeed, in 1997, Sheiner's disruptively innovative paper describing a "learn-confirm" strategy suggested we move away from thinking in terms of separate phase I/II/III studies (Sheiner, 1997). Although Sheiner's paper focused on early studies, the concept has been applied across the development continuum – namely, initial studies to learn about the drug and later studies to confirm a positive benefit–risk profile. There are, however, a number of stakeholders to be considered when designing seamless adaptive studies (patients, regulators, payers), and clearly one size does not fit all medical situations (Eichler et al., 2015; Woodcock, 2012). However, given the challenges in developing therapies for NASH, it seems reasonable to assume that seamless adaptive designs provide a "good fit" in terms of development for marketing authorizations and that these adaptive designs will continue to evolve as we learn more about this emerging epidemic.

## 14.3 Endpoints for Clinical Development in NASH

The lack of accurate, reproducible, and easily applied methods to assess NASH creates major limitations not only for drug development, but also in the clinical management of NASH patients.

Liver biopsy remains the gold standard for the diagnosis of NASH, but it has several limitations. There is always a risk that the biopsy taken might not be representative for the amount of fibrosis in the whole liver. Increasing the length of liver biopsy decreases the risk of sampling error. In general, except for cirrhosis, for which microfragments may be sufficient, a 25-mm-long biopsy is considered an optimal specimen for accurate evaluation, although 15 mm is considered sufficient. Not only the length but also the caliber of the biopsy needle is important in order to obtain a piece of liver of adequate size for histological evaluation: A 16-gauge needle is considered appropriate for percutaneous liver biopsy (Sanyal et al., 2011; Papastergiou et al., 2012). Interobserver variation is another limitation, which is related to the discordance between pathologists in biopsy interpretation. This can be as high as 25%, but the variation is less pronounced when the biopsy assessment is done by specialized liver pathologists (Papastergiou et al., 2012).

Besides the technical problems with biopsy, the procedure is unpleasant for the patient, costly, and carries a risk of rare but potentially life-threatening complications – this limits the use of liver biopsy for mass screening. Due to these limitations, the fact that NASH is usually asymptomatic and that patients usually have normal liver aminotransferases, most NASH patients are undiagnosed. It has been reported that only a minority (less than 25%) of academic gastroenterologists and hepatologists in the United States routinely perform liver biopsies in patients with presumed NASH (Lominadze et al., 2014). This level of undiagnosed patients and the need of two liver biopsies within a 48- to 72-week period establish a significant hurdle in the enrollment of patients in clinical trials.

## 14.3.1 Endpoints in Clinical Trials

In spite of all these limitations, liver histology currently offers the best short-term method not only for grading and staging but also for tracking the progression of NASH.

The primary objective of the treatment for NASH is to prevent liver-related morbidity and mortality, due mainly to the development of cirrhosis, which generally takes more than 10–20 years to develop. Due to this long natural history, there is a need of surrogate markers of avoidance of cirrhosis and thus liver-related mortality. The main predictor of disease progression is increasing fibrosis. On the other hand, patients with steatohepatitis are more likely to have a progressive disease compared to patients with isolated fatty liver. Therefore, complete resolution of NASH (i.e., absence of ballooning with no or minimal inflammation) with no worsening of fibrosis, and actual improvement in fibrosis are recommended as "surrogate endpoints, reasonably likely to predict clinical benefit in progression to cirrhosis and liver-related death" (Sanyal et al., 2014) (Table 14.1). Ideally, a co-primary endpoint of two composite endpoints, one-complete resolution of NASH with no worsening of fibrosis and, two-at least one-point improvement in the fibrosis stage with no worsening of steatohepatitis, should be demonstrated in the marketing authorization trial. However, the clinical outcome study aiming to demonstrate a reduction in progression to cirrhosis and portal hypertension/cirrhosis-related events needs to be demonstrated.

For dose-ranging trials, the histology endpoint of improvement in activity as assessed by a reduction in at least two points in NAS (including at least one point in ballooning or inflammation) is an acceptable surrogate marker of improvement. It is important to note that although NAS has proven useful for comparative analyses and interventional studies, it does not provide information about fibrosis or the location of lesions. Therefore, the reduction in *NAS must be associated with a lack of progression in fibrosis.*

**TABLE 14.1**

Endpoints and Population in Clinical Trials in NASH

| Phase | Primary Endpoint | Target Population |
|---|---|---|
| Trials to support a marketing application | Composite endpoint: Complete resolution of steatohepatitis and no worsening of fibrosis<br>Composite endpoint: At least one-point improvement in fibrosis with no worsening of steatohepatitis (no increase in steatosis, ballooning, or inflammation)<br>Clinical outcome underway by the time of submission:<br>Histopathologic progression to cirrhosis<br>MELD score change by >2 points or MELD increase to >15 in population enrolled with ≤13<br> • Death<br> • Transplant<br> • Decompensation events<br>  • Hepatic encephalopathy – West Haven ≥ grade 2<br>  • Variceal bleeding – requiring hospitalization<br>  • Ascites – requiring intervention<br>  • Spontaneous bacteria peritonitis | Biopsy-confirmed NASH patients with moderate/advanced fibrosis (F2/F3) |
| Dose ranging/phase II | Improvement in activity (NAS)/ballooning/inflammation without worsening of fibrosis can be acceptable<br>Include a subpopulation with moderate/advanced fibrosis (F2/F3) to inform phase III | Biopsy-proven NASH (NAS ≥ 4)<br>Include patients with NASH and liver fibrosis with any stage of fibrosis<br>Include patients with NASH and ≥ fibrosis stage 2 to inform phase III |
| Early-phase trials/POC | Endpoints should be based on the mechanism of drug<br>Consider using improvement in NAS (ballooning and inflammation) and/or fibrosis<br>Reduction in liver fat with a sustained improvement in transaminases | Ideal to use patients with biopsy-proven NASH, but acceptable to use patients at high risk for NASH (fatty liver + T2DM, the metabolic syndrome, and high transaminases are acceptable) |

*Source:* Filozof et al. (2017).

Biopsy-driven endpoints are, in general, not feasible in a 12- to 24-week POC trials. In short-term POC studies, which are designed mainly to assess tolerability of new drugs and to look for futility signals to direct decisions regarding further development, an improvement of hepatic steatosis, as determined by magnetic resonance technology, might be suitable since improvement in steatohepatitis is generally associated with a reduction in liver fat (Sanyal et al., 2011). Improvement in liver aminotransferase and other noninvasive biomarkers of insulin sensitivity, inflammation, apoptosis, and fibrosis could be helpful to evaluate the efficacy of the compound and support decision-making. However, it is important to note that the use of noninvasive biomarker methods is still considered experimental and there are no validated noninvasive biomarkers.

## 14.3.2 Target Population

Prevention of cirrhosis and demonstrating a positive effect on the well-defined liver outcomes are key clinical goals when considering a NASH drug development program. Therefore, for trials aiming to support marketing application, it is important that subjects with the greatest risk of progression to cirrhosis are enrolled (Table 14.1). Among individual features, liver fibrosis has proven the best independent association with liver-related mortality. Patients with NASH develop progressive fibrosis in 25%–50% of individuals over 4–6 years, while 15%–25% of individuals with NASH can progress to cirrhosis (Musso et al., 2011). In another study, with a mean follow-up of 13 years, 13.3% of NASH patients with mild to moderate fibrosis (stages 1–2) and 50% of patients with fibrosis stage 3 at inclusion developed cirrhosis (Ekstedt et al., 2006). Since in patients with NASH and advanced fibrosis (F2–F3), the probability of developing cirrhosis is much higher than in patients with early fibrosis (F1), this population is recommended for the long-term outcome trials in order to enhance the chances of demonstrating a benefit within a reasonable time frame. The enrollment of patients with moderate/advanced fibrosis for the evaluation of long-term outcomes, including progression to cirrhosis, should ensure that an expected number of events, calculated based on the progression rate for each fibrosis stage, are obtained based on the literature (see, e.g., Ekstedt et al., 2006; Argo et al., 2009; Pagadala and McCullough, 2012; Angulo et al., 2015; Singh et al., 2015). In patients with NASH and advanced fibrosis (F2–F3), this progression rate can be estimated at 8% per year for fibrosis stage 3, and 6% per year for fibrosis stage 2. Since the progression rate in some patients with mild fibrosis with additional risk factors of progression (e.g., presence of T2DM, the metabolic syndrome, high transaminases) might be fast, it is worth exploring this subgroup of patients, as an additional exploratory group.

A broad population of NASH patients, including those with mild fibrosis, is acceptable in dose-ranging (phase II) trials. However, it is recommended that a sufficient number of patients with moderate and severe fibrosis are enrolled in order to get preliminary data to inform the trial(s) to support marketing application.

Ideally, in early POC trials, the target population should also be patients with biopsy-confirmed NASH. However, patients at high risk of NASH, namely, patients with fatty liver and diabetes and/or the metabolic syndrome with or without high liver enzymes, can be acceptable at this stage. Noninvasive serum biomarkers of imaging can be used to enrich a population in a POC trial (EA, 2015).

## 14.4 Adaptive Design Approach in NASH

### 14.4.1 Statistical Considerations

One of the most commonly considered adaptive designs is probably a multiple-stage seamless adaptive design that combines several independent studies into a single study that can address the study objectives of the intended individual studies. For two-stage adaptive designs, four categories have been reported (Chow and Tu, 2008; Chow and Lin, 2015) depending upon their study objectives and study endpoints at different stages. These categories include (i) designs with the same study objectives and study endpoints at different stages, (ii) designs with the same study objectives but different study endpoints at different stages, (iii) designs with different study objectives but the same study endpoints at different stages, and (iv) designs with different study objectives and different study endpoints at different stages. Details regarding different categories of two-stage adaptive seamless designs can be found in Chapter 11. The use of the adaptive design has several advantages and some limitations (Table 14.2). Though in the 2010 FDA draft guidance (see also, FDA, 2018), the two-stage seamless adaptive design was emphasized as a less well-understood design because the valid statistical methods were not

**TABLE 14.2**

Relative Merits and Limitations of Two-Stage Adaptive Design in NASH

| Characteristic | Two Independent Trials | Two-Stage Adaptive Design |
|---|---|---|
| Power | 90% 90% (81%) | 90% |
| Sample size | $N = N + N_2$ | $N < N_1 + N_2$[a] |
| Operational bias | Less | Moderate to severe |
| Data analysis | By study analysis | Combined analysis |
| Efficiency | 6–12 months of lead time between studies | Reduced lead time between trials |
| Flexibility/ long-term follow-up | New study design based on previous data. New patients are enrolled | Adaptations based on IA: Stop one or more study arms/randomize more patients, etc. Continue follow-up |
| Regulatory aspects | Standard practice | Requires buy-in by global authorities prior to initiation |
| Statistical perspective | Valid statistical methods are well established | Evolving statistical methods |
| Operational complexity | Low | High |

*Source:* Filozof et al. (2017).
*Note:*
[a] Depends on adaptations.
$N_1$ is the sample size for trial 1.
$N_2$ is the sample size for trial 2.

fully developed at that time (Chow and Corey, 2011), the two-stage adaptive design has gradually become well understood due to the increasing number of regulatory submissions utilizing this design (see, e.g., Lu et al., 2009, 2010, 2012).

### 14.4.2 Utility Function for Endpoint Selection

We recommend a utility function be adopted to link all NASH endpoints at different stages (Chow and Lin, 2015). We propose a therapeutic index function for the analysis of two-stage adaptive designs with distinct study objectives and study endpoints at each stage under the assumption that even though the two endpoints are not the same, there is a well-established relationship between them.

A therapeutic index function is defined for each one of the endpoints. It takes different endpoints with pre-specified criteria into consideration, and it is based on a vector of therapeutic index function rather than the individual endpoints. The vector of therapeutic index model allows the investigator to accurately and reliably assess the treatment effect in a more efficient way (see Appendix: Utility Function and Statistical Tests). The direct application of standard statistical methods leads to the concern that the obtained $p$-value and confidence interval for the assessment of the treatment effect may not be correct or reliable. Most importantly, the sample size required for achieving a desired power obtained under a standard group sequential trial design may not be sufficient for achieving the study objectives under the two-stage seamless adaptive trial design, especially when the study objectives and/or study endpoints at different stages are different.

## 14.5  Clinical Development NASH Program

### 14.5.1  Marketing Authorization Trial

A single phase III–IV seamless adaptive trial design (Figure 14.1), rolling over patients from phase III into phase IV, allows for continuous exposure and long-term follow-up of NASH patients. At present, an interim analysis confirming efficacy in resolution of NASH by histology without worsening of fibrosis and improvement in fibrosis without worsening of NASH might lead to accelerated/conditional marketing authorization, but the clinical outcome study is required for full marketing authorization.

For this phase III/IV NASH trial at which the drug is intended to be approved first based on a surrogate endpoint, the sample size needs to be planned by considering the interim alpha at the end of phase III as well as the final alpha at phase IV. Although the primary endpoints for both the phase III and phase IV are supposed to be positively correlated, we recommend the split of alpha (e.g., 0.01 and 0.04) for the two separate tests to ensure

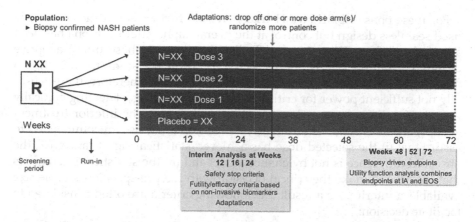

Population:
▶ Biopsy confirmed NASH patients

Adaptations: drop off one or more dose arm(s)/
randomize more patients

N XX

R

Weeks

N=XX   Dose 3
N=XX   Dose 2
N=XX   Dose 1
Placebo = XX

0    12    24    36    48    60    72

Screening period    Run-in

**Interim Analysis at Weeks 12 | 16 | 24**
Safety stop criteria
Futility/efficacy criteria based on non-invasive biomarkers
Adaptations

**Weeks 48 | 52 | 72**
Biopsy driven endpoints
Utility function analysis combines endpoints at IA and EOS

**FIGURE 14.1**
POC/dose-ranging adaptive design trial. (*Note*: EOS=end of study; IA=interim analysis; $N$=number of subjects per study arm; $R$=randomized patients. Filozof et al. (2017).)

the type I error control of the entire study as well as to take into account the assessments of other potential key secondary endpoints. Since the accelerated approval is likely to be considered on the basis of a "soft" surrogate endpoint, we strongly recommend that a much smaller alpha is allocated to the first test compared to the second one.

The total number of events in the phase IV can be estimated assuming an average progression rate of 7% in NASH patients with F2 F3 (see, e.g., Ekstedt et al., 2006; Argo et al., 2009; Pagadala and McCullough, 2012; Angulo et al., 2015; Singh et al., 2015). However, the real statistical inference will be evaluated when X number of events is reached. Additionally, an analysis of this interim safety data is recommended and should be performed by an independent unblinded data analysis group (DAG) that is not affiliated with the trial's sponsor.

## 14.5.2 Adaptive Design in Early Phases

A single POC dose-finding study utilizing seamless adaptive trial design allowing patients to continue from one phase to the other at a chosen dose/doses might also have some advantages in early phases. Figure 14.1 summarizes the POC/dose-ranging adaptive design trial. Such design would enroll patients with biopsy-confirmed NASH and allow continuous exposure rolling over those patients on the most promising doses and allowing adaptations during the interim analysis as required. The evaluation of changes in liver fat (as assessed by magnetic resonance technology) and other non-invasive biomarkers of liver function, inflammation, and fibrosis could help in decision-making during the IA after a pre-specified period of 12–24 weeks. At this point, efficacy and futility analysis will allow adaptations (i.e., drop off study arm(s), randomize more patients/stop the study if a safety concern arises). The continuous follow-up will allow the evaluation of changes in liver histology in the selected dose(s).

For these phase IIb/III NASH trials, one can further adopt the commonly used seamless design but control at the overall alpha after the alpha has been adjusted for phase III and phase IV. For this type of phase IIb/III adaptive design, the sample size is often selected to power the study for detecting a meaningful treatment effect at the end of the phase III. Therefore, there is usually not sufficient power for critical decision-making (e.g., detecting clinically meaningful difference for dose selection or dropping the inferior treatment groups) at earlier stages. In this case, a precision analysis is recommended to assure that (i) the selected dose has achieved statistical significance (i.e., the observed difference is not by chance alone) and (ii) the statistical inference is that we have to make the critical decision with the pre-specified sample size available at interim. As a result, the following criteria are often considered to facilitate decision:

i. The dose with the highest confidence level for achieving statistical significance will be selected;

ii. The doses with confidence levels for achieving statistical significance less than 75% will be dropped.

It should also be noted that precision assessment in terms of confidence interval approach is operationally equivalent to hypotheses testing for comparing means. Of note, as the interim analysis conducted at the end of phase II is usually for dose selection, not stopping for efficacy, sometimes a trend test may be sufficient, and to protect the trial integrity due to the unblinding of the interim analysis results, a tiny alpha (e.g., 0.0001) needs to be allocated.

For illustration purpose, Figures 14.2 and 14.3 summarize phase III/IV adaptive design and phase II/III/IV adaptive design, respectively, which are sometimes considered by the sponsors in NASH drug research and development.

A single phase III/IV adaptive design allows for continuous exposure and the long-term follow-up. A therapeutic index (or utility) function can be adopted to link all NASH endpoints at different stages. Furthermore, different pre-specified weights can be allocated in the function. Endpoints at the interim analysis are (i) resolution of NASH by histology without worsening of fibrosis and/or (ii) improvement in fibrosis without worsening of NASH. If positive, a long-term follow-up to confirm efficacy in reduction in clinical outcome is mandatory. It is important to ensure type I error control. At present, because marketing authorization is based on a surrogate endpoint that is reasonably likely to predict benefit on morbidity or mortality, "based on epidemiologic data" but not "surrogates that are validated by definitive studies," a smaller alpha is allocated to the first test compared to the second one (e.g., 0.01 and 0.04, respectively). Abbreviations: $N$, number of subjects per study arm; $R$, randomized patients.

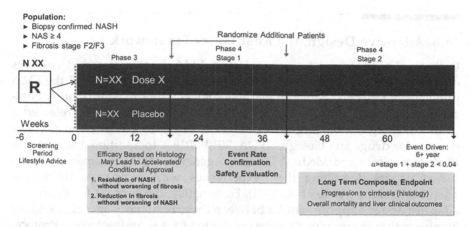

**FIGURE 14.2**
Phase III/IV adaptive design. (Filozof et al. (2017).)

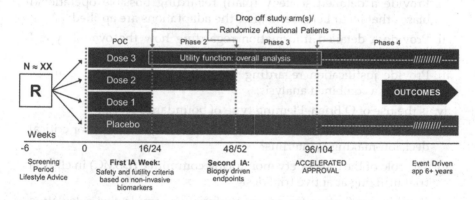

**FIGURE 14.3**
Phase II/III/IV adaptive design. (Filozof et al. (2017).)

On the other hand, a phase II/III/IV adaptive design allows adaptations, continuous exposure, and long-term follow-up. Endpoints and interim analysis are reduction of at least 2 points in NAS, resolution of NASH by histology without worsening of fibrosis, and/or improvement in fibrosis without worsening of NASH. One (the most promising dose) or two doses may continue to the next phase. A post-marketing phase IV with the demonstration of improvement in clinical outcomes will lead to final marketing authorization. Because only one trial would lead to approval, a very small overall alpha (i.e., <0.001) is recommended to ensure proper control of a type I error.

## 14.6  Adaptive Design, the Regulatory Framework

Both the US FDA and European Union EMA have released documents describing the use of adaptive designs in clinical studies. In 2007, the EMA published a reflection paper on methodological issues in confirmatory clinical trials planned with an adaptive design, and in 2010, FDA released a draft guidance, and in 2019, FDA published a revised guidance on adaptive designs for drugs and biologics (FDA, 2010, 2019c). In addition, the Japanese Pharmaceuticals and Medical Devices Agency (PMDA) in Japan has published papers describing their experiences with companies discussing adaptive designs (Ando et al., 2011). Fortunately, the concept for the use of adaptive designs appears similar between these regulatory agencies, with an emphasis that the designs must be supported by a sound rationale. Protocol should address the typical following issues:

i. Provide a detailed strategy (plan) regarding possible operational biases that incur before and after the adaptations are applied;

ii. Provide a detailed information regarding how the overall type I error rate is controlled or preserved;

iii. Provide justification regarding the validity of statistical methods used for a combined analysis;

iv. Is the use of O'Brien–Fleming type of boundaries feasible?

v. Are there any risk assessments for the precision analysis for critical decision-making at interims?

vi. The role of the data safety monitoring committee (DSMC) in clinical trial utilizing adaptive trial design;

vii. Provide justification for power analysis for sample size calculation and sample size allocation especially where the study objectives and study endpoints are different at different stages;

viii. How a blinded sample size re-estimation will be performed in the two-stage adaptive design.

## 14.7  Concluding Remarks

The prevalence of NAFLD and NASH is increasing worldwide. NAFLD is the most common cause of chronic liver disease in the Western World today. NASH is currently the second cause for liver transplantation, and it is expected to be the first one in 2020. Due to the increasing prevalence and the health burden, there is a high need to develop therapeutic strategies for patients with NASH. Unfortunately, the long and bidirectional (i.e., some patients progressing but other ones regressing spontaneously) natural history of the disease and the requirement for sequential liver biopsies create substantial challenges in clinical development. Despite considerable research and multiple clinical trials, at present there is no approved therapy.

The primary objective of the treatment for NASH is to prevent liver-related morbidity and mortality, which generally takes more than 10–20 years to develop. It is not feasible to perform such studies, and regulatory agencies like US FDA are willing to look at surrogate endpoints for approval. The use of surrogate endpoints for accelerated and conditional approval with a later confirmation of the positive impact of the compound in clinical outcomes will facilitate clinical development programs and bring therapeutic options to patients. On the other hand, adaptive seamless clinical designs in NASH provide an opportunity to shorten the overall path to registration for a new drug, and offer continuity for patients enrolled in the study. Here, we review the statistical and logistical characteristics for applying adaptive designs to clinical trials in NASH.

While not all indications will be suited to product development via the use of adaptive designs, there is a growing experience on the use of these paradigms to gain marketing authorizations on a global basis. The development of drugs for NASH appears to have a strong rationale to fit well into the current regulatory frameworks for drug development.

---

## Appendix of This Chapter: Utility Function and Statistical Tests

We recommend a utility function be adopted to link all NASH endpoints at different stages (Chow and Lin, 2015). We propose the following statistical tests for the analysis of two-stage adaptive designs with distinct study objectives and study endpoints at each stage under the assumption that even though the two endpoints are not the same, there is a well-established relationship between them.

Let $y = \{y_1, y_2, \ldots, y_m\}$ be the composite endpoint of our interest by $m$ components. At each of these endpoints, $y_i$ is a function of criteria $y_i(x)$, $x \in X$, where $X$ is a space of criteria. We can then define the following therapeutic index function for endpoints as follows:

$$U_{sk} = \sum_{j=1}^{m} w_{skj} = \sum_{j=1}^{m} w\left(y_{skj}\right), \; k = 1, \ldots, K; s = 1, \ldots, S, \qquad (14.1)$$

where $U_{sk}$ denotes the kth endpoint derived from the therapeutic index function at the sth stage of the multiple-stage adaptive design and

$w_{skj}$, $j = 1, \ldots, m; k = 1, \ldots, K$ are the pre-specified weights. As an example, if $K = 1$, $U_{sk} = U_s$, which reduces to a single composite index such as $NAS \geq 4$ and/or F2/F3 (fibrosis stage 2 and fibrosis stage 3) at the $s$th stage. When $K = 2$, the therapeutic index function for endpoints suggests co-primary endpoints, i.e., $U_{s1}$ and $U_{s2}$ at the $s$th stage. For example, we may consider the two co-primary endpoints as follows:

i. Improvement in fibrosis by at least 1 stage with no worsening of NASH defined as no increase in hepatocellular ballooning or lobular inflammation;

ii. Resolution of NASH as defined by the overall histopathological interpretation with no worsening of fibrosis.

The above therapeutic index function, which takes different endpoints with pre-specified criteria into consideration, is based on a vector of therapeutic index function rather than the individual endpoints. The vector of therapeutic index model allows the investigator to accurately and reliably assess the treatment effect in a more efficient way.

Chow and Lin (2015) suggested testing two sets of hypotheses at different stages under a two-stage adaptive design with planned interim analyses assuming that the study endpoints at different stages are different (i.e., biomarker or surrogate endpoint at the first stage and clinical endpoint at the second stage). Chow and Lin (2015) first converted the two-stage adaptive design with planned interim analyses into a multiple-stage adaptive design.

Now, suppose the intended two-stage adaptive design is to compare $k$ treatment groups, $E_1, \ldots, E_k$ with a control group $C$, and suppose a surrogate (biomarker) endpoint and a well-established clinical endpoint are available for the assessment of the treatment effect. We assume that such a biomarker or surrogate endpoint can be obtained from the utility function as described in the previous subsection. Denoted by $\theta_i$ and $\psi_i$, $i = 1, \ldots, k$ the treatment effect comparing $E_i$ with $C$ assessed by the surrogate (biomarker) endpoint and the clinical endpoint, respectively.

Similar to what described in Section 11.5.1, under the surrogate and clinical endpoints, the treatment effect can be tested by the following hypotheses:

$$H_{0,2}: \psi_1 = \ldots = \psi_k, \tag{14.2}$$

which is for the clinical endpoint, while the hypothesis

$$H_{0,1}: \theta_1 = \ldots \theta_k, \tag{14.3}$$

is for the surrogate (biomarker) endpoint. Chow and Lin (2015) assumed that $\psi_i$ is a monotone increasing function of the corresponding $\theta_i$. Their proposed tests are briefly described below. For simplicity, the variances of the surrogate (biomarker) endpoint and the clinical outcome are denoted by $\sigma^2$ and $\tau^2$, which are assumed known.

At phase II, $(k+1)n_1$ subjects are randomly assigned to receive either one of the $k$ treatments or the control at a 1:1 ratio. In this case, we have $n_1$ subjects in each group. At the first interim analysis, the most effective treatment will be selected based on the surrogate (biomarker) endpoint and proceed to subsequent phases. For pairwise comparison, consider test statistics $\hat{\theta}_{i,1}$, $i = 1,...,k$ and $S = \text{argmax}_{1 \le j \le k} \hat{\theta}_{i,1}$. Thus, if $\hat{\theta}_{S,1} \le c_1$ for some pre-specified critical value $c_1$, then the trial is stopped and we are in favor of $H_{0,1}$. On the other hand, if $\hat{\theta}_{S,1} > c_{1,1}$, then we conclude that the treatment $E_S$ is considered the most promising treatment and proceed to subsequent stages. Subjects who receive either the promising treatment or the control will be followed for the clinical endpoint. Treatment assessment for all other subjects will be terminated but will undergo necessary safety monitoring.

At phase III, $2n_2$ additional subjects will be equally randomized to receive either the treatment $E_S$ or the control $C$. An interim analysis can be scheduled when the short-term surrogate measures from some of these $2n_2$ subjects (e.g., 50% of the planned size) and the primary endpoint measures from those $2n_1$ stage 1 subjects who receive either the treatment $E_S$ or the control $C$ become available. Let $T_{1,1} = \hat{\theta}_{S,1}$ and $T_{1,2} = \hat{\psi}_{S,1}$ be the pairwise test statistics from the partial phase II data based on the surrogate endpoint and the primary endpoint, respectively, and $\hat{\theta}_{S,2}$ be the statistic from phase III based on the surrogate. If

$$T_{2,1} = \sqrt{\frac{n_1}{n_1 + n_2}}\hat{\theta}_{S,1} + \sqrt{\frac{n_2}{n_1 + n_2}}\hat{\theta}_{S,2} \le c_{2,1},$$

then stop the trial and accept $H_{0,1}$. If $T_{2,1} > c_{2,1}$, then reject $H_{0,1}$. The study will be continued till all of the planned patients obtain their final clinical outcomes. Note that to ensure the adequacy of the study power, one can consider the sample size re-estimation by comparing $T_{1,1}$ with $c_{1,1}$.

At phase IV, no additional subjects will be recruited unless the sample size re-estimation is performed and the decision is made to increase the sample size. The final analysis will be performed when all of the planned subjects complete their primary endpoints. Let

$$T_{2,2} = \sqrt{\frac{n_1}{n_1 + n_2}}\hat{\psi}_{S,1} + \sqrt{\frac{n_2}{n_1 + n_2}}\hat{\psi}_{S,2},$$

where $\hat{\psi}_{S,2}$ is the pairwise test statistic from phase IV for the patients who did not contribute any outcome data at the interim analysis. If $T_{2,2} > c_{2,2}$, then reject $H_{0,2}$ and the study drug can be claimed to be effective.

Note that the surrogate data in the first stage are used to select the most promising treatment rather than assessing $H_{0,1}$. As discussed in Section 11.5.1, under the three-stage transitional seamless design, two sets of hypotheses, namely, $H_{0,1}$ and $H_{0,2}$, are to be tested. Since the rejection of $H_{0,2}$ leads to the claim of efficacy, it is considered the hypothesis of primary interest. However, in the interest of controlling the overall type I error rate at a pre-specified level of significance, $H_{0,1}$ need to be tested following the principle of closed testing procedure to avoid any statistical penalties.

# References

Ando, Y., Hirakawa, A., and Uyama, Y. (2011). Adaptive clinical trials for new drug applications in Japan. *Eur Neuropsychopharmacol*, 21, 175–179.

Angulo, P., Kleiner, D.E., Dam-Larsen, S., Adams, L.A., Bjornsson, E.S., Charatcharoenwitthaya, P., Mills, P.R., Keach, J.C., Lafferty, H.D., Stahler, A., and Haflidadottir, S. (2015). Liver fibrosis, but no other histologic features, is associated with long-term outcomes of patients with nonalcoholic fatty liver disease. *Gastroenterology*, 149(2), 389–397.

APD (2017). Antibiotic Platform Design. https://www.berryconsultants.com/antibiotic-platform-design/.

Argo, C.K., Northup, P.G., Al-Osaimi, A.M., and Caldwell, S.H. (2009). Systematic review of risk factors for fibrosis progression in non-alcoholic steatohepatitis. *Journal of Hepatology*, 51(2), 371–379.

Bauer, P. and Kohne, K. (1994). Evaluation of experiments with adaptive interim analysis. *Biometrics*, 50, 1029–1041.

Bauer, P. and Rohmel, J. (1995). An adaptive method for establishing a dose-response relationship. *Statistics in Medicine*, 14, 1595–1607.

Berry, S.M., Petzold, E.A., Dull, P., Thielman, N.M., Cunningham, C.K., Corey, G.R., McClain. M.T., Hoover, D.L., Russell, J., Griffiss, J.M., and Woods, C.W. (2016). A response adaptive randomization platform trial for efficient evaluation of Ebola virus treatments: a model for pandemic response. *Clinical Trials*, 23(1), 22–30.

Bonferroni, C. E. (1936). Teoria statistica delle classi e calcolo delle probabilità. *Pubblicazioni del R Istituto Superiore di Scienze Economiche e Commerciali di Firenze*, 8, 1–62.

Bretz, F., Hothorn, T., and Westfall, P. (2010). *Multiple Comparisons Using R*. Chapman and Hall, CRC Press. New York.

Brunt, E.M., Janney, C.G., Di Bisceglie, A.M., Neuschwander-Tetri, B.A., and Bacon, B.R. (1999). Nonalcoholic steatohepatitis: a proposal for grading and staging the histological lesions. *American Journal of Gastroenterology*, 94(9), 2467–2474.

Brunt, E.M., Kleiner, D.E., Wilson, L.A., Belt, P., **Neuschwander-Tetri**, B.A., and NASH Clinical Research Network (CRN). (2011). Nonalcoholic fatty liver disease (NAFLD) activity score and the histopathologic diagnosis in NAFLD: distinct clinicopathologic meanings. *Hepatology*, 53(3), 810–820.

Cardoso, F., Van't Veer, L.J., Bogaerts, J., Slaets, L., Viale, G., Delaloge, S., Pierga, J.Y., Brain, E., Causeret, S., DeLorenzi, M., and Glas, A.M. (2016). 70-gene signature as an aid to treatment decisions in early-stage breast cancer. *New England Journal of Medicine*, 375(8), 717–729.

Chang, M. (2007). Adaptive design method based on sum of *p*-values. *Statistics in Medicine*, 26, 2772–2784.

Chao, A, C., Liu, C.K., Chen, C.H., Lin, H.J., Liu, C.H., Jeng, J.H., Hu, C.J., Chung, C.P., Hsu, H.Y. Sheng, W.Y., and Hu, H.H. (2014). Different doses of recombinant tissue-type plasminogen activator for acute stroke in Chinese patients. *Stroke*, 45, 2359–2365.

Chow, S.C. (2000) (Ed.). *Encyclopedia of Biopharmaceutical Statistics*. Marcel Dekker, Inc., New York.

Chow, S.C. (2011). *Controversial Issues in Clinical Trials.* Chapman and Hall/CRC, Taylor & Francis, New York.

Chow, S.C. (2013). *Biosimilars: Design and Analysis of Follow-On Biologics.* Chapman Hall/CRC Press, Taylor & Francis, New York.

Chow, S.C. (2019). Innovative Statistics in Regulatory Science. Chapman and Hall/CRC Press, Taylor & Francis, New York.

Chow, S.C. (2020). Innovative thinking for rare disease drug development. *American Journal of Biomedical Science & Research,* 7(3). doi: 10.34297/AJBSR.2020.07.001159.

Chang, M. and Chow, S.C. (2005). A hybrid Bayesian adaptive design for dose response trials. *Journal of Biopharmaceutical Statistics,* 15, 677–691.

Chow, S.C. and Chang, M. (2011). *Adaptive Design Methods in Clinical Trials,* Second Edition. Chapman Hall/CRC Press, Taylor & Francis, New York.

Chow, S.C., Chang, M., and Pong, A. (2005). Statistical consideration of adaptive methods in clinical development. *Journal of Biopharmaceutical Statistics,* 15, 575–591.

Chow, S.C. and Chang, Y.W. (2019). Statistical considerations for rare diseases drug development. *Journal of Biopharmaceutical Statistics,* 29, 874–886.

Chow, S.C. and Corey, R. (2011). Benefits, challenges and obstacles of adaptive clinical trial designs. *Orphanet Journal of Rare Disease,* 6, 79–89.

Chow, S.C. and Huang, Z. (2019a). Demonstrating effectiveness or demonstrating not ineffectiveness – A potential solution for rare disease drug development. *Journal of Biopharmaceutical Statistics,* 29, 897–907.

Chow, S.C. and Huang, Z. (2019b). Innovative thinking on endpoint selection in clinical trials. *Journal of Biopharmaceutical Statistics,* 29, 941–951.

Chow, S.C. and Huang, Z. (2020). Innovative approach for rare disease drug development. *Journal of Biopharmaceutical Statistics,* 30, 537–539. doi; 10.1080/10543406.2020.1726371.

Chow, S.C. and Lin, M. (2015). Analysis of two-stage adaptive seamless trial design. *Pharmaceutica Analytica Acta,* 6(3), 1–10. doi: 10.4172/2153-2435.1000341.

Chow, S.C. and Liu, J.P. (2008a). *Design and Analysis of Bioavailability and Bioequivalence Studies,* Third Edition. Taylor & Francis, New York, vol. 25, pp. 237–241.

Chow, S. C., and Liu, J.P. (2008b). *Design and Analysis of Clinical Trials: Concepts and Methodologies.* John Wiley & Sons, New York.

Chow, S.C., Lu, Q.S., and Tse, S.K. (2007). Statistical analysis for two-stage seamless design with different study endpoints. *Journal of Biopharmacutical Statistics,* 17, 1163–1176.

Chow, S.C. and Shao, J. (2002). *Statistics in Drug Research – Methodologies and Recent Development.* Marcel Dekker, Inc., New York.

Chow, S.C. and Shao. J. (2005). Inference for clinical trials with some protocol amendments. *Journal of Biopharmaceutical Statistics,* 15, 659–666.

Chow, S.C., Shao, J., Wang, H., and Lokhnygina, Y. (2017). *Sample Size Calculations in Clinical Research,* Third Edition. Chapman and Hall/CRC Press, Taylor & Francis, New York.

Chow, S.C., Song, F.Y., and Cui, C. (2017). On hybrid parallel-crossover designs for assessing drug interchangeability of biosimilar products. *Journal of Biopharmaceutical Statistics,* 27, 265–271.

Chow, S.C. and Tu, Y.H. (2008). On two-stage seamless adaptive design in clinical trials. *Journal of Formosa Medical Association,* 107, S52–S60.

Chow, S.C., Xu, H., Endrenyi, L., and Song, F.Y. (2015). A new scaled criterion for drug interchangeability, *Chinese Journal of Pharmaceutical Analysis*, 35(5), 844–848.

Coors, M., Bauer, L., Edwards, K., Erickson, K., Goldenberg, A., Goodale, J., Goodman, K., Grady, C., Mannino, D., Wanner, A., Wilson, T., Yarborough, M., and Zirkle, M. (2017). Ethical issues related to clinical research and rare diseases. *Translational Science of Rare Diseases*, 2, 175–194.

Corrigan-Curay, J. (2018). Real-World Evidence – FDA Update. Presented at RWE Collaborative Advisory Group Meeting, Duke-Margolis Center, Washington, DC, October 1, 2018.

Cui, L., Hung, H.M.J., and Wang, S.J. (1999). Modification of sample size in group sequential clinical trials. *Biometrics*, 55, 853–857.

Dai, B., Ding, S., and Wahba, G. (2013). Multivariate Bernoulli distribution. *Bernoulli*, 19(4), 1465–1483.

Dmitrienko, A., Tamhane, A.C., and Bretz, F. (2010). *Multiple Testing Problems in Pharmaceutical Statistics*. Chapman & Hall, CRC Press, New York.

Duan, Y., Ye, K., and Smith, E.P. (2006). Evaluating water quality using power priors to incorporate historical information. *Environmetrics*, 17, 95–106.

EA (2015). European Association for Study of, L. and H. Asociacion Latinoamericana para el Estudio del, EASL-ALEH clinical practice guidelines: non-invasive tests for evaluation of liver disease severity and prognosis. *Journal of Hepatology*, 63(1), 237–264.

Efron, B., and Tibshirani, R. J. (1994). *An Introduction to the Bootstrap*. CRC Press. New York.

Ekstedt, M., Franzén, L.E., Mathiesen, U.L., Thorelius, L., Holmqvist, M., Bodemar, G., and Kechagias, S. (2006). Long-term follow-up of patients with NAFLD and elevated liver enzymes. *Hepatology*, 44(4), 865–873.

EMA (2014). *Pilot Project on Adaptive Licensing*. European Medicines Agency, London. https://www.ema.europa.eu/en/documents/other/pilot-project-adaptive-licensing_en.pdf.

Endrenyi, L., Declerck, P., and Chow, S.C. (2017). *Biosimilar Product Development*. Taylor & Francis, London.

EPAD (2018). European Prevention of Alzheimer's Dementia Consortium. http://ep-ad.org/.

Esserman, L., Hylton, N., Asare, S., Yau, C., Yee, D., DeMichele, A., Perlmutter, J., Symmans, F., van't Veer, L., Matthews, J., Berry, D.A., and Barker, A. (2019). I-SPY2 unlocking the potential of the platform trial. In *Platform Trials in Drug Development*, Eds. Antonijevic, Z. and Beckman, R.A. Chapman and Hall/CRC Press, Taylor & Francis, New York, pp. 3–21.

FDA (1998). *Guidance for Industry – Providing Clinical Evidence of Effectiveness for Human Drug and Biologic Products*, The United States Food and Drug Administration, Silver Spring, MD. https://www.fda.gov/downloads/Drugs/GuidanceComplianceRegulatoryInformation/Guidances/UCM072008.pdf.

FDA (2001). *Guidance for Industry – Statistical Approaches for Evaluation of Bioequivalence*. Center for Drug Evaluation and Research, the United States Food and Drug Administration, Rockville, MD.

FDA (2003). *Guidance for Industry – Bioavailability and Bioequivalence Studies for Orally Administrated Drug Products – General Considerations*. Center for Drug Evaluation and Research, the United States Food and Drug Administration, Rockville, MD.

FDA (2010). *Draft Guidance for Industry – Adaptive Design Clinical Trials for Drugs and Biologics.* The United States Food and Drug Administration, Silver Spring, MD.

FDA (2014a). *Guidance for Industry – Expedited Programs for Serious Conditions – Drugs and Biologics.* The United States Food and Drug Administration, Silver Spring, MD.

FDA (2014b). *Guidance for Industry – Immunogenicity Assessment for Therapeutic Protein Products.* The United States Food and Drug Administration, Silver Spring, MD. https://www.fda.gov/downloads/drugs/guidances/ucm338856.pdf.

FDA (2015a). *Guidance for Industry – Scientific Considerations in Demonstrating Biosimilarity to a Reference Product.* The United States Food and Drug Administration, Silver Spring, MD.

FDA (2015b). *Guidance for Industry – Rare Diseases: Common Issues in Drug Development.* Center for Drug Evaluation and Research, the United States Food and Drug Administration, Silver Spring, MD.

FDA (2015c). *Guidance for Industry – Considerations for the Design of Early-Phase Clinical Trials of Cellular and Gene Therapy Products.* Center for Drug Evaluation and Research, the United States Food and Drug Administration, Silver Spring, MD. https://www.fda.gov/downloads/ BiologicsBloodVaccines/GuidanceComplianceRegulatoryInformation/ Guidances/CellularandGeneTherapy/UCM564952.pdf.

FDA (2016). *Guidance for Industry – Non-Inferiority Clinical Trials to Establish Effectiveness.* The United States Food and Drug Administration, Silver Spring, MD.

FDA (2017a). *Guidance for Industry – Considerations in Demonstrating Interchangeability with a Reference Product.* The United States Food and Drug Administration, Silver Spring, MD.

FDA (2017b). *Use of Real-World Evidence to Support Regulatory Decision-Making for Medical Device.* Guidance for Industry and Food and Drug Administration staff, US Food and Drug Administration, Silver Spring, MD.

FDA (2018a). *Draft Guidance for Industry – Long Term Follow-Up After Administration of Human Gene Therapy Products.* The United States Food and Drug Administration, Silver Spring, MD.

FDA (2018b). *Guidance for Industry – Human Gene Therapy for Rare Diseases.* The United States Food and Drug Administration, Silver Spring, MD.

FDA (2018c). *Guidance for Industry – Master Protocols: Efficient Clinical Trial Design Strategies to Expedite Development of Oncology Drugs and Biologics.* The United States Food and Drug Administration, Silver Spring, MD.

FDA (2018d). *Guidance for Industry – Noncirrhotic Nonalcoholic Steatohepatitis with Liver Fibrosis: Developing Drugs for Treatment.* The United States Food and Drug Administration, Silver Spring, MD.

FDA (2019a). *Guidance for Industry – Rare Diseases: Common Issues in Drug Development.* The United States Food and Drug Administration, Silver Spring, MD.

FDA (2019b). *Framework for FDA's Real-World Evidence Program.* The United States Food and Drug Administration, Silver Spring, MD.

FDA (2019c). *Guidance for Industry – Adaptive Designs for Clinical Trials of Drugs and Biologics.* The United States Food and Drug Administration, Silver Spring, MD.

Filozof, C., Chow, S.C., Dimick-Santos, L., Chen, Y.F., Williams, R.N., Goldstein B.J., and Sanyal, A. (2017). Clinical endpoints and adaptive clinical trials in precirrhotic nonalcohotic steatohepatitis: facilitating development approaches for an emerging epidemic. *Hepatology Communications*, 1, 577–585.

Finner, H. and Strassburger, K. (2002). The partitioning principle: a powerful tool in multiple decision theory. *Annals of Statistics*, 30(4), 1194–1213.

Grady, C. (2017). Ethics ad IRBs, proposed changes to the common rule, and rare disease research. *Translational Science of Rare Diseases*, 2, 176–178.

Gravestock, I. and Held, L. on behalf of the COMACTE-Net consortium (2017). Adaptive power priors with empirical Bayes for clinical trials. *Pharmaceutical Statistics*, 16, 349–360.

Haidar, S.H., Davit, B., Chen, M.L., Conner, D., Lee, L., Li, Q.H., Lionberger, R., Makhlouf, F., Patel, D., Schuirmann, D.J., and Yu, L.X. (2008), Bioequivalence approaches for highly variable drugs and drug products. *Pharmaceutical Research*, 25, 237–241.

He, J., Kang, Q., Hu, J., Song, P., and Jin, C. (2018). China has officially released its first national list of rare diseases. *Intractable & Rare Diseases Research*, 7(2), 145–147.

Herbst, R.S., Gandara, D.R., Hirsch, F.R., Redman, M.W., LeBlanc, M., Mack, P.C., Schwartz, L.H., Vokes, E., Ramalingam, S.S., Bradley, J.D., and Sparks, D. (2015). Lung master protocol (Lung-MAP) – a biomarker-driven protocol for accelerating development of therapies for squamous cell lung cancer: SWOG S1400. *Clinical Cancer Research*, 21(7), 1514–1524.

Hochberg, Y. (1988). A sharper Bonferroni's procedure for multiple tests of significance. *Biometrika*, 75, 800–803.

Holm, S. (1979). A sharper Bonferroni procedure for multiple tests of significance. *Scandinavian Journal of Statistics*, 6, 65–70.

Hommel, G. (2001). Adaptive modifications of hypotheses after an interim analysis. *Biometrical Journal*, 43, 581–589.

Hommel, G., Lindig, V., and Faldum, A. (2005). Two-stage adaptive designs with correlated test statistics. *Journal of Biopharmaceutical Statistics*, 15, 613–623.

Huang, Z. and Chow, S.C. (2019). Probability monitoring procedure for sample size determination. *Journal of Biopharmaceutical Statistics*, 29, 887–896.

Ibrahim, J.G. and Chen, M.H. (2000). Power prior distributions for regression models. *Statistical Science*, 15, 46–60.

ICH (2000). International Conference on Harmonization Harmonized Tripartite Guideline. *Choice of Control Group and Related Issues in Clinical Trials*, E10. Geneva, Switzerland.

Jennison, C. and Turnbull, B.W. (2006). Adaptive and nonadaptive group sequential tests. *Biometrika*, 93, 1–21.

Khatri, C.G. and Shah, K.R. (1974). Estimation of location of parameters from two linear models under normality. *Communications in Statistics – Theory and Methods*, 3, 647–663.

Kleiner, D.E., Brunt, E.M., Van Natta, M., Behling, C., Contos, M.J., Cummings, O.W., Ferrell, L.D., Liu, Y.C., Torbenson, M.S., Unalp-Arida, A., and Yeh, M. (2005). Design and validation of a histological scoring system for nonalcoholic fatty liver disease. *Hepatology*, 41(6), 1313–1321.

Lehmacher, W. and Wassmer, G. (1999). Adaptive sample size calculations in group sequential trials. *Biometrics*, 55, 1286–1290.

Leisch, F., Weingessel, A., and Hornik, K. (1998). On the Generation of Correlated Artificial Binary Data. Working Paper Series No. 13, *Adaptive Information Systems and Modeling in Economics and Management Science*. Vienna University of Economics and Business Administration, Vienna.

Li, G., Shih, W.C.J., and Wang, Y.N. (2005). Two-stage adaptive design for clinical trials with survival data. *Journal of Biopharmaceutical Statistics*, 15, 707–718.

Lilford, R. J., Thornton, J. G., and Braunholtz, D. (1995). Clinical trials and rare diseases: a way out of a conundrum. *British Medical Journal*, 311(7020), 1621–1625.

Liu, G.F. (2017). A dynamic power prior for borrowing historical data in noninferiority trials with binary endpoint. *Pharmaceutical Statistics*, 17, 61–73.

Liu, Q. and Chi, G.Y.H. (2001). On sample size and inference for two-stage adaptive designs. *Biometrics*, 57, 172–177.

Lominadze, Z., Harrison, S., Charlton, M., Loomba, R., Neuschwander-Tetri, B., Cald-well, S., Kowdley, K., and Rinella, M. (2014). Survey of Diagnostic and Treatment Patterns of NAFLD and NASH in the United States: Real Life Practices Differ from Published Guidelines. *Program and Abstracts of the 65th Annual Meeting of the American Association for the Study of Liver Diseases (AASLD)*, November 7–11, 2014, Boston, MA. Abstract 838.

Lu, Q., Chow, S.C., Tse, S.K., Chi, Y., and Yang, L.Y. (2009). Sample size estimation based on event data for a two-stage survival adaptive trial with different durations. *Journal of Biopharmaceutical Statistics*, 19(2), 311–23.

Lu, Q., Tse, S.K., and S.C. Chow (2010). Analysis of time-to-event data under a two-stage survival adaptive design in clinical trials. *Journal of Biopharmaceutical Statistics*, 20(4), 705–719.

Lu, Q., Tse, S.K., Chow, S.C., and Lin, M. (2012). Analysis of time-to-event data with nonuniform patient entry and loss to follow-up under a two-stage seamless adaptive design with weibull distribution. *Journal of Biopharmaceutical Statistics*, 22(4), 773–784.

Lu, Y., Kong, Y., and Chow, S.C. (2017). Analysis of sensitivity index for assessing generalizability in clinical research. *Jacobs Journal of Biostatistics*, 2(1), 9–19.

Ma, R., Li, D.G., Zhang, X., and He, L. (2011). Opportunities for and challenges regarding prevention and treatment of rare diseases in China. *Chinese Journal of Evidence-based Pediatrics*, 6, 81–82. (in Chinese)

McPherson, S., Hardy, T., Henderson, E., Burt, A.D., Day, C.P., and Anstee, Q.M. (2015). Evidence of NAFLD progression from steatosis to fibrosing-steatohepatitis using paired biopsies: implications for prognosis and clinical management. *Journal of Hepatology*, 62(5), 1148–1155.

Meier, P. (1953). Variance of a weighted mean. *Biometrics*, 9, 59–73.

Miller, F., Zohar, S., Stallard, N., Madan, J., Posch, M., Hee, S.W., Pearce, M., Vågerö, M. and Day, S., (2018). Approaches to sample size calculation for clinical trials in rare diseases. *Pharmaceutical Statistics*, 17(3), 214–230.

Misra, V.L., Khashab, M., and Chalasani, N. (2009). Nonalcoholic fatty liver disease and cardiovascular risk. *Current Gastroenterology Reports*, 11(1), 50–55.

Motzer, R.J., Penkov, K., Haanen, J., Rini, B., Albiges, L., Campbell, M.T., Venugopal, B., Kollmannsberger, C., Negrier, S., Uemura, M., and Lee, J.L. (2019). Avelumab plus axitinib versus sunitinib for advanced renal-cell carcinoma. *New England Journal of Medicine*, 380(12), 1103–1115. doi: 10.1056/NEJMoa1816047. Epub 2019 February 16.

Muller, H.H. and Schafer, H. (2001). Adaptive group sequential designs for clinical trials: combining the advantages of adaptive and of classical group sequential approaches. *Biometrics*, 57, 886–891.

Musso, G., Gambino, R., Cassader, M., and Pagano, G. (2011). Meta-analysis: natural history of non-alcoholic fatty liver disease (NAFLD) and diagnostic accuracy of non-invasive tests for liver disease severity. *Annals of Medicine*, 43(8), 617–649.

Nie, L, Niu, Y., Yuan, M., Gwise, T., Levin, G., and Chow, S.C. (2020). Strategy for similarity margin selection in comparative clinical biosimilar studies. Currently under clearance at Office of Biostatistics, Center for Drug Evaluation and Research, Food and Drug Administration.

ODA (1983). Orphan Drug Act of 1983. Pub L. No. 97-414, 96 Stat. 2049.

Pagadala, M.R. and McCullough, A.J. (2012). The relevance of liver histology to predicting clinically meaningful outcomes in nonalcoholic steatohepatitis. *Clinics in Liver Disease*, 16(3), 487–504.

Pan, H., Yuan, Y., and Xia, J. (2017). A calibrated power prior approach to borrow information from historical data with application to biosimilar clinical trials. *Applied Statistics*, 66(5), 979–996.

Papastergiou, V., Tsochatzis, E., and Burroughs, A.K. (2012). Non-invasive assessment of liver fibrosis. *Annals of Gastroenterology*, 25(3), 218–231.

Park, J.W., Liu, M.C., Yee, D., Yau, C., van't Veer, L.J., Symmans, W.F., Paoloni, M., Perlmutter, J., Hylton, N.M., Hogarth, M., and DeMichele, A. (2016). Adaptive randomization of neratinib in early breast cancer. *New England Journal of Medicine*, 375(1), 11–22.

PDUFA VI (2018). *Prescription Drug User Fee Act (PDUFA) – PUDFA VI: Fiscal Years 2018–2022*. US Food and Drug Administration, Silver Spring, MD. Retrieved September 27, 2018 from http://www.fda.gov/ForIndustry/UserFees/PrescriptionDrugUserFee/ucm446608.htm.

Posch, M. and Bauer, P. (2000). Interim analysis and sample size reassessment. *Biometrics*, 56, 1170–1176.

Proschan, M.A. and Hunsberger, S.A. (1995). Designed extension of studies based on conditional power. *Biometrics*, 51, 1315–1324.

Proschan, M.A., Lan, G.K.K., and Wittes, J.T. (2006). *Statistical Monitoring of Clinical Trials: A Unified Approach*. Springer Science Business Media, LLC, New York.

Redman, M.W. and Allegra, C.J. (2015). The master protocol concept. *Seminars in Oncology*, 42(5), 724–730.

Rider, P. R. (1960). Variance of the median of small samples from several special populations. *Journal of the American Statistical Association*, 55(289), 148–150.

Rosenberger, W.F. and Lachin, J.M. (2003). *Randomization in Clinical Trials*. John Wiley & Sons, Inc., New York.

Rugo, H.S., Olopade, O.I., DeMichele, A., Yau, C., van't Veer, L.J., Buxton, M.B., Hogarth, M., Hylton, N.M., Paoloni, M., Perlmutter, J., and Symmans, W.F. (2016). Adaptive randomization of veliparib-carboplatin treatment in breast cancer. *New England Journal of Medicine*, 375(1), 23–34.

Rzany, B., Bayerl, C., Bodokh, I., Boineau, D., Dirschka, T., Queille-Roussel, C., Sebastian, M., Sommer, B., Edwartz, C., and Podda, M. (2017). An 18-month follow-up, randomized comparison of effectiveness and safety of two hyaluronic acid fillers for treatment of moderate nasolabial folds. *Dermatologic Surgery*, 43(1), 58–65. doi: 10.1097/DSS.0000000000000923.

Sanyal, A.J. (2011). NASH: a global health problem. *Hepatology Research*, 41(7), 670–674.

Sanyal, A.J., Brunt, E.M., Kleiner, D.E., Kowdley, K.V., Chalasani, N., Lavine, J.E., Ratziu, V., and McCullough, A. (2011). Endpoints and clinical trial design for nonalcoholic steatohepatitis. *Hepatology*, 54(1), 344–353.

Sanyal AJ, Friedman SL, McCullough AJ, Dimick-Santos L; American Association for the Study of Liver Diseases; United States Food and Drug Administration. (2015). Challenges and opportunities in drug and biomarker development for nonalcoholic steatohepatitis: findings and recommendations from an American Association for the Study of Liver Diseases-U.S. Food and Drug Administration joint workshop. Hepatology 2015;61:1392–1405.

Saville, B.R. and Berry, S.M. (2016). Efficiencies of platform clinical trials: a vision of the future. *Clinical Trials*, 13(3), 358–366.

Schemper, M., Kaider, A., Wakounig, S., and Heinze, G. (2013). Estimating the correlation of bivariate failure times under censoring. *Statistics in Medicine*, 32(27), 4781–4790.

Schuirmann, D.J. (1987). A comparison of the two one-sided tests procedure and the power approach for assessing the equivalence of average bioavailability. *Journal of Pharmacokinetics and Biopharmaceutics*, 15(6), 657–680.

Senn, S. and Bretz, F. (2007). Power and sample size when multiple endpoints are considered. *Pharmaceutical Statistics*, 6(3), 161–170.

Shao, J. and Chow, S.C. (2002). Reproducibility probability in clinical trials. *Statistics in Medicine*, 21, 1727–1742.

Sheiner, L. (1997). Learning versus confirming in clinical drug development. *Clin Pharmacol Ther*, 61, 275–291.

Simon, R. (1989). Optimal two-stage designs for phase II clinical trials. *Contemporary Clinical Trials*, 10(1), 1–10.

Singh, S., Allen, A.M., Wang, Z., Prokop, L.J., Murad, M.H., and Loomba, R. (2015). Fibrosis progression in nonalcoholic fatty liver vs. nonalcoholic steatohepatitis: a systematic review and meta-analysis of paired-biopsy studies. *Clinical Gastroenterology and Hepatology*, 13(4), 643–654, e1–9; quiz e39–40.

Song, F.Y. and Chow, S.C. (2015). A case study for radiation therapy dose finding utilizing Bayesian sequential trial design. *Journal of Case Studies*, 4(6), 78–83.

Sorrentino, P., Tarantino, G., Conca, P., Perrella, A., Terracciano, M.L., Vecchione, R., Gargiulo, G., Gennarelli, N., and Lobello, R. (2004). Silent non-alcoholic fatty liver disease-a clinical-histological study. *Journal of Hepatology*, 41(5), 751–757.

Todd, S. (2003). An adaptive approach to implementing bivariate group sequential clinical trial designs. *Journal of Biopharmaceutical Statistics*, 13, 605–619.

Tsiatis, A.A. and Mehta, C. (2003). On the inefficiency of the adaptive design for monitoring clinical trials. *Biometrika*, 90, 367–378.

Viele, K., Berry, S., Neuenschwancler, B., Amzal, B., Chen, F., Enas, N., Hobbs, B., Ibrahim, J.G., Kinnersley, N., Lindborg, S., and Micallef, S. (2014). Use of historical control data for assessing treatment effects in clinical trials. *Pharmaceutical Statistics*, 13 (1), 41–54.

Wasserstein, R. L., and Lazar, N. A. (2016). The ASA's statement on p-values: context, process, and purpose. *The American Statistician*, 70(2), 129–133.

Westfall, P. and Bretz, F. (2010). Multiplicity in clinical trials. In *Encyclopedia of Biopharmaceutical Statistics*, Ed. Chow, S.C., 3rd Edition, Taylor & Francis, New York.

Williams, G., Pazdur, R., and Temple, R. (2004). Assessing tumor-related signs and symptoms to support cancer drug approval. *Journal of Biopharmaceutical Statistics*, 14, 5–21.

Woodcock, J. and LaVange, L.M. (2017). Master protocols to study multiple therapies, multiple diseases, or both. *The New England Journal of Medicine*, 377, 62–70.

Younossi, Z.M., Abdelatif, D., Fazel, Y., Henry, L., and Wymer, M. (2016). Global epidemiology of non-alcoholic fatty liver disease-meta-analytic assessment of prevalence, incidence and outcomes. *Hepatology*, 64, 73–84.

Zheng, J. and Chow, S.C. (2019). Criteria for dose-finding in two-stage seamless adaptive design. *Journal of Biopharmaceutical Statistics*, 29, 908–919.

Zhou, J., Vallejo, J., Kluetz, P., Pazdur, R., Kim, T., Keegan, P., Farrell, A., Beaver, J.A., and Sridhara, R. (2019). Overview of oncology and hematology drug approvals at US Food and Drug Administration between 2008 and 2016. *Journal of the National Cancer Institute*, 111(5), 449–458. doi: 10.1093/jnci/djy130.

# Index

Note: **Bold** page numbers refer to tables and *italic* page numbers refer to figures.

Printed in the United States
by Baker & Taylor Publisher Services

Printed in the United States
by Baker & Taylor Publisher Services